Notice to the Reader

August 2017

As part of the 100th year commemoration of the birth of the 4th Infantry Division, on 10 December 1917, I have taken it as a personal objective to make as many of the out of print 4th Infantry Division books as I am aware of, and can get rights to, available. By doing that, the 4th Infantry Division Family, and anyone else with an interest in our long and illustrious history, can have their personal reprinted copies of these older books. We have done our best to make the books we reprint as close to the original version as possible, but in most cases with a new cover and with a unique ISBN for easy ordering.

Among the books that will be reprinted, and made available in print and e-book format, will be:

History of the 4th Division in the World War by Christian Bach and Henry Noble Hall, 1920

History of the 12th Infantry Regiment in World War II by Colonel Gerden Johnson, 1947

History of the 22nd Infantry Regiment in World War II by Chaplain Bill Boice, 1956

Five *World War II 4th Infantry Division Yearbooks* – 8th, 12th, 22nd Infantry Regiments, Division Artillery, and HQ & Special Troops, 1946

Operation Iraqi Freedom I: A Year in the Sunni Triangle by Robert O. Babcock, 2005

Operation Iraqi Freedom 07-09: Dispatches from Baghdad by Robert O. Babcock, 2010

Any or all of these books can be ordered through the 4th Infantry Division Association website at www.4thinfantry.org or from the publisher, www.deedspublishing.com. The books will be available in printed and in Kindle format e-books.

All profits from the sale of the above books will be used by me, in my role as 4ID historian and archivist, to make other 4ID written history available in coming months and years. All the new publications we can make available can be ordered from either of the above websites.

It is an honor to make the proud history of the 4th Infantry Division and our assigned units over the first 100 years of our existence, from 1917 to 2017, available to a much wider audience at an affordable price.

Steadfast and Loyal,

Bob Babcock
Past-President, Historian, Archivist – 4ID Association
www.4thinfantry.org
CEO/Founder, Deeds Publishing LLC
www.deedspublishing.com

To the 4th Division and its Commander with sincere appreciation of their efficient and loyal service coupled with my very best wishes.

June 19, 1919, John J. Pershing.

THE FOURTH DIVISION

ITS SERVICES AND ACHIEVEMENTS IN THE WORLD WAR

Gathered from the Records of the Division

BY

CHRISTIAN A. BACH

COLONEL, GENERAL STAFF, CHIEF OF STAFF
FOURTH DIVISION

AND

HENRY NOBLE HALL

WAR CORRESPONDENT LONDON "TIMES"
ACCREDITED TO THE AMERICAN ARMY

WITH FIVE ILLUSTRATIONS
PAINTED FROM LIFE
BY
CAPTAIN LEON DABO, A.D.C.
AND
NUMEROUS PHOTOGRAPHS AND FIVE MAPS

Maps are not included in this reprint but are available online at www.4thinfantry.org and www.deedspublishing.com

ISSUED BY THE DIVISION

1920

TO THE COUNTLESS THOUSANDS
MANY YET UNBORN
WHO WILL READ THE HISTORY OF THIS GREAT WAR
AND
DRAW INSPIRATION FROM THE HEROISM
OF THOSE WHO DIED
THAT OTHERS MIGHT LIVE IN FREEDOM
THIS BOOK IS DEDICATED

Foreword

THIS is the story of a Division which was organized, trained, brought overseas, fought in the final great offensives of the World War and which saw the signing of the Armistice—all within a period of eleven months. It is a feat not paralleled in the history of our Army. This achievement was made possible only because, earlier in its career, officers and men alike became imbued with a high sense of loyalty to their division and to each other. It was this steadfast and loyal spirit, gradually developed in the school of discipline, made perfect in the school of war, which rendered all things possible.

What soul is to the individual, spirit is to the Division. It is something intangible and hard to define but we all know its animating power. It lives, it hovers about and over the Division. It inspires every officer and man, no matter how long or short his service. It is greater than morale, more personal than patriotism, more exacting than duty. It makes men hold their heads high and face the world proudly in the joy of good work well done, in the calm self-reliance of disciplined efficiency, in the mutual confidence of true comradeship—that harmony of line and staff without which victory is impossible. It is a spirit, not of pride, but of service—of service to one another, to our great country, to mankind.

Born of the ideals of the Regular Army, nurtured in the trying conditions of the training camp, this spirit strengthened men when their muscles ached during the long marches as they staggered through mud and darkness under heavy loads; steeled them in the great test of battle when they lay, wet and hungry, under shell-fire which racked their nerves and tore their bodies, and, when death stalked on all sides, carried them irresistibly forward without thought of themselves or heed of danger.

In the following pages will be found the tangible things which have produced the intangible, but no less masterful, force:

THE SPIRIT OF THE FOURTH DIVISION

Contents

Contents

Illustrations

PAINTINGS BY CAPTAIN LEON DABO

MAPS

Introduction

It is proper that the deeds of a Division that fought in the World War be recorded with painstaking care. The reader of this book may feel satisfied that no effort has been spared to insure the historical accuracy of the facts stated. For most of them the original documents, substantiating the statements, are on file in the Historical Branch, War Plans Division, General Staff, Washington, D. C.; others are based upon the oral or written statements of two or more reliable eye-witnesses; still others are taken from the personal knowledge of one of the authors.

This book is not a tactical study of the operations of a combat division; nor is it merely a colorful account of a series of actions, such as might be written by a non-military observer. A sincere effort has been made to produce, in readable form, a bird's-eye view of the whole war, an intimate history of the operations of the Division and a correct record of facts that embrace sufficient data to make the book of some military value.

The manuscript has been revised by the commanders of regiments and of independent battalions of the Division. As published it bears the stamp of their official approval.

The material contained herein has been gathered from a variety of sources. Each organization, by means of its history and the stories of incidents that it furnished, has borne its part in the preparation of this volume.

The thanks of the Division are extended to Captain Charles L. Bolté, 58th Infantry, for his labors in the preparation of the data contained in the maps and in the appendix, for intelligent proofreading of manuscript and printed matter, and for the excellent business judgment displayed by him in handling many of the final details of publication; to Captain Leon Dabo, Aide-de-Camp to Major General Hersey, for five paintings that appear as illustrations in the book. Captain Dabo visited the Aisne-Marne and

Meuse-Argonne battlefields and made sketches, on the spot, for four of the paintings, the fifth—that of the sinking of the *Moldavia*—being based on descriptions by eye-witnesses. The Division acknowledges its gratitude to Master Engineer Charles W. Linberger, 4th Engineers, for several of the photographs that appear in the book.

C. A. B.
H. N. H.

Viola L. Norton

THE FOURTH DIVISION

THE HISTORY OF THE FOURTH DIVISION

CHAPTER I

WHY AMERICA ENTERED THE WAR

"WHERE do we go from here?" "When do we eat?" These were the thoughts uppermost, "over there," in the minds of the men who crossed the sea to fight. The rank and file saw little of the stupendous drama in which they played so essential a part. Yet, to borrow a figure of speech from a great French poet, the leaders would have been as naught but for the myriad tiny letters which go to make up the page of history. So it is only right that the part played in the final victory by the men who fought should be made clear.

1 Nearly eighty thousand young Americans lie dead in
France. When fighting ceased more than three and a half
million other Americans were in arms to insure the triumph
2 of the great cause for which those men had died. A nation,
vast in numbers, greatest in wealth, and greatest in the purpose of its people, had abandoned the path of peace and devoted its every energy to a war which, in the last analysis, was as truly a war of self-defense as that which France had sustained since the invader set foot upon her soil in 1914. In taking up arms, in throwing its decisive weight into the fight for civilization, American democracy was repelling an invasion of its sovereign rights which was the slow, steady, logical outcome of the invasion of Belgium, just as the latter was the slow, steady, logical outcome of the growth of Prussian

Note: The marginal index numbers refer to the list of authorities, which follows Chapter XVII.

military autocracy which for centuries had served the political fortunes of the Hohenzollerns.

The Great War for Civilization was fundamentally a war between Autocracy and Democracy. The Kaiser, it is true, dreamed of a World Empire—as others have dreamed before him—but his triumph would have been more than the triumph of one dynasty or one nation; it would have been the triumph of a system of government which for more than seven hundred years had been slowly crumbling beneath the pressure of human rights and human liberty. Those who fought to overthrow German power were fighting not only for their own freedom; they were fighting to fix the future of all generations to come. It was early apparent that the Prussian military autocracy must be crushed or European democracy would be obliterated. Later, when "the high temper of the great affair" was fully discovered, it was seen that the aim of America was to secure the freedom of mankind and to unite the whole world in a brotherhood of lasting peace based on Liberty and Justice. In the final struggle America played a leading part, but to understand its full bearing upon history, we must examine the origin of the cataclysm and recall the events leading up to America's entry into the war.

In the first quarter of the thirteenth century the Teutonic Knights, returning from the Crusades, undertook the conquest of Prussia, which two centuries previously had renounced Chris-
tianity. Berlin, which had been located among the barren 3
and unproductive sand heaps of Brandenburg by Albert the
the Bear (1163), was taken and a military colony established. 4
Two hundred years later, Friedrich of Hohenzollern, Burggraf
of Nürnberg, obtained, by purchase from Sigismund the
Great, the Margravate of Brandenburg. 5

From then on the history of Prussia is the history of the House of Hohenzollern. It is not the national development of a people but the territorial expansion of a dynasty based upon military power. In 1525, the Margravate of Branden-

6 burg was raised to a dukedom as a fief of Poland; and on
January 18, 1701, Frederick III, Elector of Brandenburg, in
an assembly of his knights and notables, placed a crown upon
his own head and upon that of his consort and proclaimed
7 himself King of Prussia. Louis XIV, King of France, refused
to recognize the new kingdom and that is why, in 1871, after
the Franco-Prussian War, William of Prussia chose the 18th
8 of January to have himself crowned German Emperor in the
Hall of Mirrors in the great Palace of Versailles where Louis
XIV had held court and where, in justice to France, on June
28, 1919, the Treaty of Peace was signed.

So, in 1701, Prussia became a kingdom; and the head of
the House of Hohenzollern, who had ranked beneath the Elec-
tors of Bavaria and Saxony and had been a vassal of the Kings
of Poland, elevated himself to royal rank. From the first
the Kingdom of Prussia was based upon the existence of a
military autocracy, a sacrosanct class superior to any other,
and upon the subservience of civil to military authority.
This was the system of Frederick the Great, who brought his
Prussian army to such a state of perfection that he was able
9 to withstand the rest of Europe. It continued until the re-
volution in Germany which forced the military to lay down its
10 arms in 1918. Thrice was this power fortified; by the Con-
gress of Vienna, by the failure of the German revolution in
1848, and by the genius of Bismarck, who placed the scepter
of the German Empire in the hand of the Kings of Prussia.

The Congress of Vienna (1815), whose main purpose was to
reduce Republican France to helplessness, either for attack
11 or defense, gave Prussia the guardianship of the Rhine. The
Kingdom of the Netherlands was formed out of Holland and
Belgium to permit easy invasion of France by both Great
12 Britain and Germany, but this careful structure crumbled
at the Belgian revolution of 1830, and the Powers granted Bel-
gium her independence only on condition of her neutrality,
13 which they guaranteed.

The failure of the German revolution, in 1848, one of the
greatest tragedies of the human race, had made militarism
more secure in Prussia than anywhere else, and enabled Prus-
sia to obtain such an ascendency over the other German States
that when Bismarck, in 1862, declared Germany could never
become great as a nation of poets and philosophers but that she
must adopt a policy of blood and iron, there was little oppo- 14
sition. In 1864 Denmark was despoiled of Schleswig-Hol-
stein. Austria was next attacked and the supremacy of the 15
Hohenzollerns over the Hapsburgs fully established, the latter
being forced to consent to a reorganization of Germany in
which they had no participation. Hanover, Hesse-Cassel,
and Nassau were annexed to Prussia. 16

Then came the War of 1870. Bismarck, by the falsification
of the Ems dispatch, let loose hostilities at the moment he 17
judged most favorable. The Prussian war machine was placed
in motion by Von Moltke. France was unprepared and, after
a brief and bitter war, the Germans invested Paris. Through-
out the winter the French Capital withstood the siege but was
forced by starvation to capitulate. The German Empire
was founded in the palace of the Kings of France at Versailles,
and the Hohenzollerns became the dominant dynasty in
Central Europe. Alsace-Lorraine was wrested from France.
The Kingdoms of Bavaria and Würtemberg, the Grand Duchy
of Baden and the Southern Provinces of the Grand Duchy of
Hesse were added to the North German Federation. All the
German States which had survived the struggle of 1866, with
the exception of Austria, the tiny Grand Duchy of Luxemburg
and the Principality of Lichtenstein, were incorporated as a
federated nation under the hereditary leadership of the King of
Prussia, with the title of German Emperor. In 1870 Belgium 18
narrowly escaped German domination, as only England's
threat that she would fight on the side of France if Belgium's
neutrality were violated caused Bismarck to hesitate.

In the next forty-four years—from 1871 to 1914—the

Prussian autocracy, which had now made itself the military
master of all Germany, devoted its full strength to relentless
19 preparation for world dominion. Every German child was
taught at its mother's knee that the God to which it prayed
was a German God. In the schools, at the universities, from
press, platform, and pulpit writers, teachers, philosophers
never tired of repeating that the Germans were a race of
supermen, that German "Kultur" was the height of civiliza-
tion, that the destiny of Germany was to rule mankind.
Force was extolled as the basis of world empire with such a
disregard for moral restraint that very few outside Germany
20 took this Prussian philosophy seriously. Hatred of, and con-
tempt for, democracy were the natural concomitants of these
teachings and yet, so cleverly did the German Government
conceal its military preparations, that it almost succeeded in
lulling into a false security the great nations which it planned
to conquer. All the time the mind of the German people was
being slowly and scientifically prepared for "Der Tag"—
the day when the Hohenzollern dynasty should make its bid
21 for World Empire.

Meanwhile, Germany's progress was uninterrupted. Cen-
tralized authority, efficient organization, scientific manage-
ment were necessary for the creation of the mighty engine with
which Prussian autocracy intended to subdue the world. They
brought an extraordinary degree of material prosperity to the
hardest working and most highly disciplined nation in Europe.
Germany amassed great wealth; her population increased to
more than 70,000,000; her exports were approaching in volume
those of England; she founded a vast colonial empire; her ship-
ping grew a hundredfold; she built a great navy and organized
an army which was the wonder and admiration of all profes-
22 sional soldiers. There was no secret about Germany's readi-
ness for war, although the full extent of her preparations was
unknown due to the secrecy with which she had accumulated
vast supplies of ammunition and built large quantities of heavy

artillery far more powerful than any then in existence. But
the blessings that prolonged peace had brought to the German
people were so great that even statesmen hesitated to believe
war possible. The nations of Europe continued to bear the 23
burden of armament without realizing that the military mas-
ters of Germany had the power to plunge the whole world into
war, and that they would do so the day that it served their pur-
pose best.

The assassination of the Archduke Ferdinand, heir-apparent
of Austria-Hungary, at Sarajevo, the capital of Bosnia, on
June 28, 1914, provided the pretext. Germany forced
Austria to adopt, and supported her in, a policy which could
have but one result—war. We now know that Germany,
seeing England almost on the verge of civil war in Ireland and
faced with other serious domestic problems, believed she would
keep out of the war and that the conflict could be restricted to
a struggle for the supremacy of continental Europe between
Germany and her allies on one side, Russia and France on the
other. The result Germany never doubted. Her military 24
leaders felt sure they could crush France suddenly before
Russia could mobilize, and then turn eastward and destroy the
Czar's forces. After the Kaiser had been made supreme in
Europe and had wrested from France her colonies, Germany
would be able to attack England and make her bid for that
world dominion which for a generation had been the dream of
Prussia.

Not only had France strongly fortified the Franco-German frontier at all vulnerable points but the nature of the French terrain, facing Germany, was, in itself, a defense against invasion. Between the basin of Paris and the German frontier is a series of slowly rising plateaus which end in sharp cliffs facing eastward, at the foot of which generally meanders a river. On the other side, the ground again rises gradually, to end in another abrupt drop facing toward Germany. The last rise terminates in the crest of the Vosges Mountains, which form a final barrier. So to the natural defenses of the Vosges Mountains and the Moselle River are added successively the sharp scarps of the Woëvre Cliffs, those west of the Meuse River and those fringing the eastern edge of the Champagne Plateau. These, with the strongly fortified positions of Verdun-Toul and Épinal-Belfort, and the preparations for the concentration of the French behind the *Trouée de Charmes*, known in English as the Gap of Toul, constituted an almost impregnable barrier, one which, in any case, could not be quickly broken through. But what Germany wanted above all, was to strike quickly at the heart of France. This could not be done through the eastern frontier.

The northeastern frontier of France facing Belgium was weakly protected, in comparison. There were a few fortified places, such as Lille and Maubeuge, beyond which the road to Paris lay open and without natural obstacles. The only real protection lay in the neutrality of peaceful and industrious Belgium. This, however, did not deter Germany. The Kaiser's forces were thrown into the little country whose inviolability Germany had joined with the other Great Powers in guaranteeing. England at once entered the war in defense of Belgian neutrality, as she was pledged to do by the Treaty
13 of 1839—the "scrap of paper" contemptuously referred to by the German Chancellor, who, however, a few days later in the Reichstag, pleaded the law of necessity in extenuation of "the
2 wrongs we are committing."

With an army equal to only about three American divisions, King Albert prepared to resist the German invasion. While the bulk of the Kaiser's military machine was held for seven precious days by the forts of Liège, the Belgians, more or less effectively, destroyed their bridges, culverts, and railroads. This quick action foiled the German plan for a lightning blow at France. England threw her small but well trained regular army across the Channel. The French, quickly modifying their plans, rushed large bodies of troops to the Belgian frontier. But gallant as was the resistance of their combined forces—the French at Charleroi, the British at Mons—the German horde swept with irresistible force across northern France. The seat of the French Government moved to Bordeaux. The retreat of the Allies carried them back more than 125 miles in twelve days until they reached the shelter of the Grand Morin under the guns of the outer defenses of Paris. The original plan of the French General Staff was to withdraw as far as the Seine, where the army would have menaced the enemy flank if the main German forces had advanced to Paris. But, obeying the strategic plan prepared by Von Moltke in 1859,* the Germans turned from Paris and struck southeast to annihilate the Franco-British armies then gathered on the Marne. The forces defending Paris attacked the German flank commanded by Von Kluck; Foch drove in the German center; and after seven days of bitter struggle Joffre won the Battle of the Marne, forcing the Germans back to the Aisne.

The Battle of the Marne did more than save Paris. It sealed the fate of the Kaiser's bid for world dominion. Germany, with her war machine at its highest point of efficiency, after forty-four years of preparation, had not only failed to destroy the French army and reach Paris, but had been beaten back by inferior forces incompletely mobilized. Then, while England was raising and training her armies, the French held the Germans in check. For nearly three years the battle line remained virtually unchanged on the western front, while on the eastern it swayed to and fro until Russia finally succumbed.

It is not our purpose to tell the story of the great struggle which was watched with breathless interest by the American

---

*In the report given by Von Moltke at Boehm, in February, 1859, he foresaw an invasion of France, through Belgium, by a strong army which would descend the banks of the Oise. The conclusion of this report is as follows:

"If we should find the French army assembled around Rheims, we should have to turn away from Paris. We should attack the French behind the Aisne and, profiting by our superior forces, would beat them and throw them back over the Marne, the Seine, the Yonne and finally beyond the Loire. Our road would then lie open to Paris."

people. How the British and the Belgians on the Yser and at the second battle of Ypres in the Spring of 1915 foiled the attempt of the Germans to reach the Channel Ports; the splendid offensive of the French in Champagne, which nearly broke the enemy's lines in the Autumn of the same year; and then, in 1916, the heroic defense of Verdun, where Germany spent her strength in vain efforts to capture the French stronghold but retained power sufficient to withstand the terrific onslaughts of combined British and French in the bloody battle of the Somme.

Vast and complicated systems of narrow trenches and deep dugouts were built, defended by hundreds of thousands of miles of barbed wire; high explosives reinforced shrapnel; poisonous gases and liquid fire were introduced; gas masks and steel helmets were supplied to all the troops; the use of artillery developed on a scale hitherto undreamed of. Aviation passed from the experimental stage and became the eyes of the contending armies; aviators fought duels in the air and rained bombs on the earth below. New problems of transport and communication were solved. The whole civilian populations of the nations at war were mobilized to the enormous economic effort needed for the war which affected all. Everywhere industry and commerce, food production and consumption were placed under national control.

True to the Prussian philosophy, Germany waged war with the utmost ruthlessness. She vainly sought by "frightfulness" to bend and break the morale of the Allied nations. She swept aside all restraints dictated by considerations of humanity and the rights of non-combatants. Unspeakable out-
27 rages were committed in Belgium, in northern France, Russian Poland, and Serbia. Civilians were massacred, women abused and children slain. The *Lusitania* was sent to the bottom of the sea without challenge or warning and 1154 non-combatants, men, women and children, including 114 Americans, were drowned. The priceless library of Louvain

was burned without the shadow of military necessity. The
Cathedral of Rheims, one of the greatest heirlooms of mankind,
was bombarded and hundreds of other churches and public
buildings destroyed. Defenseless cities were bombed by night
and peaceful inhabitants killed in their beds. During Hin-
denburg's so-called strategic retreat in March, 1917, hundreds
of square miles of French territory were devastated so com-
pletely that every highway, town, and well, every trace of ani-
mal and vegetable life was obliterated. Tens of thousands of 26
French and Belgian civilians were deported to Germany, the 28
men to enforced servitude in German industry, the women
and girls often to a fate worse than slavery. All these things
aroused the conscience of America and crystalized, against
Germany, public opinion which at the outset had been con-
fused as to the merits and issues of the war.

For nearly three years the United States maintained an
attitude of neutrality but, from the beginning, was unable to
avoid controversy with Germany, first over the sinking of the
Lusitania and other submarine outrages, then over the sup- 29
ply of munitions to the Allies and the open invasion of 29
American sovereignty by paid agents of the Kaiser's govern-
ment. The submarine controversy, which had been brought 30
to a climax by the sinking of the *Sussex* without warning, was
apparently settled by Germany's pledge henceforth to conduct
her submarine warfare in accordance with international law. 30
On the other hand German intrigues against the United States
were increasing. German officials violated American laws, in-
cited labor troubles in munitions plants and paid for the plac-
ing of bombs on vessels thus destroying American lives and
property on the high seas. The German government fanned
anti-American feeling in Mexico and fomented Irish revolution-
ary plots, against Great Britain, in the United States. It filled
unsuspecting communities and even Government offices with
spies. As early as the end of 1915 the United States had
found it necessary to dismiss the German attachés Von Papen

and Boy-Ed on clear proof of guilt, but no apologies to America
30 or reprimands to them were issued by their government.

In December, 1916, it had become apparent that unless
peace in Europe could speedily be restored, the United States
would inevitably be drawn into the struggle. In answer to
President Wilson's overtures of December 18, 1916, the Allies
stated their readiness to make peace on a basis of Reparation,
Restoration and Security, but Germany refused to disclose her
terms. On January 22, 1917, President Wilson outlined to the
Senate such a peace as the United States could join in guar-
anteeing. His proposal met with a mixed reception both from
the Entente countries and the Central Powers and was un-
productive of results. Shortly thereafter the "Zimmermann
Note" fell into the hands of the United States Government.
In this the German Secretary of Foreign Affairs secretly in-
formed the German Minister to Mexico of the German inten-
tion to repudiate the *Sussex* pledge and instructed him to offer
to the Mexican Government three American States—New
Mexico, Texas, and Arizona—if Mexico would join with
31 Japan in attacking the United States.

Then came the official notification from the German Govern-
28 ment that ruthless submarine warfare would be resumed.
The German Chancellor openly announced that the only reason
why the intensified submarine warfare had been so long delayed
was that sufficient submarines had not been built to make
effective the attack on the commerce of America and the
world. Thus it was revealed that the promises Germany had
made to the United States were never intended to be kept and
were made only in order to gain time to build more sub-
32 marines. On February 3rd President Wilson dismissed the
German Ambassador and severed diplomatic relations with
28 the Imperial German Government. On March 12th orders
28 were issued to arm all American merchant ships and, on April
2nd, the President urged upon Congress the recognition of a
state of war with Germany. The declaration of war was form-

ally passed in the Senate on April 4th, in the House on April 6th, and at once signed by the President. It was Good Friday.

In his address to Congress, urging the declaration of war, the President had pointed out that this would involve the utmost practical coöperation in counsel and action with the governments already at war with Germany; that this would mean not only furnishing financial credits but the organization and mobilization of all material resources of the country to supply the fighting forces and serve the nation in a satisfactory manner. At once the Army and Navy were placed upon a war footing and within a few weeks French and British War Missions landed in the United States. Joffre, in person, came as military head of the French Mission and made an appeal to the President to send, immediately, American troops to Europe to take their place in the battle line. The War Department had already set itself to the work in hand. The Allied experts found tentative plans in process of elaboration. Conferences were held and arrangements made for sending over large numbers of British and French officers and instructors for the armies that America was about to raise.

Meanwhile General Pershing and his staff were preparing to sail. The American Commander-in-Chief's instructions were to proceed with his staff to Europe, there "to command all the land forces of the United States operating in continental Europe and in the United Kingdom of Great Britain and Ireland" and "to establish, after consultation with the French War Office, all necessary bases, lines of communication, etc., and make all incidental arrangements essential to active participation at the front."

General Pershing sailed in May on board the S. S. *Baltic* and landed at Liverpool. He at once proceeded via Boulogne to Paris and established headquarters in two houses in the Rue Constantine. The General himself had offices at No. 31, and the windows of his study looked out on the gilt dome of Napoleon's Tomb. From the very first the American

Commander grasped the magnitude of the undertaking. He saw big, very big. He recommended to Washington the sending of at least twenty combat divisions to France for action and told the War Department a million men was "the smallest unit which in modern war could be considered as a complete fighting organization." He urged that preparations be made in the United States for training at least three million men in two years. Realizing the important effect that the arrival of American troops in France would have on the morale of the Allied armies, he hastened the coming of the first detachment by every possible means, and, at the end of June, 1917, went to St. Nazaire to meet the first few thousand regulars and marines, barely half a division but the nucleus of the great army that was to be.

Then, refreshed and strengthened, General Pershing went back to Paris and completed his plans. The harbors of France on the western coast were to be rebuilt on a scale commensurate with the task of handling millions of men and all that they would need to live and fight, three thousand miles from home. Great railroads, stretching across the whole width of France were to be created, huge warehouses built, storage plants erected and millions of tons of supplies brought over to fill them. As this stupendous undertaking was to be carried out in a land drained by three years of war, the United States was to bring over everything, from the huge freight locomotives to the smallest detail of the medical supply. Not only did the ships to bridge the Atlantic have to be built but the very shipyards for their construction had to be created, as most of the existing facilities were taken up with the needs of the Navy.

Meanwhile, in America, great training camps had been built. Sixteen wooden cities had sprung into being, as if by magic, in less than ninety days, some with streets laid out, their water mains and sewage systems installed, their hospitals and lecture halls ready. By the beginning of September, the

youth of America, chosen under the draft law, was being taken in hand and trained for battle. Democracy was answering the call to arms at home. In France, General Pershing had moved his headquarters nearer to the front chosen for American troops. The small houses in Paris were exchanged for huge barracks at Chaumont, the chief city of the Department of the Haute Marne. Here the American Commander was close to the Lorraine front, where the United States was to make its entry into the World War. The first shot was fired on October
23rd and, a few days later, on November 4, 1917, the first 34
American soldiers to fall in the fight for Liberty and Justice
were buried with military honors at Bathlemont. 35

As the year drew to its close American troops continued to arrive. The 1st Division was succeeded by the 2nd and this by the 26th from New England. The 42nd or Rainbow Division came next. America's force was taking form. It was at this stage that, at Camp Greene, North Carolina, the 4th Division was born.

CHAPTER II

ORGANIZING THE FOURTH DIVISION

ON DECEMBER 3, 1917, the order creating the 4th
36 Division was issued by the War Department. Major General George H. Cameron was placed in command, and Camp Greene, North Carolina, designated as the point of assembly and organization. At that time Camp Greene, located among a succession of small but well-wooded hills and valleys in the neighborhood of Charlotte, was blanketed with a foot of snow. In addition to the 3rd Division, in training there, the elements from which the 4th Division was to be formed—four infantry and three artillery regiments, the infantry in skeletonized form partly filled out with voluntary enlistments—were already in camp, the men housed in pyramidal tents heated by Sibley stoves.

Immediately following the declaration of war on Germany, Congress had provided for an increase in the military forces of the United States on a scale without precedent. The Army Act, approved May 18th, authorized the raising of all organizations of the Regular Army to the maximum enlisted strength sanctioned by law, the incorporation into the Federal Service of the National Guard and the National Guard reserves, and the drafting of a force of 1,000,000 men upon the principle of universal liability to service. In order to raise the Regular Army to maximum strength each of the old regiments was divided into three groups. One of these was designated to be the nucleus around which the parent regiment would be continued, the others being used to constitute the framework of two new regiments. Thus each had for its foundation a strong element of men who brought to the new regiment the

splendid spirit and traditions of the old army in which they
had served. The regiments selected for service in the 4th
Division had been created out of some of the most famous
regiments of the old army. The 58th and 59th Infantry were
formed from the old 4th Regiment, which dated back al- 37
most to the Revolutionary War and carried on its regimental
colors no less than one hundred and eleven silver bands, each
engraved with the name of a battle in which the regiment had
participated. The 47th Infantry was an offspring of the old
9th Regiment, formed just before the Civil War and fam- 37
ous for distinguished service in Indian Campaigns, in Cuba, in
the Philippines and during the Boxer Rebellion. The 39th
Infantry was formed from the 30th which, after fighting in 37
the Philippines, had been returned to the Pacific Coast and
rendered valuable service during the San Francisco earth-
quake and fire. The 77th Field Artillery, originally the 19th
Cavalry, could trace a proud descent from the 2nd Cavalry,
(the old 2nd Dragoons) which, organized for the Seminole 37
Campaign in Florida, had run the gamut of army history
through the Mexican and Civil Wars, the Indian Campaigns,
and the Spanish-American War. The 13th Field Artillery
was formed out of the old 5th Artillery and the 16th Field
Artillery out of the old 8th Artillery. 37

All of the units had been created and partly trained in camps scattered over the United States. The infantry came from Gettysburg, Pennsylvania, and from Syracuse, New York, the artillery from camps and training stations in Texas, Wisconsin and Vermont. There the skeleton organizations had been filled out to some extent by the voluntary enlistment of men anxious to go to the front as quickly as possible. The volunteer recruits, imbued with the spirit of ardent patriotism, were the very pick of the nation's youth and quickly absorbed the spirit of the old regular units in which they were to train and fight. An incident will show how keen that spirit was. The 19th Cavalry received word at 10 A. M. on

July 20th that it was transformed into the 77th Field Artillery. Before nightfall of the same day the batteries had logs mounted on the running gear of escort wagons, in lieu of cannon, and the school of the cannoneer was started.

This was the material which General Cameron found awaiting him when he arrived at Camp Greene on December 10th, from Camp Kearny, California. For such men he was a worthy commander. Born in 1861 and graduated from West Point in the Class of 1883, General Cameron brought with him a wealth of experience and the highest traditions of the United States Army. To him more than to any other man must be given credit for instilling in the 4th Division the magnificent spirit which has distinguished it since its birth. By his tact, his understanding of men and their motives, his military knowledge and ability, his justice, his human sympathy, his willingness to give credit, he commanded not only the respect of his officers and men but their affection as well. He created a soul in the Division; he breathed life into what might have been otherwise only a military machine. Above all he was loyal to the Division, and the entire personnel responded with an equal loyalty. An earnest student of his profession he had a keen appreciation of a military situation and the courage of his convictions.

When the General arrived at Camp Greene he found that, under orders from the War Department, the 47th and 39th Regiments of Infantry with the 11th Machine Gun Battalion, then in process of organization, formed the 7th Infantry Brigade and the 58th and 59th Infantry with the 12th Machine Gun Battalion, also in process of organization, formed the 8th Infantry Brigade. Colonel Leon S. Roudiez of the 47th Infantry, as senior officer of the 7th Brigade, had begun to organize the 11th Machine Gun Battalion, and Colonel Benjamin W. Atkinson, 59th Infantry, as senior officer of the 8th Brigade had begun to organize the 12th Machine Gun Battalion and was also directed to form the 10th Machine Gun

Battalion with men taken from the four infantry regiments.

The companies of the regiments were very weak, some having only thirty or thirty-five men and none exceeding forty in strength. In some companies there was not a single officer. The organization of the Artillery Brigade was further advanced. Some of the batteries had received their guns as far back as June and had nearly their full complement of enlistments before they reached Camp Greene.

After making a rapid inspection of the camp and the personnel of the Division, General Cameron left for Washington to receive detailed instructions and to expedite the formation of his staff. He returned to Camp Greene on December 13th, having accomplished what he desired. The same day 430 recruits arrived from Camp Meade. On the following day a detachment of the Headquarters Troop arrived from Fort Myer, Virginia. On the 14th General Cameron was joined by his Chief of Staff and the duties of divisional organization were begun. 38

On Christmas day the people of Charlotte presented General Cameron with a silver loving cup. A field meet, with all kinds of athletic events, and a Christmas tree celebration were held for the men of the Division near the camp-site of the 16th Field
Artillery. The people of Charlotte attended in numbers. 38
The 4th Division was popular in Charlotte—perhaps more popular than Charlotte was with the 4th Division. Although many of the officers and men enjoyed that lavish hospitality for which the Carolinas are famous, the commercial enterprise of the city could not resist the temptation to reap a golden harvest. The cost of everything soared. As one doughboy put it, "the price of a meal would pay for a week's board at home." Even the "movies" trebled their prices. Furnished rooms were rented for the rates charged at Broadway hotels. But the Christmas season was a great success.

The weather had, so far, alternated between snow and rain, making anything beyond the indoor work of organization

impossible. On December 30th and 31st, the 4th Engineers
38 arrived from Vancouver Barracks, Washington. They were
well organized, trained and equipped and had developed a
confidence in their ability to surmount obstacles that never
left them. At the time there was not a single drafted man in
the regiment. With the exception of two hundred and thir-
teen old enlisted men from Companies "E" and "F" of the 2nd
Engineers, the men were all volunteers—mechanics, students
and skilled workmen—who had joined the engineers because
of their experience in some branch of construction or engineer-
ing work. Two field hospitals and two ambulance companies,
which had reached Camp Greene in November, 1917, formed
the nucleus of the Sanitary Train, later completed by the addi-
tion of units from Fort Riley, Kansas. With the arrival of
the 8th Field Signal Battalion from Monterey, California, on
January 2nd in the midst of a blinding snowstorm, the unit
organization of the Division was nearly complete. The am-
38 munition and supply trains still remained to be organized.

Arms, equipment, and clothing were secured as fast as they
could be furnished by the War Department but, repeatedly,
equipment and arms were taken from units of the 4th Division,
to fit out units of the 3rd Division and hasten its departure,
which, however, did not actually take place before the middle
of March. Supplies came in satisfactory quantities but re-
cruits were slower in arriving. In January the War Depart-
ment ordered all voluntary enlistments throughout the United
States to be sent to the 4th Division which thus received men
from every State in the Union. These men began to arrive in
38 the latter part of February.

Training at Camp Greene was made very difficult by weather conditions. Most of the men had gone south patting themselves on the back for their good fortune. They pictured themselves drilling in shirts, playing baseball or basking in the sunshine of the Southland. What they found was far different. It was cold, very cold, and the only fuel available to offset the raw, penetrating dampness was green timber. Camp Greene took on the aspect of a northern lumber camp in midwinter. The men had to cut their own firewood from pine trees growing on the reservation. More land was cleared around Charlotte in a few months than in all the years since the Civil War. Even the oldest inhabitants could recall no such disagreeable weather in North Carolina. It rained and froze and snowed and rained. The cold raw wind swept through the trees and over the camp. Morning would find the ground covered with driven snow. Then this would thaw and turn the camp into a vast sea of reddish-yellow mud. It was a thick viscous clay, of the consistency of library paste and about the color of a new saddle, and made progress of any sort almost impossible. An article dropped in it was immediately engulfed. It defied the efforts of the motor truck company which was supposed to serve the camp. The mud on several roads was up to the floor of the truck bodies. So the Camp Commander, General Dickman, applied for two pack trains to transport rations from the railroad station to the organizations. Be it said to the everlasting credit of the lowly army mule that he labored faithfully and well, although at times his work had to be supplemented by fatigue parties, the men carrying meat and other foodstuffs on their backs.

With the exception of the paved road to Charlotte, the camp had only ordinary dirt roads and these soon became almost impassable. Officers and men all lived under canvas with serious overcrowding, the standard pyramidal tents housing ten to twelve men instead of the authorized eight. No artificial drainage existed and the pits frequently overflowed. Had it not been for the stoves and the fact that timber could be had for the cutting there would doubtless have been a high record of sickness. As it was Camp Greene had one of
the lowest sick rates of all the camps in the United States, 39
and, at the end of the winter, its men were healthier than they had ever been and were thoroughly hardened to severe

weather conditions. Perhaps the lack of comforts helped to impart that independence and self-reliance so characteristic of the regular army.

At the end of the second officers' training camps the full quota of line officers was assigned to the Division, and reported
38 in the latter part of December. The rolling character of the land provided few parade or drill grounds; and then there was the mud. The troops simply could not work out of doors. It is a physical impossibility to get men to go through drill, with snap and precision, half way to their knees in soft clay. A system of trenches had been dug but was swallowed up by the mud. Bad as conditions were they at least had the advantage of giving the men a good idea of what the front was like. Indeed, some of the hardships of life at Camp Greene were shock-absorbers which prepared the troops for what was in store for them in France. Both in Flanders and in the Argonne, mud at times offered at least as much resistance as the enemy to the advance of the Allied forces under conditions not dissimilar to those of Camp Greene.

The month of January brought continuous bad weather but, notwithstanding the difficulties caused thereby, the organization of special schools of instruction went on apace under the direction of British and French instructors who lectured to selected groups of officers and non-commissioned officers from all organizations. Each of these foreign officers was thoroughly competent and had seen active service at the front. Most of them wore wound chevrons. The school of Heavy Machine Gun firing was under Captain F. W. Petrie-Hay of the Gordon Highlanders, and that of Field Fortification under Lieut. Auguste Arrighi of the 285th French Infantry. Trench Mortar firing was taught by Lieut. Ivor Macfarlane of the London Regiment, and Automatic Rifle firing by Lieut. Réné Dael of the 277th French Infantry. Captain George Borthwick of the 9th Royal Scots taught Bayonet Fighting, and Grenade Warfare instruction was under Captain Edouard

Dupont of the 295th French Infantry. The School for Scouting and Sniping was under Captain F. B. Money and that in Gas Defense under Captain Harry Kemp, both of the North Lancashire Regiment. To Lieut. Maxim Bucher, French Field Artillery, was assigned the duty of acquainting the artillery regiments with the difficult problems of trench warfare. The student officers and men, upon completing these courses, were returned to their own units as instructors in turn, and imparted their newly gained knowledge to the men.

A review of all troops in camp was ordered for January 15th but had to be cancelled on account of the weather. Two days later Mr. Crowell, Assistant Secretary of War, and Major General Biddle, Assistant Chief of Staff, arrived from Washington and inspected the camp and all troops, under arms, on their organization parade grounds, and General Cameron submitted a detailed report on the existing conditions and on the needs of the camp. On January 22nd another effort was made to hold a general review but a combination of rain, sleet, and snow made this impossible. To add to the discomfort an outbreak of spinal meningitis in Charlotte caused the entire camp to be placed in quarantine, and no officer or man was allowed to leave the camp and visit the city except on official business. Quarantine barriers were also established around the area occupied by each organization of the camp. 38

During the whole of February the bad weather kept up and made outside instruction impossible except at rare intervals. Indeed, from December 10th, when General Cameron arrived, until March 4th, there were only sixteen working days during which outdoor training was possible. In February the 4th Supply Train was called upon to convoy trucks from Buffalo to Baltimore and made several trips with credit to itself and to the Division. The road conditions around Camp Greene became so serious that each organization was assigned a certain road area to maintain, to the exclusion of regular instruction and duties, a situation which lasted for more than a week. No rock was available and the soil was a sticky red clay that had no bottom. All that could be done was to fill up ruts, crown the roads and build side ditches to provide drainage. The 4th Engineers corduroyed an unpaved stretch of the main road between Charlotte and the camp and installed turnouts to allow two-way traffic.

Between March 1st and 21st some 10,000 men were received by the 4th Division from Camps Custer, Grant, Lewis, Travis, and Pike but this was not yet sufficient to fill up the Division. Voluntarily inducted recruits continued to pour in, however, in detachments of from ten to fifty men. They usually arrived at Camp Greene in their citizens' clothing and presented a very nondescript appearance. The majority were totally untrained. The men from the training camps came in uniform, and had received several weeks or months of instruction. All were willing workers and anxious to learn. They formed a plastic mass of human material which was quickly molded around the regular army nucleus of old soldiers, gradually filling out to authorized strength each organization of
38 the Division.

It was found that many of the men drafted under the regulations of the Selective Service Act were totally unfit, mentally and physically, to be sent overseas. These men were transferred out of the Division to a Casual Camp in Camp Greene to be retained in service for such other work as they could perform. This led to an extension of the work of the Division neuro-psychiatrist who made an examination of each man sent to the Division and recommended him not only for acceptance or rejection but also for the special duty to which he was best adapted. This work proved of importance in utilizing most efficiently the man power furnished to the Division.

Among the schools which were started was one for the study of Signal Corps work, which was attended by the signal platoons of all infantry regiments and by detachments from other organizations. A telephone system was installed and operated throughout the camp, all the work being done by the students under the supervision of officers and non-commissioned officers of the 8th Field Signal Battalion. This system was, for the camp, a marked improvement over the Charlotte city system then in use. It enabled General Cameron to assemble all organization commanders on the telephone and talk to them at one time.

The approach of Spring found the camp fairly well cleared of timber and filled to capacity with troops. Early in March the weather moderated and the improvement in climatic conditions gave an opportunity for the development of the work in field fortification. In each of the four infantry regiments a platoon of 54 pioneers had been formed but it was impossible for the foreign instructor to handle all these men together. So, in addition to his class in field fortification, attended only by officers and non-commissioned officers, the 4th Engineers were entrusted with the duty of training the privates of the four infantry regiments in this work. Fire and communication trenches were dug, cheveaux-de-frise built, all kinds of barbed wire defenses put up, mines, saps, trench accessories, deep dugouts, machine-gun emplacements, obstacles, revetments, observation posts and posts of command constructed. Under French and British instructors hand grenade throwing and bayonet practice were carried on vigorously. In the former it took much persuasion and many a sore arm to discourage the natural inclination of the men to pitch a grenade like a baseball but, once they had grasped the advantages of the overhead lob, which enabled them to throw almost as far with one tenth the strain, they quickly became proficient in it. The same can hardly be said about the use of the bayonet. The primary object of bayonet drill, as of boxing, is to instil in the soldier the determination to close with his opponent. Many of the men drafted had never struck a blow in anger in their lives. The bayonet drill instilled a fighting spirit and gave them individual aggressiveness, but it was never really popular. The rifle is the national arm of the American people, and they do not take kindly to the use of cutting or thrusting weapons. But, although the men of the 4th Division had few occasions to use their bayonets in hand-to-hand fighting, the training received was of real value and had a distinct psychological effect.

On March 5th and 6th Colonel Applin, British Army, arrived in Charlotte and delivered two lectures, the first on "Discipline and Training," the other on "The tactical handling of machine guns," to 1,700 officers and 1,100 sergeants at the Charlotte auditorium. These lectures created a great impression; they were entertaining as well as instructive and were not only enjoyed but remembered by all who heard them.

Weather conditions at Camp Greene were rapidly improving. On February 26th General Dickman, commanding the 3rd Division, had left and turned over the camp command to General Cameron. In the middle of March the 3rd Division began to leave and this permitted the 4th Division to expand and rendered the situation less cramped and crowded. Recruits, however, kept arriving in large numbers, and it became necessary to establish two casual camps, from which, after examination by the personnel section and the termination of the prescribed period of quarantine, the men were assigned to the various organizations of the Division. The personnel section did excellent work in this connection, putting each new man where he would be of most use. Many of the men in the draft had been born in foreign lands, and the records of the 4th Division show that some of the best stuff in America is in the men who are naturalized citizens of the United States.

The departure of the 3rd Division enabled the 4th to begin its target practice on the rifle range. All organizations except the 39th Infantry and one battalion of the 58th completed the prescribed course of firing before entraining for Camp Mills.

The Artillery Brigade had suffered, perhaps more than the infantry, from the weather conditions but, as the roads finally became passable for the guns, field exercises replaced the schools, indoor work, and dismounted drills. Gun crews were whipped into shape, drivers and horses conditioned and trained, special details perfected, a range chosen and acquired in the wooded area between King's Mountain and Gastonia, 28 miles from Charlotte. Here Camp Chronicle was established with accommodations for one half of the Brigade at a time. In thirty days, commencing March 18th, a full year's allowance of ammunition was fired away, the sharp bark of the 3-inch guns of the 16th and 77th Artillery being accompanied by the more sonorous tones of the 13th Artillery's 4.7 howitzers. It was here that the men learned to handle rapidly and accurately live ammunition, the junior officers to observe and sense shell-bursts and to manipulate a sheaf, and the horses what it meant to pull a loaded American caisson. The booming of the guns and the bursting of the shells gave the natives of the country round about an idea that the war was getting rather close to them after all.

April found the Division straining every nerve to get into shape for overseas service. Better weather and the gradual drying of the mud made it possible to speed up outdoor training, and work at camp became severe and exacting. Schools of all kinds were in progress. The British and French instructors rounded out their courses with practical field demonstrations. Discipline became iron-handed. The men drilled longer hours and more intensively. There were frequent road marches which gradually became longer and longer. From a three-mile hike the men were brought up to marching twelve miles with full pack. Road discipline was strictly insisted upon. The men were taught to keep well closed up in marching, to stick to the right of the road and, equally important, to fall out on the right side. Special attention was devoted to teaching them how to get the maximum amount of rest during the halts that were made, and instruction was given in the handling of transportation. All this was found exceedingly useful in France. It was a splendid sight at the end of a long march to see the men swing by in cadence. The 4th Division began to acquire that cohesion and soldierly aspect which distinguished it later. One of the final touches was given by Colonel Giles of the British General Staff who arrived and delivered a five-day series of lectures on staff work to the field officers of the Division, giving them the benefit of observation and experience acquired in actual combat.

The 4th Division was now up to the strength prescribed by the new Tables of Organization and comprised the 7th Infantry Brigade, the 8th Infantry Brigade, the 4th Artillery Brigade, the 4th Engineers, the 10th Machine Gun Battalion, the 8th Field Signal Battalion, the 4th Train Headquarters and Military Police, the 4th Engineer Train, the 4th Ammunition Train, the 4th Sanitary Train, the 4th Supply Train and the Headquarters Troop.

General George H. Cameron, in command of the Division had Lieut. Col. Christian A. Bach as his Chief of Staff. The other Divisional Staff Officers were Major Edward Canfield, Jr.,who as Division Quartermaster performed the duties of G-1, to which post he was later appointed; Captain Max S. Murray, who acted as G-2; Major Charles H. Rice, G-3; Major Jesse D. Elliott, Adjutant; Lieut. Col. James M. Petty, who had succeeded Colonel Marcellus G. Spinks, as Inspector; Lieut. Col. Robert L. Carswell, Surgeon; Major Lyman N. Hine, Ordnance Officer; Major Charles C. Teare, Jr., Judge Advocate; Lieut. Col. George E. Kumpe, Signal Officer, and Major Robert McAuslin, Veterinarian.

Brigadier General Benjamin A. Poore was in command of the 7th Infantry Brigade, having succeeded Brigadier General John S. Mallory on April 3rd. The 39th Infantry was commanded by Colonel Frank C. Bolles, who had arrived on April 9th from the 2nd Infantry to take the place of Colonel

William C. Bennett. Colonel Leon S. Roudiez commanded the 47th Infantry. The 11th Machine Gun Battalion was under Major William G. Murchison. Brigadier General Frank D. Webster commanded the 8th Brigade, having succeeded Brigadier General Samuel W. Miller on March 24th. Colonel Carroll F. Armistead had taken over command of the 58th Infantry on March 16th. It had previously been under Colonel Lutz Wahl and, after the latter's promotion on February 3rd, his place had been temporarily taken by Captain George T. Everett. The 59th Infantry was commanded by Colonel Benjamin W. Atkinson, and the 12th Machine Gun Battalion by Major Max B. Garber.

Brigadier General Edwin B. Babbitt commanded the 4th Field Artillery Brigade. The 13th Field Artillery was commanded by Colonel Wright Smith, who had organized it from the 5th Field Artillery at El Paso, Texas. Colonel H. T. Newbold commanded the 16th Field Artillery and Colonel Guy H. Preston the "Silent" 77th. The 4th Trench Mortar Battery was under Captain Robert H. Ennis.

The 4th Engineer Regiment was commanded by Colonel Jay J. Morrow, the 10th Machine Gun Battalion by Captain Franklin C. Sibert, and the 8th Field Signal Battalion by Major Girard L. McEntee. Colonel James J. Hornbrook commanded Train Headquarters and Military Police; Lieut. Hardy J. Story, the Engineer Train; Lieut. Col. Alden Trotter, the Ammunition Train; Major William E. Wilmerding, the Sanitary Train; Captain Jack L. Meyer, the Supply Train, and Captain H. J. M. Smith, the Headquarters Troop.

On April 6th, the anniversary of the entry of the United States into the war, the Division, commanded by Major General Cameron, paraded through the city of Charlotte to inaugurate the drive for the 3rd Liberty Loan. It was the Division's first opportunity to appear as a unit and although many of the men had never before marched in a formal parade, they presented a fine appearance and well deserved the cheers that greeted them. Everyone felt that this was a farewell review and that the Division would soon be on its way to France. That same day President Wilson, speaking in Baltimore, declared: "Germany has once more said that force and force alone shall decide whether justice and peace shall reign in the affairs of men; whether right as we conceive it or dominion as she conceives it shall determine the destiny of mankind. There is therefore but one response possible from us—force, force to the utmost, force without stint or limit, the righteous and triumphant force that shall make Right the law of the world and cast every selfish dominion down in the dust."

CHAPTER III

CROSSING THE ATLANTIC

MARCH, 1918, was, for the Allies, a critical period of the war. For more than three and a half years the great struggle had been waged with varying fortunes, but all efforts of both contestants definitely to break the front had failed. The battles of the Marne and the Yser had stabilized the lines and practically ended movement, because neither the French nor the Germans were strong enough to take the initiative of further operations on a large scale. Trenches had been dug from Switzerland to the sea, from Altkirch to Nieuport. All these trenches the French held while the British were raising their armies and organizing all the resources of their Empire. The French "poilu" revealed to the world the stoic endurance of which he was capable, through mud of winter and heat of summer, beneath a never ending storm of high explosives.

On the western front 1915 was a year of defense, of waiting and of organization. Germany, having failed to crush France, had thrown her weight against Russia. In order to relieve the pressure there the Franco-British forces twice took the offensive, the first time in Artois on May 9th and again on September 25th in Champagne. The second winter in the trenches passed and every day saw increasing numbers of British
troops in France. Germany had been victorious in Russia and 40
had helped her allies to repulse the attack on the Dardanelles,
to vanquish Serbia and Montenegro. She determined once 41
again to throw her whole weight against the French, to an- 42
nihilate France in single combat, before the new British armies could be fully trained and ready to bring their strength to bear.

Major General George H. Cameron
Commanding 4th Division

Brigadier General Edwin B. Babbitt
Commanding 4th Artillery Brigade

On February 21, 1916, the Germans struck at Verdun. Their artillery concentration surpassed anything military experts had conceived possible. In a day the French front positions were wiped out and behind a heavy curtain of high explosives the German infantry advanced in mass formation. The colossal effort, which lasted until the middle of July, was in vain although, after the fall of Fort Douaumont on February 26th, the German General Staff sent wireless messages throughout the world to announce that "the keystone of the French front had fallen." But beyond the slopes of Douaumont no German ever went. The famous motto of Verdun "*On ne passe pas*" echoed over the valley of the Meuse. The Kaiser, determined to take Verdun at any cost, extended his attacks, at the beginning of March, from the right to the left bank of the river and then began a series of the most furious alternating assaults. All the divisions in the German army on the western front were thrown into the attack at one time, but their efforts were frustrated by Pétain, Nivelle, and Mangin, and the sixty-two French divisions which participated in the most famous defense of history. The French losses were terribly heavy, 350,000 men, but at Verdun German
43 military power wasted itself. On October 24th, the soldiers of France retook Forts Vaux and Douaumont and won back the positions from which they had been driven in the first
44 German onrush. The victorious resistance of Verdun saved not only France but the whole of the great cause for which America was soon to fight.

The sacrifices of the French at Verdun during the first four months of the defense made possible the battle of the Somme where, for the first time, the new armies of Great Britain were to be used on a vast scale. Twenty-three months had elapsed since the beginning of the war and during all that time France had borne the brunt of the German attacks, had held by far the largest portion of the front. Now, in July, 1916, her troops were once again thrown forward to coöperate

with the hundreds of thousands of men which the British
Empire was at last able to place in line. This victorious
offensive, which stopped only when the French were before
Peronne and the British before Bapaume, is now known to
have brought Germany to the very verge of defeat. Saved 45
from final disaster by the breaking of the weather, the Ger-
mans decided to retreat to other positions rather than fight
another such costly battle. So the Hindenburg line was
constructed. After a third winter in the trenches, before
the Allies could renew their offensive, Hindenburg withdrew
in March, 1917, to his prepared positions. Skilfully con-
cealing his plans, the German commander was able to carry
out the first part of the operation without being discovered. 46
At the same time he was gradually able to strengthen his posi-
tion, thanks to the situation on the eastern front and in
Russia, which permitted the Germans to transfer more and
more new divisions to the Franco-British front.

At the beginning of 1917 the French and British armies
were in the fullness of their power. Despite their formidable
losses at Verdun and on the Somme, the Allies were able to
accumulate more than three and a half million men and an
extraordinary amount of all kinds of artillery and munitions
with hundreds of tanks, enormous fleets of aeroplanes, millions
of trench weapons and hand grenades and vast supplies of
materiel. The general plan was for the British to attack 47
from west to east in the direction of Cambrai while the 48
French, moving from south to north, struck at the Aisne
front. But there was no unity of command. Instead of 46
one great battle which might have been successful, several
battles were fought, none of which finally achieved the de-
sired result. The French attack was stopped after sus-
taining heavy losses. For a time offensives on a huge scale
were abandoned and the doctrine of "active defense" became
paramount; economy of human lives was the first considera-
tion, and the complete destruction of a position by heavy

artillery was aimed at before infantry was sent forward to
occupy it.

America had entered the war in April, 1917, and during this
trying period when—waiting for the American forces to mater-
ialize—only small operations were undertaken, the Allied
soldiers, though tired and worn, maintained their confidence
in final victory. Franco-British forces were sent into Italy
49 to stabilize the front there after the disaster of Caporetto,
but despite the sacrifices of the French in the Battle of the
Aisne, the heroic British offensives at Vimy and la Bassée, a
local offensive at Verdun, the slow wearing down of German
forces before Ypres, and the splendid offensive of the Mal-
maison, no decision was had and another winter—the fourth—
was spent in the trenches.

Just as Germany had tried to crush France at Verdun in
1916, before the full power of Great Britain could be felt, so she
opened the campaign in 1918 with a desperate effort to divide
the British and French armies, destroy them separately and
take Paris and the Channel Ports before the arrival of the
50 Americans. The utter collapse of the Russian front permit-
ted Germany to concentrate virtually all her available forces
51 —almost two hundred divisions—on the western front, and
on March 21, 1918, sixty-four German divisions attacked the
British front which was held by nineteen divisions and had
51 but thirteen in reserve. In twenty-four hours the enemy
had broken through on a front of fifty miles; the Fifth British
Army was compelled to give ground, involving the Third Army
in its retreat. Noyon was lost, as was Peronne, and the enemy
reached a point within 12 kilometers of Amiens. Paris was
uncovered. On the evening of March 26th, the Germans had
all but succeeded in separating the British and French armies.
The breach was filled by engineers, auxiliary forces, camp fol-
lowers, drivers, cooks, anything that could be found, until the
Australians and the French cavalry and a few batteries of
52 75 mm. came up. It was a critical moment. The French

had barely managed to stop the gap. The British were fighting "with their backs to the sea." Only 300,000 American troops were at that moment in Europe. Paris was under 53
bombardment by a long-range gun. The outlook was indeed a gloomy one.

The Allies sent a pressing appeal for help to America. This took the form of a cable sent by Mr. Lloyd George, the British Prime Minister, to Lord Reading, the British Ambassador in Washington, on March 27th. At once Lord Reading went to the White House and laid the facts before the President, who immediately acceded to his request that every available man should be sent at once or as soon as transportation could be found. That same night, at the Lotus Club in New York, Lord Reading read Mr. Lloyd George's cable, which was as follows:

> "We are at the crisis of the war. Attacked by an immense superiority of German troops, our army has been forced to retire. The retirement has been carried out methodically before the pressure of a steady succession of fresh German reserves, which are suffering enormous losses. The situation is being faced with splendid courage and resolution. The dogged pluck of our troops has for the moment checked the ceaseless onrush of the enemy, and the French have now joined in the struggle. But this battle, the greatest and most momentous in the history of the world, is only just beginning. Throughout it the French and British are buoyed up with the knowledge that the great Republic of the West will neglect no effort which can hasten its troops and its ships to Europe.
>
> "In war, time is vital. It is impossible to exaggerate the importance of getting American reinforcements across the Atlantic in the
> shortest possible space of time." 54

The gravity of the situation was such that the Allied Governments, for the first time during the war, decided to entrust the supreme command of all their forces to one man. General Foch was chosen. He assumed command on March 27th. On the morning of March 28th, General Pershing called upon the new Commander-in-Chief, whom he found

walking in his garden. Informally, speaking in French, he
offered Foch all the American troops and American materiel
55 in France to use as he thought best. One American division
was sent to the front and three others to reserve sectors where
they replaced veteran French divisions. In six days liaison
between the British and French fighting forces was fully re-
established and the German offensive in Picardy brought to
a standstill.

In America the response to the Allies' appeal for help was
immediate and whole hearted. Orders were at once issued
to move every available division to the seaboard ready to be
56 embarked for Europe. At Camp Greene the utmost activity
reigned. The 3rd Division began to move to the Port of Em-
57 barkation on March 15th. The 3rd Division gone, every-
body in the 4th "snapped into it"—more training, more
enthusiasm, more soldiering. When the orders finally came,
57 on April 15th, to move to Camp Mills and Camp Merritt
all knew that the time had come and "Over There" was
filled with a new and more personal meaning. Between
April 21st and 28th the Ivy Division, less its artillery,
was moved by rail from Camp Greene to the New Jersey
57 and New York camps. Some of the units had their journey
broken at Washington and enjoyed a two-hour period of
marching, for exercise, through the outlying streets of the
capital.

Some realization of what moving a division means, even
under what were virtually peace time conditions, may be
gathered from the fact that the 4th Division took with it
some 24,000 pieces of freight, 35,000 of heavy baggage and
6,000 of light baggage, without counting the hundreds of
vehicles, which are the indispensable paraphernalia of the mod-
58 ern fighting unit. Before leaving Camp Greene the Ivy
Leaf design, prepared by General Cameron, had been adopted
as the insignia of the Division. It was derived from the Ro-
man numeral IV, and was especially appropriate, as in the

language of flowers, the ivy means "steadfast and loyal," two qualities which have ever distinguished the 4th Division.

As soon as the different units of the Division arrived at the embarkation camp a thorough inspection was made of the men's individual equipment, and although in some units the clothing was virtually new, the inspecting officer directed that everything showing the slightest sign of wear should be discarded. This was done. The inspection was made with painstaking care, all the articles of clothing were minutely examined and each man was required to bend over and expose his rear center to the sun, while the inspector—the president of one of America's principal railroads—gazed intently at the point of greatest resistance to make certain that no signs of wear made rips or holes imminent in his nether garments.

At Camp Mills, 400 men from Camp Devens were received to take the place of enlisted men who had failed to pass the physical and mental tests, and the Ivy Division finally embarked with its full quota of 29,180 officers and men. As sailing dates approached more orders, circulars, memoranda, bulletins and regulations piled in, all bearing dire warnings to any organization that should reach the shore of France even so much as a button short.

The men who were at Camp Mills saw more of New York during their stay than those quartered at Camp Merritt. Camp Mills had a system of 24-hour passes and each morning 20 per cent. of the men were allowed leave until reveille the following morning. This enabled every man to enjoy at least one day in New York and no train on the Long Island Railroad was as popular as the "Reveille Flyer" which, leaving the Pennsylvania Station, arrives at Garden City and Hempstead just in time to allow its happy passengers to fall in for reveille.

The 4th Engineers were the first troops of the Ivy Division

to sail. The regiment embarked on April 29th at New York and landed at Bordeaux on May 12th. On the same day a battalion of the 59th Infantry reached Liverpool on the *Olympic* and forty-eight hours later the Machine Gun Company of the regiment, and the 12th Machine Gun Battalion, made the same port in the *Aquitania*, both of these great liners having made the dash across the Atlantic, without convoy, in seven days. The Division Headquarters, and the remainder of the 8th Brigade, with the exception of the 58th Infantry, left New York on May 3rd and disembarked at Liverpool on May 16th. They were escorted out of New York by the U. S. S. *San Diego*, sent to the bottom by a mine a few days later. The 7th Brigade, with some of the artillery, arrived at Brest on May 23rd as part of a convoy of sixteen ships, carrying 35,000 men, which had left Hoboken and Newport News thirteen days previously. The rest of the artillery followed, the last battery leaving on May 22nd, and landing at Brest on the 30th. On the following day the 77th Artillery landed at Liverpool, and by June 5th the entire 4th Division was in France with the exception of the Sanitary Train, part of the 58th Infantry, and the 77th Artillery which stayed for a few days in camp near Winchester, England.

The absolute necessity of getting large American forces to Europe as quickly as possible had taxed the shipping resources of Great Britain and the Allies to the utmost. Instead of transporting thirty thousand men a month across the Atlantic which, prior to March 18th, had been done almost exclusively in American bottoms, ten times that number of troops were to be sent over if possible. In the effort to achieve this, everything that could float was pressed into service either as a transport or to take the place of some freighter which could be released for troop carrying purposes. Alongside of the finest liners with their palatial fittings ruthlessly torn out, there lay in New York Harbor weird looking "tramps" for
56 which Great Britain had combed the seven seas. The crews

came from every seaport in the world. The 39th Infantry sailed on an Italian ship manned by a mixed crew, chartered by the British Government to carry American troops to France.

Few of the men had ever been to sea before, but whether they sailed in one of the huge ocean greyhounds, or embarked in some obscure vessel as part of a convoy, their experiences were very much the same. There was the march down to the docks in the early morning through deserted streets, the minute examination on the dock, the distribution of hot coffee, sandwiches, candy, and cigarettes by willing women workers of the Red Cross—and then a few steps up the gangway and they found themselves in their floating camp. Company after company joined on board and when the last available inch of space was filled up the order to cast off was given, and one more ship, big or little, slid slowly and silently from its pier to the lower harbor. The embarkation formalities were the same in all cases. Each man was identified by his company commander and by his first sergeant as he stepped upon the gang plank. Immediately on board ship the men were shown to their quarters on the troop decks, packs were removed and they set about examining their new home. It was a strange sensation to many, and those on the lower decks cast apprehensive glances at the narrow hatchways which offered the only exit to the decks above. The men crowded to the portholes to get a last look at the skyline of the great city, as every one had to be below decks on the regular troopships while in the upper harbor. The Statue of Liberty was soon passed and their troop ship quickly found itself lying in the lower bay, one of a number of curiously camouflaged vessels awaiting their convoy. It was a sight which may never be seen again and will certainly never fade from the memory of any man who saw it. Large ships were curiously painted with huge stripes and curves of black and white and vivid colors, until even at a short distance their own builders would not have known them. There, lying against the island are two small vessels, one behind the other, looking so insignificant that no one would ever waste a torpedo on them, and yet closer inspection reveals the optical illusion; it is really a fair sized troopship cleverly disguised. Seen at close quarters, however, the camouflage looked as if it would attract attention rather than reduce visibility, and the most complimentary remark from the doughboys on board the closely packed ships was that "the camouflage man had had a nightmare."

Only those who have sailed on a transport under conditions which obtained on the Atlantic, in the early spring of 1918, can know the full measure of the old expression "packed like sardines in a box." Men literally slept on top of one another. Although on some of the transports men could eat their meals sitting at tables, on others it was necessary to eat them sitting on deck, with the possibility of some soldier finding another soldier's hob-nailed foot in his messpan. Those who were lucky enough to cross on big boats had none of these discomforts. In spite of everything, the average doughboy managed to enjoy himself on the way across. For one thing there was remarkably little seasickness among the men. They were all in good physical condition, the sea was quite smooth, and such sickness as they experienced was quickly overcome.

The convoy carrying General Cameron and the bulk of the 8th Brigade was one of the largest that had sailed for France. More than 45,000 troops were embarked on fifteen vessels, which were escorted into the mid-Atlantic by the U. S. S. *San Diego*. The convoy was arranged in three lines of five ships each, covered a front of five or six miles and extended about three miles in depth. All the ships carried at least one gun mounted on the stern. The troop ships were preceded by an auxiliary cruiser, as flagship, and followed by a fast and heavily armed vessel. During the daytime the men were allowed on deck. At dark all smoking on board ceased, all lights were put out on the upper deck and all portholes covered. During the night no soldier was allowed to leave his troop deck or to open a porthole. No light was shown except a tiny red one on the stern of each transport. During the daytime the ships kept in communication by means of wigwag signals; at night a system of low whistle signals was used.

Lifebelts had to be worn constantly by officers and men alike. Everyone, from general to private, carried his lifebelt to mess and to bed with him. It was a constant and necessary companion.

The men of the 59th Regiment on the *Olympic* had a thrilling experience with submarines. She is a mammoth vessel and on the trip carried some 7,500 men. She left Hoboken on Sunday, May 5th, unconvoyed. The *Olympic* is very fast and well able to take care of herself, carrying eight 6-inch guns. On the morning of the 12th, about 4:30, loud reports and a scraping along the ship's bottom brought the troops out of their bunks at a bound. The captain had sighted a submarine, swung squarely at it, at full speed, and beaten the enemy at his own game. The big hooks on the bow caught the submarine, ripped it open, and, passing over it, the huge *Olympic* rolled the submarine over and over as though it were a log. A destroyer which had come on the scene soon wirelessed that it had picked up the crew, 4 officers and 28 men. A few hours later were heard more loud reports. The *Olympic* quivered and rolled. Destroyers had sighted a second submarine and dropped depth bombs. Soon pools of oil and strewn wreckage told unmistakably that another "baby killer" had been accounted for. On this same day an observation balloon, towed by a destroyer, and a seaplane joined the *Olympic* and accompanied it to port at Southampton.

The first casualties in the Division, as the result of an enemy act, occurred at sea when the *Moldavia,* a refitted British liner with Companies "A" and "B" of the 58th Infantry on board, was torpedoed and sunk by a German submarine with

the loss of 56 men, all but one being Company "B" men. This occurred at 2:40 on the morning of May 23, 1918, at a point
59 about midway between Land's End and the Isle of Wight.

The *Moldavia* was armed with six light naval guns and two anti-aircraft guns. She was leading her convoy of five ships, "mothering" them, so to speak, while five British destroyers, which had joined them the previous day, darted around and between them. The sky was overcast with clouds and the night was very dark—ordinarily an ideal condition for outwitting the lurking U-boats. A sudden rift in the clouds permitted the moon to peep out for perhaps five minutes during which short time the invisible periscope enabled the U-boat commander to glimpse his prey and send the fatal torpedo crashing into her.

The explosion tore a gaping hole in the port side of the hull and shattered the compartment where the "B" Company men were sleeping and the ladders leading to the deck above. Nearly all the men in the compartment were killed outright by the explosion. The listing of the ship to port undoubtedly saved those who survived the explosion, enabling them to utilize the uneven surface of the sloping compartment walls in climbing to the next deck above, and from there to the boat deck and rescue.

Fortunately, the engines were not injured by the explosion, and the bursting of a starboard bulkhead caused the ship to resume an even keel. An attempt was made to reach shore, but the water gained rapidly and stilled the ship's throbbing engines within an hour.

The explosion had awakened all on board. Contrary to orders, many of the men had undressed before going to bed in the belief that the submarine danger was practically over, as land had been visible during the entire day. In the darkness and confusion many of them were unable to find their clothes and were forced to go to their boat stations in little more than their underwear—some wore even less.

The dying down of the engines found nearly all the men at their proper boat stations. A decided list to starboard had developed, however, which made the launching of the boats at the port side a very precarious matter. A number of men received a chilly bath during the process, but, eventually, the boats were launched. One of the British destroyers came alongside, and made fast to the starboard side of the *Moldavia*, taking all the men she could find room for direct from one ship to the other. Another destroyer circled around, picking up those who were in boats and hanging to the life-rafts.

Only two men were lost by drowning, although many unwelcome cold baths were taken by those who tried to climb upon the life-rafts only to overbalance the unwieldy affairs and slide, headlong, back into the sea. One soldier performed this stunt three times before he became convinced of the error of his ways.

After dropping depth bombs over the area in which the *Moldavia* was struck, three of the destroyer escort and the other five ships of the convoy had scurried away to avoid possibility of further disaster. This left only the two rescuing destroyers with their shivering but thankful burden, to watch the death throes of the stricken vessel. She had settled forward rapidly, her stern slowly rising higher and higher in the air.

Picture, if you will, a glassy, calm sea in the early dawn of that season of the year; a heavily clouded sky through which the sun had not yet made its way, but light enough to make each detail of the tragedy visible to the spectators aboard the little rescue ships; abandoned life boats and rafts bobbing about; here and there a dead body; on the decks of the two waiting destroyers a motley, shivering, awestruck crowd in various stages of dress and undress; a long minute of breathless silence; a lurch; a gurgle; a ponderous gathering of that immense mass of steel, as if for physical effort, followed by the rearing of the stern high in the air; a sickening dive; a seething cauldron in which are tossed bits of wooden wreckage, and the *Moldavia* was no more. The rescued soldiers, packed on the two destroyers, greeted the final plunge of their ship with a tumult of cheers.

CHAPTER IV

TRAINING WITH THE BRITISH

IT WAS on the short march from the docks at Liverpool to the trainshed of the Lancashire and Yorkshire Railroad that the men of the 59th Infantry, who led the invasion of the 8th Brigade into England, enjoyed their first view of the city. They gazed with natural and pardonable curiosity on the narrow, winding streets and the rows of small houses, dark with the grime of age, from which, possibly, some of their grandfathers had gone forth to populate the New World. The people of Liverpool who lined the streets cheered them heartily.

Events moved rapidly in those days. The troops were hurried into second-class coaches with compartments built to carry eight passengers, four sitting on each side. This was another novelty to the Americans, who were soon being carried, whither they knew not, through the smiling fields and neat hedgerows of old England. Mile after mile of meadowlands rolled past their eyes. They saw miles of red-brick houses. They saw women and children working amid the growing crops and crowded gardens, taking the places of men gone to war. England appeared to them to have been measured out inch by inch before being apportioned among its present owners.

Soon the falling shadows of evening hid the countryside, and when night came the train was steaming into London. From every window appeared American heads, struggling to get a glimpse of the wonderful old city. It was a strange sight. Myriads of lights were shaded from above to prevent their beams piercing the pall of darkness and attracting the unwelcome German night bombers. Nothing else could be seen.

Suddenly at nine-thirty every light disappeared. The "What the hell?" of the men was promptly answered by a giant policeman, appearing from out of the gloom, who approached the door of a compartment and told the sergeant to put out the light. "Who are you?" demanded the sergeant. "Never mind; put it out; Jerry's up," came the reply. "And who is Jerry?" "Jerry is a German bomber," explained the patient policeman. The lights went out.

About midnight the train reached Dover. The men were awakened, ordered to sling packs and fall in. They marched up the steep hill to Dover Castle overlooking the white cliffs of England which, before Christ was born, had led the Roman Conquerors to give Britain the name of Albion.

Here the soldiers were quartered for the night. In the morning they breakfasted on an English ration of tea, jam, oatmeal and dark war bread. A few hours later they bade farewell to England and were crossing the Channel from Dover to Calais in the wake of hundreds of thousands of British troops.

The sea trip occupied only a couple of hours, but the many determined attempts, not only of German submarines but also of enemy destroyers, to sever that vital line of communication gave the journey a thrill in keeping with the glamour of adventure with which the men had come to regard their going to war. As the troops debarked at Calais all eyes were turned to the debris of wrecked shops and houses along the waterfront and to the multitude of French signs which the the men could not understand.

They marched to a rest camp a mile or so outside the port. Just as they were stacking their arms, an aeroplane appeared and attracted the rattling anti-aircraft fire of the British. It brought a new thrill. A British sentry explained with a grin that it was a German who had "come to take your picture." So it proved to be. At night when all were seeking a rest, a flock of German aeroplanes came over and let fly their bombs.

Here was a taste of real war—utter darkness, the dull drone of the motors, the roar of bombs as they crashed down on city and camp shaking the stoutest buildings; the sound of falling masonry and breaking timbers; the rattle of anti-aircraft guns, like frantic pneumatic riveters; then a short silence followed by the tinkle of breaking glass as an occasional anti-aircraft bullet fell through a window or skylight. The droning of the motors grew fainter and fainter. The raid was over. Every doughboy wanted to see it. Some came out from the doubtful safety of a wine cellar and peered aloft but could see nothing except the fitful bursts of shrapnel in the sky, the long fiery lines of tracer bullets and the probing fingers of searchlights seeking to find the invader. They stood and listened to the menacing hum of the aeroplane motors overhead. Others more reckless, disregarding the advice of the British soldiers to "get under cover," stood out in the open and fired rifles at imaginary objects in the air. Few of the troops of the 8th Brigade who passed through Calais missed this German welcome, and few of the men who experienced it failed to write long accounts to the folks at home about this first adventure with death in France.

Next morning, May 18th, the serious work of war commenced. All barrack bags were confiscated and many cherished personal belongings, never to be seen again, were dumped in a nondescript pile. Then came the sudden realization that a rifle, ammunition, and a shovel were the essential implements of war, particularly the shovel. The heaviest blow was yet to come. The weapon which they knew and loved, the Springfield rifle, was exchanged for the British Lee-Enfield. This was almost more than the old regulars who formed the nucleus of the Division could stand, but protests, firm and respectful, passed up through the non-commissioned officers to the company commanders, were of no avail. The British rifle, which is heavier and clumsier than the Springfield, was issued to all units and the men set to work to learn how to care for and

Brigadier General Benjamin A. Poore
Commanding 7th Infantry Brigade

The sinking of the *Moldavia*

use it. Later, when orders were received for the 4th Division to leave the British area and support the French in their resistance to the drive on Paris, the Lee-Enfields were withdrawn and the American rifles re-issued to the troops.

The men of the 4th Division did not take kindly to the British Army ration. It was not their "chow" and neither appealed to their taste nor satisfied their stomachs. The lack of coffee "got the doughboy where he lived" as one officer put it; but it was something more than the mere substitution of tea for the fragrant black beverage dear to the American Army. The British ration is lighter than the American. The men ate very sparingly of oatmeal. They sorely missed their plentiful sugar allowance and the jam served instead failed to take its place. They missed also the bacon and corn-willy to which they were accustomed. But they took it all in good part, although it was perhaps fortunate that the stolid British mess sergeants did not always catch the drift of the remarks hurled in their direction at meal time.

Three days were spent at the Calais camp. Between May 18th and 20th the 59th Regiment boarded French military
75 trains en route for its training ground in the Samer area. Companies "C," "D," and "I," and the 2nd Battalion, 58th Infantry, which had sailed from Philadelphia, May 7th, and landed at London, May 24th, followed immediately behind and went into camp at Calais after crossing from Dover. The remainder of the 58th sailed from New York via Halifax and did not reach London until June 5th and Calais until June 9th.

Meanwhile the 7th Brigade was invading France from the French seaboard. The 47th Regiment, together with two companies of the 39th and the 11th Machine Gun Battalion, had landed at Brest on May 23rd, and the remainder of the 39th at Bordeaux. Those who landed at Brest went into camp in the vicinity of the Pontanezen Barracks built by Napoleon, but they were more concerned with the present difficulties and the peculiarities of the French language than with the mighty deeds of the past. It was the middle of May when the lightness and charm of a French Spring suffused itself throughout the land. The men slept in shelter tents, happy to be near the war and wondering, "Where do we go from here?"

Only a few days in Brest and then a procession of units of the 7th Brigade were introduced to a friend with whom they were to become only too familiar, the "Hommes 40 Chevaux 8"—the French box car. These "side door Pullmans" are about a third of the size up and down and back and forth of the American box car, and more nearly resemble in capacity a six-mule wagon. The inscription quoted above is their estimated capacity for transport in time of war. Dirt is their chief characteristic and discomfort their main recommendation. But, as the men of the 4th Division were quickly to discover, they play a highly important part in the great game of war. With a smile and a joke they swung into the dinky little cars and were soon absorbed in the beauty of the country of Normandy as they made their way to the British rest camp at Calais and forgot the discomforts of their mode of travel.

One detachment was fortunate enough to rest in the warm beach sand of the French port listening to the French girls sing. At night, however, they made up for it. They had the thunderous music of the air raids to stir their emotions to the depths. The next day they listened to an eloquent sermon by the gas instructor beginning with the text "For God's sake" and closing with a snappy salute.

The experience of the 7th Brigade followed that of the 8th at Calais even to the loss of the American rifles and the acquisition of the British ration. Early in June its units followed the 8th Brigade to Samer for intensive training under British instructors before they got any taste of the real thing.

Arriving at Samer, after passing through Boulogne, the men detrained and marched in the heat of a June afternoon to the little villages throughout the training area. They were on historic ground. The names of Crêcy and Agincourt recalled scenes in the history of England of great and gallant days when the bow was the chief arm of the infantry. In fact the area on which they stood was that of the famous Field of the Cloth of Gold where Henry VIII of England and Francis I of France had, at their meeting, striven each to outshine the other in the magnificence and panoply of kings.

The rest of the 39th Regiment and the 10th Machine Gun Battalion came to the Samer area via Bordeaux and Le Havre and by the 5th of June the 4th Engineers, the 47th and 39th Regiments, 11th Machine Gun Battalion, the 59th Regiment and Companies "C," "D," and "I," and the 2nd Battalion of the 58th, together with the 10th and 12th Machine Gun Battalions were assembled at Samer. The 77th Field Artillery, which had passed through England, was camped at Le Havre; later it joined the other artillery units at Camp de Souge, Bordeaux. Headquarters and the Headquarters, Supply and Machine Gun Companies and Companies "A," "B," "K," "L," and "M" of the 58th Infantry were still in England. The Sanitary Train and the Signal Battalion were on the Atlantic; the former nearing Liverpool, the latter Brest. Division Headquarters was established at Samer on the 19th.

In the little villages of the area the men had their first experience with French billets. Many of them discovered that a barn, even one in which it was rumored Napoleon had stabled his horses, was worse than no barn. There were, in fact, many requests to be allowed to bivouac in the open fields warm with the heat of the June sun, requests that were not refused. In such little places as Desvres, Deaudeauville, and Berniéville, the French people saw, for the first time, the careless smiles of American soldiers who, in the midst of searching for water drawn from wells four hundred feet deep, found time to play with the children and to endeavor to learn the simplest terms of the French language.

Here the men of the Division heard for the first time the rumble of heavy artillery firing on the front and many experienced the then novel sensation, later to become the rule of life, of training on the edge of woods in order to conceal themselves from the observation of enemy aeroplanes.

The Division was attached for training and supply to the 16th British Infantry Division, which had its headquarters at Samer. Thus consultation with the British officers was simplified and many questions were asked. The British officers and men were a fine lot. Their division had suffered heavily in the German attack of April 4th and was recuperating and awaiting replacements. For administrative purposes the 4th Division was assigned on May 18th to the II American Corps, which had Headquarters at Fruges; for tactical disposition it was under the First British Army.

Relations with the British were most cordial. The men learned the game of war from their instructors, attended the various schools which they conducted, absorbed their machine gun, bayonet, and combat "dope," ate their rations which often left them hungry, drew equipment and engineering supplies, gazed on the horses, interested by the extraordinary care lavished upon them, and puzzled themselves about the peculiar differences of English as spoken by the English and as spoken by themselves. When Corporal Oswald Olsen was asked by a British instructor who had picked up a wrench, what it was, he replied "A monkey wrench," only to be told, "No, that is a combination extension spanner wrench."

No incident that occurred during this period got "under the skins" of the men of the 4th Division quite so deeply as did the bombing of the British hospital at Étaples. This hospital was situated only a short distance outside the divisional area. It was a large one located on the seashore where the wounded might have the benefit of the sea air. It was some distance from any town and marked by large, easily visible red crosses placed both on the ground and on the roofs of a number of the Adrian barracks used to house the patients. One bright, moonlight night the distinctive drone of the German air-motors was heard approaching the area from the northeast. Everyone expected another bombardment. Instead of that the planes passed over the divisional area and continued southwest. Soon the dull, heavy explosions of about a dozen large bombs were heard. In the morning the men listened in cold fury to the report that "Fritz" had demolished two of the barracks, killing several score of patients in their beds as well as a number of Red Cross nurses. Another mark was added to the score the Germans would have to pay.

The 4th Engineers drew on the Royal Engineers for supplies, studied their methods at a large dump and built rifle ranges. Vickers guns were served out to the machine gunners, much to their delight. But the great day came when the men of the various units were ordered to march to Samer to exchange their Lee-Enfield rifles. The distance in some cases was eight miles, but a round trip of sixteen miles on a hot afternoon was more than compensated in the minds of the men by the fact that the Springfield rifles were to be returned to them. Then came another change. The distinctive American campaign hats were exchanged for overseas caps and steel helmets, and the canvas leggings were discarded in favor of the wrapped puttees. To the men of long service these changes were revolutionary but it did not take long for everyone to realize the merits of the steel helmet.

Meanwhile, the 4th Artillery Brigade and the Ammunition Train assembled at Camp de Souge, near Bordeaux, early in
60 June. Here the 16th and the 77th Field Artillery Regiments devoted themselves to mastering the intricacies of the mechanism of the French "75's," the difficulties of the French sight, the working of the recoil device and the principles governing the elaborate assortment of ammunition and fuses. The 13th Field Artillery wrestled with the 155 mm. howitzers,

the novel firing mechanism, the totally unfixed ammunition, and the difficulties of loading. Before leaving the United States all arrangements had been made for the motorization of the 13th, and a large number of trained drivers had been transferred out of the regiment in exchange for mechanics. On arrival in France it was found that this plan for motorizing the heavy artillery was not to be carried out, and many difficulties were met in training new drivers, particularly as there was delay in supplying new horses. All the officers went through a thorough course of instruction under French and American instructors. The specialists attended schools in their particular branches—materiel, munitions, telephone, radio, or orientation. Similar courses were given to enlisted specialists. Three full crews for every piece were trained. In a short time they were handling French materiel as familiarly as they had handled the American materiel at Gastonia and service practice was commenced. Progress was rapid as the men were enthusiastic to learn all that was to be learned. Soon barrages were being fired, day and night, while officers in the towers were making observations and solving firing problems. It was not of course foreseen how little opportunity there would be to use a great many of the principles, that were so laboriously acquired at this time. Little by little the brigade obtained its animals, large French farm horses, but very soft and difficult to train as they were accustomed to being led instead of driven. The surroundings amid which all this training was strenuously pursued were far from comfortable. In fact, many of the artillerymen of the Division will never forget Camp de Souge, which was really a collection of long one-story brick barracks in the middle of a desert of heat, sand, scrub brush and sandflies, which made even the banks of the Rio Grande seem a beautiful home. The camp had none of the conveniences of those in the States; no bathing facilities were provided, and a small, slimy creek a short distance away had to suffice for the few baths that could be taken. Vin rouge et blanc were easier to obtain than water.

The men picked up various French phrases and some settled down, with commendable initiative, to master the French language. They all inquired more intimately into the sacrifices that France had made and the weight which the war had thrust upon the French people. The tremendous realization of what it all meant became fixed in the minds of the soldiers and when the fears of the French people as to the future of the war, and particularly the German drive to the Marne, began to be more openly expressed, every man yearned to be at the front to render the service which he had crossed the ocean to perform.

The Americans, young, sturdy, eager and willing to learn all about the great game of war, were to the French people the one bright spot on a dark sky. In those parts of France where the 4th Division found a temporary home its presence infused a cheerfulness into the people which was of inestimable value in the struggle. The menace of possible German success made the men work harder and harder in the school of training. There was in fact at that moment no division in France which could show a finer fighting spirit, a keener desire to meet the enemy in combat, and a greater determination to prove its worth on the battlefields. It was the spirit of the old regular army inspiring the mingled elements of the 4th Division.

CHAPTER V

TRAINING WITH THE FRENCH

THE March drive of the Germans along the Valley of the
Somme had been stopped by the British before Amiens
and had reduced itself to a series of desultory local attacks, the
inevitable aftermath of all great drives. Then the enemy,
discovering a weak spot in the Allied front, quickly concen-
trated large forces, under excellent concealment, and on
May 27, 1918, launched a new attack against the French posi-
tion on the Chemin des Dames. The French were totally 61
unprepared for this onslaught. Their position was weakly
held, and a number of divisions were resting in the back area. 62
They were soon driven from the shell-swept heights of the
Aisne, across the Vesle to the Valley of the Marne. Oppo- 61
sition to the Germans had almost faded away in the face of
the tremendous onslaught and the drive gained such impetus
that the enemy moved forward more with the regularity of a
procession than like an army fighting its way over the ground.
Soissons fell. On June 1st Château-Thierry was occupied. 63
The news quickly spread over the whole front. The 2nd and
3rd Divisions were rushed to the vicinity of Château-Thierry
to assist in stopping the drive. The men of the 4th Division 64
waited, in tense eagerness, for the order that would place
them in the battle line. The French people were greatly
dispirited.

Orders were issued from Headquarters 4th Division on June
8th directing the various units of the Division, in the region
of Samer, to hold themselves in readiness to move, and these
were welcomed with all the thrill of anticipated battle. The 65
"Where do we go from here" inquiry of the soldiers was

mingled with the fear that the war might be called off before they actually settled down to the solemn and mechanical business of killing Germans. Broad smiles adorned the faces of the men as they left behind them British rations and British equipment, and headed toward Hesdin and the south.

Again it was necessary for officers and men to strip themselves of everything not absolutely essential for sustaining them in combat. This was a real "strip." Officers from Division Commander down were allowed to transport only fifty pounds of baggage, including bed, and during the campaign no officer had any baggage save his bed roll. The baggage thus discarded was collected from all over the area, and stored in an unused brewery near Samer under guard furnished by the Division. Later it was taken to Gievres and stored there. The Division never saw it again until hostilities had ceased. Much of it was lost or stolen.

The officers were, of course, the ones who suffered most in this reduction of personal equipment. It is unfortunate that there was not a better understanding between the War Department and the A.E.F. as to the character and amount of baggage which an officer should bring with him to France. In America the Division received lists, from the office of the Inspector General of the Army, of the articles which each officer must have in his possession before he would be permitted to sail. As every officer in the Division would, if necessary, have sacrificed cheerfully at least one of his fingers for the opportunity of getting to France with the Division, there were no shortages of any articles which could possibly be obtained. Each officer thus expended several hundred dollars for equipment which he had to store immediately upon arrival in France and which, in many cases, he never saw again.

No fault was found by any one with the A.E.F. order requiring every officer to reduce his baggage to fifty pounds. The justice and necessity for this was apparent to all. The reasons, however, that prompted the War Department to demand the purchase, in America, of equipment which must be discarded in France were by no means apparent.

On June 9th and 10th the Division marched to Beaurainville, Montreuil, Maresquel, and Hesdin, where the various units were entrained on June 11th. The march to the entraining points was carefully observed by staff officers from G. H. Q., who later complimented the Division Commander on the excellent march discipline and on the fact that there were no stragglers. The wisdom of discarding every non-essential article and thus lightening the packs of the men had proved itself. At each entraining point a company of engineers was designated to make all preparations for and supervise the loading of property and to do all things necessary to expedite the orderly entrainment of troops. The system proved so satisfactory that it was used for all subsequent movements by rail.

There was much wild speculation among officers and men as to the destination. The 59th Infantry reached Paris in the darkness of night, staying sufficiently long to view the darkness and nothing else. The men of the 47th, on the other hand, considered themselves highly unfortunate in that they reached St. Denis to the north of Paris and then swung around the city without a glimpse of the French capital.

The whole of the Division traveled by the most direct railroad route to the area of Meaux and the Marne River, detraining at Meaux and Trilport on June 12th. The 2nd Division was still fighting the action at Belleau Wood, and American wounded were streaming back to the hospitals in the region where the 4th Division had detrained.

For a day the Division was grouped in the area just north of Meaux while its future movements were being determined by the French Sixth Army. Then, on the evening of June 13th, orders were received placing the 7th Brigade at the disposal of the 4th French Infantry Division, with Brigade Headquarters at Rosoy-en-Multien. Here the 39th Infantry 66 was given the opportunity of completing its long-delayed and as yet unfinished target practice. These men had heard the thunder of the guns at the front but had never fired a rifle.

The remainder of the Division was attached to the 164th French Infantry Division and directed to march, on June 14th, to an area just south and west of La Ferté-sous-Jouarre, on the Marne River. Division Headquarters was established at
66 La Ferté-sous-Jouarre. It was a step nearer to the plunge
67 into action.

The sturdy, smart appearance of the men of the 8th Brigade as they marched along the main highway from Meaux to La Ferté, a section of the famous Paris-Metz road, brought encouragement to the hearts of the French people. Tears of joy were in the eyes of many of the women who saw them pass.

Throughout the little villages lying in the junction of the Marne and Ourcq valleys the men settled down to a period of intensive training. The green fields, broken here and there by patches of woods, were specially adapted for this purpose. At night the German aeroplanes bombed the area but there were no serious daylight raids.

It was here that the infantry handled live grenades for the first time and received their Chauchats, facetiously termed "Hot cats," otherwise automatic rifles. They underwent a systematic course of training under the French, and learned their methods of attack; how best to use their automatic rifles and machine guns; how to pass through barrages with a minimum of loss; as well as all the latest wrinkles in digging trenches and stringing barbed wire. While ranges were being constructed they practised rifle firing, putting holes in flattened corned beef and tomato cans nailed on sticks. To advance in combat order was part of each day's work. On June 20th, the 7th Brigade spent the forenoon in field maneuvers simulating open warfare. The nearness of the front line gave to all this a reality impossible to find in the training camps in the States. The continual expectation of another drive, bursting upon them from the Marne, was the decisive factor in speeding up.

Hotchkiss guns were issued and the machine gunners made the day merry with the rounds they fired.

The 8th Field Signal Battalion and the Supply Train had arrived at Brest on June 8th. Here the Signal Battalion spent ten days at the Pontanezen Barracks before entraining to join the Division. The battalion was treated to an air raid, in Paris, on its way to the detraining point at Lizy-sur-Ourcq and was finally billeted at Jouarre. It had had its troubles of movement, but they were trifling compared with the difficulties of procuring materiel. Its property upon arrival at Jouarre consisted solely of rations, company records, and what the men had on their backs. The one hundred per cent. equipment, laboriously acquired at Campe Greene, had faded into thin air. Most of it had been, of necessity, left behind at Brest. After a time a side car was unearthed at a salvage dump in La Ferté and attached to one of the motorcycles, this constituting the sole transportation of the battalion until it went into action. Some of the side cars later supplied saw such hard service and were so often repaired that, at the close of the war, it would have been difficult to find any part that belonged to the original machine. The Supply Train, after procuring its motor transportation proceeded, overland, to join the Division. It arrived at La Ferté on June 17th.

On June 16th the French Army Commander called upon the Americans to construct an important second line of defense running from a point near Crouttes, on the Marne River, northwest to Crouy-sur-Ourcq. On June 17th the 4th Engineers moved up from La Ferté to Crouttes, a distance of eleven miles, leaving behind two companies to complete the construction of the rifle ranges. These companies rejoined the regiment seven days later.

Crouttes is a small village situated on a bend of the Marne southwest of Château-Thierry. Within six miles of the German batteries, it was often under fire. On the morning of June 18th the engineers first heard the whistle of a coming shell as several six-inch missiles landed among them. No damage was done except to a wagon wheel, but the men hurriedly moved to a more sheltered location up the river.

On June 18th, Colonel Morrow was relieved from duty with the 4th Engineers and soon afterward promoted to the rank of Brigadier General and made Chief Engineer of the First American Army. Lieut. Col. Acher came into the command of the regiment and was later, on August 1st, made a full Colonel.

The Engineers remained at Crouttes for three weeks working on the reserve lines on the heights overlooking the river. A complete system of trenches was laid out and partially excavated. Woods were cleared, wire entanglements constructed and machine gun emplacements located. They were preparing for the next German drive.

Each day a battalion of infantry was sent up to assist in the digging of the trench system and nearly all came under enemy fire. The 59th Regiment suffered some casualties in this way. It seemed strange to the men that death and destruction should be wrought amid such a smiling country as the Marne Valley. However, those who passed under fire came away with the proud feeling of veterans, although not unmindful of the value of a shovel. The story is told by a sergeant of the 59th that he saw another sergeant, two days after enemy fire had caused some casualties, still throwing up dirt from a deep hole. "What are you doing?" he exclaimed. "I'm digging in," came the reply.

Meanwhile the men of the Division were beginning to be at home in France. With about five words of French they could make their way about more easily. The first time they received their pay in francs they called it "circus money" but they found that after all it would buy at the French shops or at the commissary, the things they craved.

Under the supervision of officers from Headquarters, I American Corps, the Division engaged in maneuvers twice in the La Ferté area and once in the area near Lizy-sur-Ourcq. The primary purpose of these maneuvers was to instruct the troops in liaison, which involves communication between the various units of an organization, between the organizations and their brigade headquarters and between the headquarters of the brigades and of the division; also between troops on the ground and observers in the air who give information to the infantry or direct artillery fire from aeroplanes and balloons. The Division needed this instruction.

July 4th brought the men the first real holiday since they had left the shores of America. Paris, still under the fire of "Big Bertha" and attentively visited by German air raiders, nevertheless was at her best on that day. She arose early in the morning to garb herself exquisitely, in honor of the Fourth, in a blending of the Stars and Stripes and the Tricolor. Her people massed the streets to see, to congratulate, to cheer and to entertain the Americans. The whirlwind of enthusiasm swept all before it and many an American spent what will ever remain the happiest and proudest "Fourth" of his life.

The 4th Division took part in all this. Two battalions, one each of the 39th and 58th Regiments, commanded by Colonel Carroll F. Armistead, were sent to Paris for this event. There they marched with one battalion of Marines, fresh from its exploits in Belleau Wood, through the thronged streets of Paris to receive the salute of the great Republic. Very few of the men had ever visited the French capital before and to win a first glimpse of such a city as a representative of America marching, to be honored by a deeply grateful people, was a memorable experience. To the men the necessity of keeping in step found an overpowering distraction in the revelation of streets and monuments, of squares and shops as they passed on, mile after mile.

While these two battalions were winning praise for the Division by their excellent bearing and fine behavior, the rest of the men were celebrating behind the front with a series of field events, boxing matches, and ball games. It was a big day, overflowing with fun. When evening came the French instructors and non-commissioned officers banqueted the different units of the Division with a charm and enthusiasm that made soldiering lose its harshness and retain only the joy of fraternal comradeship.

The battalions of the 39th and 58th which had enjoyed the hospitality of the French people in Paris, and survived the battle of "Garçon le Check," left the capital on the afternoon of July 5th. As the train sped onward to the front the rumor started that the 39th Regiment had gone into action. Upon arrival at Lizy-sur-Ourcq orders were received by the battalion to join the regiment in the "alert" position not far behind the front line. The men detrained in the evening and started a forced march through their old billets at Acy-en-Multien, to the second line positions. In the dark of night they ploughed their way over fields and through woods, up and down hills, with an occasional German shell falling along the road. Dawn found them in position. That day the order came from the Commanding General, II Army Corps (French), to which the brigade was attached, "the alert position will be discontinued at six P. M. this date and the normal order of affairs will be resumed. Troops will be returned to their stations. Marches to begin after dark." All the regiment, except the 3rd Battalion, which remained in the trenches until the 7th of July, then marched back to Acy-en-Multien that night.

The other regiment of the 7th Brigade, the 47th, and the 11th Machine Gun Battalion also moved up to positions in reserve trenches on July 5th, marching to Chenévière farm, Boullarre and to Neufchelles on the Ourcq front and returning the next day to their regular stations. This "alert" maneuver was a necessary part of the training for the difficult task of placing a unit in line at night.

The 8th Field Signal Battalion, whose first section of the outpost company was ordered to report to the 47th Regiment on the night of July 5th, long remembered the experience. Revolvers had just been received by the supply officer who crawled out of bed, in answer to the alarm, to make the issue of these weapons to the section. Unfortunately, holsters had previously been issued which proved too small for the weapons and the men set out looking like old-fashioned pirates with the revolvers draped on their persons in any place they seemed likely to stick during the march. But they were on the job.

The Division, less the 7th and Artillery Brigades, moved, on July 5th, to the region of the Ourcq canal which was to be
the new area. Division Headquarters was established at 69
Lizy-sur-Ourcq. The 59th Regiment marched to May-en-Multien, where it went into shelter camp in the woods near the town and just above the canal. The men found themselves in the midst of the French heavy artillery concealed in the woods, and became accustomed to the noise and reverberation, day and night, of the large guns. The 12th Machine Gun Battalion, too, camped in this area, while the 58th Infantry was distributed through Vendrest, Villers-les-Rigault, Lizy-sur-Ourcq and Rademont. The 10th Machine Gun Battalion found a temporary home at Ocquerre. On June 21st Companies "A" and "C," 4th Ammunition Train, with their complete motorized equipment, joined the Division.

With the Division now in support position, visits to the front lines on tours of inspection became more frequent. During one of these visits two members of the Division were killed, a lieutenant and sergeant. The occupation of second line trenches also began at this time. All the daylight hours were devoted to rigorous training.

The severe training showed its effects in the orderly maneuvers of the whole Division, in the proficiency which the men were rapidly gaining on the target ranges, in the "snap" of their bayonet work, in their well set-up figures and their alert soldierly deportment, and in intelligent execution of open warfare movements. The whole Division was ready for any call.

The engineers at this time were busy in actual preparations for what was believed to be an impending German attack. On the night of July 6th-7th they left the Crouttes area, where they had built the reserve defenses for the Marne, and moved by camions to the Ourcq area, where, around Vernelle, Crouy, Rademont, and Certigny they commenced to throw up a defensive position, digging trenches, stringing barbed wire, and building strong points in anticipation of a German attack.

The artillery meanwhile remained at Camp de Souge, Bordeaux, acquiring that technical knowledge necessary for the effective utilization of their weapons.

The Fourteenth of July dawned, the great day when France celebrates the taking of the Bastille. Paris, apprehensive for her immediate future, in expectation of another German drive toward her gates, cheered heartily for the American and other Allied troops who marched through her streets. Rumors of intended evacuation of the capital and the proclamation of a state of siege went from mouth to mouth as the people lined the streets to welcome the marching soldiery. The men of the 4th Division did not take part. Waiting behind a front that might at any moment blaze into the frenzied activity of a drive they returned the gracious hospitality of their friends and gave the French officers and non-commissioned officers a banquet in honor of the Fourteenth. It was to be the last banquet for a long time.

CHAPTER VI

THE AISNE-MARNE OFFENSIVE

ON THE night of July 14th-15th the people of Paris rose from their beds to hear the dull, threatening roar that came from the direction of the Marne. They guessed, if they did not actually know, what was happening. It was the great German drive for Paris which the Crown Prince had launched on the evening of the National Fête day of France, the last drive the Germans were to make in the war. This offensive contemplated two operations; a strong one east of Rheims launched against the Fourth French Army, and a somewhat smaller one between Château-Thierry and Rheims against a combined force of French and Americans. 70

On the morning of July 15th Big Bertha opened fire on Paris and a large siege gun mounted on a turntable in the Bois du Châtelet, near Brécy, shelled Meaux, Coulommiers and La Ferté-sous-Jouarre with 380-millimeter shells. The Crown Prince threw his First and Third Armies, containing 29 combat divisions, against the Fourth French Army, commanded by General Gouraud, east of Rheims. At the same time the Seventh German Army of 21 divisions attacked between Château-Thierry and the River Ardre, southwest of Rheims. 71

The German bombardment which preceded the attack in Champagne, on the morning of the 15th, was particularly intense and lasted more than ten hours, but Gouraud was ready. A daring raid, on the night before the attack, had provided him with such valuable information that he was enabled almost to paralyze the drive before it started. The prisoners captured in the night raid revealed how, in two hours, the German artillery preparation would commence and how,

at that moment, the trenches from which the Germans were
to depart were the scene of thousands of men struggling along
in the dark finding their places for the attack. Gouraud
made his plan rapidly. The French artillery was quickly
informed and, half an hour before the German artillery
opened, the French anticipated it with a concentrated fire on the
front and rear areas of the enemy that took a heavy toll of
life and brought great confusion to the German army. Never-
theless the Germans launched their drive, but three hours
passed before they actually reached the combat line of the
French, for Gouraud had withdrawn his forces from the front
line to his principal second line of defense and an intermediate
line was held only by small groups of delaying detachments.
The German attack was met with an avalanche of machine
gun and artillery fire but it was nevertheless pushed with
great vigor and met with desperate resistance. All day long
the battle raged. At the end of the day the German hope
of taking Châlons and Epernay had faded and the drive was
72 definitely stopped.

Meanwhile, between Château-Thierry and Dormans, the
Germans had crossed the Marne but as soon as the fail-
ure of the drive east of Rheims was realized they were
73 ordered to retire. Before this, however, they had been
stopped on the left by the American troops of the 3rd Divi-
74 sion. This second German operation had been preceded by
an intense bombardment, 84 German batteries playing
all their fury into the 31 batteries of French and American
artillery. The crossing of the Marne was effected on steel
75 platforms held up by cables, on canvas boats and pontoons.
The French on the right were driven in by the weight of the
attack but on their left one of the American regiments—the
74 38th U. S. Infantry—gallantly held. Then came the order
to the Germans to retire, and at eight o'clock on the morning
of the 16th, the enemy were scrambling back across the river,
retaining a hold on only a small strip of the south bank. The

last German elements recrossed on the night of July 20th-21st.

This crossing of the Marne, on such a narrow front, was so hazardous an undertaking that the Germans must have felt supremely confident that the great attack, east of Rheims, would succeed.

The 4th Division was in the Ourcq sector, northwest of Château-Thierry, ready to occupy the reserve positions in case of a possible drive on that front, but although a few casualties fell to the Division it did not then have the opportunity to measure its strength with the enemy. The turn of the Ivy Division was soon to come.

True to the military maxim that only by attacking can final victory be secured, Foch had made preparations at the beginning of July for a Franco-American offensive on a twenty-five mile front from the Aisne to the Marne, with the object of relieving the menace to the capital. It was to be launched by the armies of Generals Mangin and Degoutte, reinforced by American divisions. Although the 4th Division had heard rumors of a French army being secretly massed behind the Forêt de Compiégne, nothing was known with certainty, and the Germans themselves never penetrated the secret of the Allied Commander-in-Chief. Before the Crown Prince began his drive the offensive had been fixed for July 18th. This drive fitted in admirably with Marshal Foch's plans, as it lengthened the German flank. When Gouraud stopped the Germans on July 15th-16th, Foch ordered the great counter offensive, which had been in preparation for more than two weeks, to begin, exactly as planned, on the morning of July 18th, but additional divisions were rushed up to add all 76
possible force to the blow directed against the German flank. General Mangin commanding the Tenth French Army, to which were assigned the 1st and 2nd American Divisions, was to attack due east and drive in the German flank south of Soissons—to form the neck of the bottle. The two American divisions, fresh, eager to demonstrate their fighting prowess on

the European battlefields, were selected together with the Moroccan shock troops, to form the spearhead of the attack.

South of Mangin's Tenth Army and supporting it in the great drive was the Sixth Army of General Degoutte, under whom the 4th Division was to receive its baptism of fire.

During the action of July 18th-22nd the 4th Division did not function as a unit. The Division Commander exercised no command. The units of the Division were brigaded with French troops; the 7th Brigade with the II French Corps and the rest of the Division with the VII French Corps. The sectors of the two corps were adjacent and extended in their full length from Ancienville to Hautevesnes and lay west of the main highway from Soissons to Château-Thierry, the spine of the German positions which were to be attacked. The II French Corps, to which the 7th Brigade was attached, occupied the northern sector.

It was a terrible night that preceded the dawn of the attack. The roads became almost impassable under the torrents of rain. The darkness and insecure foothold were formidable obstacles in the speedy movement of troops, artillery, and supplies, which nothing but the splendid endurance of the men and the enlivening prospect of a fight could overcome.

The men of the 4th Division were on the move in the valley of the Ourcq. In the 7th Brigade the 39th Regiment had left Acy on July 16th for the Forêt de Retz under the impression that it was destined for another tour in the trenches. But on the afternoon of July 16th the 33rd French Division, to which it had been attached, was ordered to relieve French troops then in the front line. The next morning the 39th, with companies "A" and "C" of the 11th Machine Gun Battalion, received orders to move forward that night to relieve the 11th French Infantry, 33rd French Division, in the front line, from a point just north of the Ourcq, at the eastern edge of the village of Troësnes, to a point two kilometers due north,

on the northern bank of the Saviéres River. Each battalion was to have two companies in the front line and two in support, the 2nd Battalion on the right, the 3rd in the center and the 1st on the left. The relief was made with great diffi- 77
culty. Staggering along the roads in the inky darkness, their eyes filled with rain and perspiration, swinging in and out of the lines of transport, holding on to each other's packs so as to keep the chain intact, falling down shell holes, catching, in the murky darkness of the night, a glimpse of a flash of red which told of the firing of a big gun, straining their ears to find out just where a German shell hissing through the air was going to land, the men moved slowly on until they were all in the trenches and settled down to snatch a fugitive bit of sleep. That night the French issued orders to the regiment for attack in the morning. It was to capture the Buisson de Cresnes immediately in its front, attacking an hour after the French had gone forward. It was the intention of the French to pinch out the ground between Ancienville and Noroy and thus leave little opposition to face the 39th. Zero hour for the French was 4:30 A. M.; for the Americans 5:30.

At half past four the French opened a tremendous barrage with seventy batteries. The men of the 39th, new to warfare, listened with bated breath to the incessant whistling of the shells overhead. All the fury of war's tumult was let loose. There would be a second's lull and then the roar of passing shells would recommence. As dawn was breaking the men looked over the top to see the German trenches four hundred or five hundred yards distant wreathed in a mass of bursting shells and concealed in the flying dirt and smoke of the explosion. They wondered how any one could live through such a fire and congratulated themselves that it was the enemy and not they who were crouching beneath it.

The French had already gone over when the Germans opened a counter-barrage on the American lines. Trench mortars and artillery joined in malicious unison and the maintenance of communication with the rear became more and more difficult. The orders having been changed at the last moment, Major Henry Terrell, Jr., commanding the 1st Battalion did not move it forward until about eight o'clock in the morning. The 3rd started at nine and the 2nd at four in the afternoon.

The Buisson de Cresnes, a large thicket on a hill, its trees stripped of their foliage by shell fire, would have proved formidable for any frontal attack had it been at all strongly organized. Except for strands of barbed wire, however, the Germans had not built any real protection on the western face of the woods.

In the immediate front of the 1st Battalion lay the Saviéres River, a narrow but deep stream with a bottom of soft mud. The banks on either side were marshes. Some of the men waded but American ingenuity came to the rescue and, on poles thrown across the water, the whole Battalion reached the other side. It then entered the Buisson de Cresnes from its western edge, meeting with little opposition. The Germans thought that no one would be foolish enough to attack over the swampy ground, and had their machine guns pointing in the direction of the Ourcq and the south. Ascending the hill from the river the 39th took its first machine gun. Two German machine gunners camouflaged in a wood pile opened fire on the left flank of Company "A." A sergeant of this company, who was later killed in the Forêt de Fére, discovered the nest, rushed it, and threw in a hand grenade. The wood pile fell over revealing two startled gunners. Before they could throw up their hands they were riddled with bullets. Another isolated machine gun met early in the day was in a glass house with sliding windows, built in the fork of a large oak tree.

The men moved steadily eastward into the northern half of the Buisson de Cresnes, meeting only with promiscuous rifle and machine gun fire from the supports of the troops which held the southern face of the woods. As the 3rd Battalion was not to move forward until some time later, the 1st sent two platoons to cover its right flank. The whole battalion then advanced, through the woods, up the hill in front of it. In about two hours, more than one hundred Germans were prisoners and several of their machine guns were prizes in the hands of the 1st Battalion.

When, at nine o'clock, the 3rd Battalion advanced under the command of Lieut. Col. Robert H. Peck, it found that the 1st had already cleared the front and all that the companies could do was to crowd into the sector of the 2nd Battalion farther to the right. Then at 4:00 P. M. came the 2nd Battalion commanded by Major Manton C. Mitchell. It had suffered losses by the artillery of the Germans while it remained in the trenches. When it advanced practically no opposition was met and only one machine gun was taken. Unfortunately in their eagerness to get to the enemy all the units by this time had become intermixed and overcrowded and reorganization became immediately necessary.

Meanwhile the French had not progressed as rapidly as they had expected and, in the afternoon, no French troops had entered Noroy which was one of their objectives. Indeed, as the day went on, the Germans had recovered from their surprise and stiffened their resistance. The French, expecting a heavy counter-attack, sent a request to Colonel Bolles, commanding the 39th Regiment, for help in the region of Noroy. While the reorganization of the battalions was still in progress, the colonel sent word down the line that a glorious Allied victory had been won, and, cheered and enthused by this, the 2nd and 3rd Battalions, with the 1st in support, moved to the help of the French. The 3rd Battalion advanced on Noroy, which is a small village at the head of a road running in a northeasterly direction through the southern part of the Buisson de Cresnes. One company of the battalion passed through the village without serious resistance and emerged on the other side. Here it was stopped by machine-gun and direct artillery fire from a gully about 300 yards distant. The company maintained its position. When darkness fell another company came up in support and was sent to the left of the village. There it came under a vicious bombardment by small enemy cannon but held its ground. The rest of the companies remained at the edge of the woods. The 2nd Battalion, after making certain that the 3rd had taken Noroy, advanced through the woods, to the south of the village, and along both banks of the Ourcq River. Here the battalion commander sent out patrols and outposts to protect his position which was in liaison with the French on the right.

The 39th dug shallow ditches and "fox-holes" to crawl into for the night, and the men, tired with their exertions and the nervous strain, lay down to reflect on their first day of battle. They slept with German shells falling around them during the night.

Companies "A" and "C," 11th Machine Gun Battalion, which went into action with the 39th at Noroy, fought with particular distinction and won the generous praise of the entire regiment. The 47th Infantry did not participate in the fighting. It was held in reserve in the vicinity of Boullarre Rouvres, and Etavigny, and later occupied the old front line position east of La Ferté-Milon, between La Loge aux Boeufs and Mosloy. Its turn was to come later.

During the night the Germans reorganized their forces and prepared new defenses. This the 39th discovered when, on the morning of July 19, the men moved out of the fox-holes where they had spent the night. The 2nd Battalion was on the right along the Ourcq River, the 3rd in the center, on the eastern edge of Noroy, and the 1st on the left facing Chouy. At 4:00 A. M. the 3rd Battalion moved forward from Noroy, without any preparatory fire and, overcoming determined resistance, captured a battery of artillery in the gully from which the battalion had been stopped the day before. But a tragedy intervened to mock its success. About 5:15 A. M. a message reached the regimental commander that the French had changed the time of the attack to 5:30 and that a barrage was to precede it. Hurriedly information was sent back by runners to tell the French that the position was already taken by the 39th and to advance the barrage. But it was too late. Soon the French shells came thick and fast. They passed through the American line and beyond, causing a considerable number of casualties. Small rockets were fired but all to no avail. The battalion nevertheless moved forward, undaunted by such an unnerving experience. Though opposed by heavy machine-gun fire, the whole line of the 39th made steady progress. The 2nd Battalion on the right had by far the most difficult terrain to cover. In its front lay the marshes, hills, woods and winding valley of the Ourcq. As soon as the assault line reached the road running south from the eastern edge of Noroy the Germans commenced shelling heavily. The right of the battalion had to pass through small but dense patches of woods where the difficulties of advancing made the men fall behind. The left, on the other hand, became immersed in a swamp and was soon suffering from the dual horrors of German machine-gun fire in front and friendly artillery fire from the rear. It was necessary to move out of the swamp. A platoon was sent forward which captured the machine guns immediately in front. Then the whole line advanced once more to a point nearly south of Chouy without further difficulty.

Toward evening the 1st Battalion without opposition entered the small village of Chouy, which was now on fire. The Germans had withdrawn when the French passed it on the left and the 3rd Battalion, 39th on the right.

By dusk the 39th was established on a general north and south line just east of Chouy, where the men dug in for the night. Companies "F" and "H," however, never came up. They became separated from the front line by the barrier of a French barrage and were led back to the rear by a French officer. During the night the 39th was relieved and returned to the Buisson de Borny for a rest. It had lost 2 officers and 54 men killed and 9 officers and 218 men wounded.

The regiment, although never under fire before, performed its task with a coolness and assurance that made its reputation as a fighting unit. The officers of the regiment are entitled to all praise. But to the men, who endured the fatigue of physical and nervous exhaustion, who asked nothing but that they should defeat the enemy, the credit for the fine achievement of the 39th is undoubtedly due. The regiment was cited for its conduct by General Tanant, commanding the 33rd French Division, under whom it fought. The 39th with its Colonel received the Croix de Guerre and the felicitations of General Philipot of the II French Corps.

Meanwhile the 8th Brigade was fighting with the French to the south. The 58th and 59th Infantry and the 12th Machine Gun Battalion were attached to various units of the 164th French Infantry Division. The infantry battalions were used as tactical units and the machine gun companies so distributed that each battalion had a machine gun company attached to it. Four companies of the 4th Engineers were attached to the attacking troops for the organization of the conquered terrain. The 10th Machine Gun Battalion was placed in reserve for the French Division. The 8th Field Signal Battalion actively aided in maintaining lines of communication. The trains were busily employed under the orders of the French.

The march to the front line on the night of July 17th was trying beyond description. A strong wind drove the rain through the soldiers' clothing and blinded their eyes. The night was pitch black; the roads were cluttered with transport—wagons, trucks, automobiles, mounted men, foot troops. The mass of this transport was moving to the front; occasionally a truck or wagon would be moving to the rear. Troops would be made aware of the presence of wagons or trucks in their front only by bumping into them. Progress was necessarily at a snail's pace. The frequent flickering of the lightning and the livid flashes of the guns, which now and then illuminated the darkness and indicated the road, were all that troops had to march by. The guns of the Allies were firing intermittently, the usual desultory night shooting, and although the Germans, who had observed the artillery concentration, did not fail to bestow upon them a full measure of attention there was little to indicate to experienced troops that a great counter-offensive was about to commence. But the men of the 1st and 2nd Battalions of the 58th and of the 12th Machine Gun Battalion were far from experienced and the roar and confusion of the whole scene, just as dawn was rising and little could be seen but mist, led many of them to think that the actual battle had begun.

The plan of battle, as drawn by General Gaucher, commanding the 164th Division, contemplated a partition of his forces into three large groups, a right group, a left group and a reserve group. Each of these groups contained three French battalions and one American battalion. Between the right and the left group was a liaison battalion. The right group was in command of Lieut. Col. Kiffer, of the 133rd French Infantry Regiment, the left group in command of Lieut. Col.
Dussauge, of the 13th Group, Chasseurs à Pied. The three 78
American battalions thus attached to the three French groups were the battalions of the 58th Infantry. The 1st Battalion, under Major Samuel H. Houston, afterward killed in the fighting on the Vesle River, was attached to the right group, the 2nd Battalion, under Major Gilbert R. Cook, to the left group, the 3rd Battalion, under Major Charles C. Drake, to the reserve group. With each battalion was a company of machine guns—the regimental machine gun company and companies "A" and "C" of the 12th Machine Gun Battalion. General Gaucher drew no dividing line between the area of operations of the right and of the left group. The commander of each was to take ground or give ground on his inside flank, in favor of the other group, as might be made necessary by circum-

stances. The 59th Infantry was held as an additional reserve. The French artillery laid down a heavy barrage, and at 4:35 A. M. the 58th Infantry climbed out of the holes that were dignified with the name of trenches and, flitting like ghosts in the heavy mist of the early dawn, sought for the enemy they had come thousands of miles to meet. The previous roar passed into utter insignificance before the thunder of that barrage, and the earth seemed to quake as shell after shell poured into the German lines. But the French had no monopoly of artillery fire. Within a minute or two after the Allied troops had left their trenches, down came the German barrage all along the line—a perfect hail of shells through which the troops were forced to pass. It was here that the French liaison officers attached to the battalions rendered valuable aid. Marching with the battalion commanders, these experienced veterans indicated to the majors the best methods for passing through the German barrage with a minimum of loss.

The 1st Battalion, as support with the right group consisting of the 133rd French Infantry Regiment, assisted in the taking, at 5:00 A. M. and after a severe local fight, of the town of Hautevesnes which crowned the crest of a high hill. Later in the day Courchamps was taken, in spite of determined resistance, after a short barrage had been laid down on the village. That night the Battalion bivouacked in the Ravine
79 de la Folie, on the western edge of Courchamps.

The 2nd Battalion, reinforced by the Machine Gun and Headquarters Companies, moved forward at 4:35 A. M. and took Chevillon at 5:30 A. M. The Battalion was the front-line element in the center of the left group, commanded by Lieut. Col. Dussauge. Chevillon is only a small and very obscure village lying in a valley amid the low hills and clumps of woods that form the landscape in that part of France. East of Chézy-en-Orxois and just east of the railroad that runs in front of that village, the 2nd Battalion started its acquaintance with real war. Chevillon lay right in the path of its advance. Just west of the valley in which Chevillon lies is the crest of a long, high ridge running north and south. Over this crest the battalion passed in its advance on the village, taking two lines of trenches, and it was here that it was met by a heavy concentrated machine gun and artillery fire that caused many casualties. The Germans fell back to the next crest—Hill 172. Undismayed by its losses the battalion pressed on to the village. There was a brief, decisive fight and Chevillon was captured, together with 75 prisoners. General Gaucher, commanding the 164th French Division, reported the village as taken "in a splendid dash."

Leaving the village behind, the battalion, now somewhat weakened in strength, pushed on eastward through minenwerfer and machine-gun fire, up the slope of Hill 172. On its 80
right was the 41st Battalion, Chasseurs à Pied. The 43rd Battalion, Chasseurs à Pied, which had started on its left, was still far in the rear. The Germans had entrenched Hill 172 very strongly with three lines of trenches, which converged on the southern crest of the hill and formed a single line running northwest and southeast. Against this system the battalion threw its strength. Over the trenches they went, passing through a hail of machine-gun bullets, bayonetting the enemy when resistance was offered. Twenty-five Germans were captured in the first trench. The battalion now divided, the right half moving slightly southeast, the left half more to the northeast in the direction of the Sept Bois.

Between Hill 172 and the Sept Bois, the ground was covered with wheat, waist-high. When the left half of the battalion approached the woods, the Germans, in strength, emerged and counter-attacked, capturing one officer and several enlisted men. The remainder retreated to Hill 172, reorganized and, in the afternoon, with the assistance of machine guns, advanced again and gained a foothold in the scattered clumps of trees that form the Sept Bois. There was no contact with French units on the left; these had not come up.

An officer of the Machine Gun Company, following 600 meters behind the rear wave of this portion of the battalion, saw what he took to be a French medical soldier standing in the wheat some 250 meters to his left rear. To secure the assistance of this man for his wounded, the officer went over to him, only to find that the "Frenchman" was a German wearing a French helmet, blouse and Red Cross brassard and that there was a machine gun, manned by four more Germans, hidden in the wheat at his feet. The camouflaged German, poking his automatic into the officer's abdomen, informed him, in perfect English, that he was a prisoner and would be shot with the other captured Americans after they had divulged what information they could. Fortunately a party of Americans made their appearance at this juncture, and opening fire, afforded the officer an opportunity to escape. As he crawled off into the wheat one of the enemy, before retreating, jumped toward him, knocked him over the head with the butt of his
81 rifle and left him unconscious.

The right half of the battalion, inclining toward the southeast, was exposed to the view of the enemy's observation posts on the high ground in the Bois de l'Orme. In a few minutes it came under heavy artillery and minenwerfer fire.

Almost deafened by the constant detonations, horrified by the vision of mangled bodies, the men hurried forward over the rough ground. Gasping for breath, they reached the unimproved road 300 yards west of the Bois de l'Orme, where they threw themselves on the ground to catch a moment's rest. Part of this road is sunken, part cut from the high bank on its eastern side. Before them was the Bois de l'Orme. Now the small arms projectiles began to come, not from the ground but apparently from the air. The mystery was explained as they drew near the wood. The Germans, with their usual ingenuity in machine gun fighting, at which it must be confessed they excelled, had constructed platforms amid the pine trees, on which they had placed machine guns.

Then came the deluge. As the troops approached the wood the Germans turned their artillery and machine guns on the Americans with extreme force. Exhausted by their previous efforts, stunned by the storm of projectiles, their ranks depleted, the command became disorganized. Suffering severe casualties, the men fell back to the road they had just crossed. Here, under the shelter of the sunken road and the high bank, they halted and were reorganized.

This occurred about 10:00 A. M. During the remainder of the day this group prepared its position for defense and established communication with the French battalion on its right. During the evening the 2nd Battalion was relieved by the 3rd Battalion, 59th Infantry, Major Louis Farrell commanding, and at 7:00 A. M., July 19th, marched back to 82
Chézy. The front line that night extended along the north and south road west of the Bois de l'Orme.

The initiation of the 2nd Battalion, 58th Infantry, and of the Machine Gun Company and elements of the Headquarters Company attached to it, had been a severe one. Two officers and 93 men had been killed, and 11 officers and 436 men wounded. One officer and 66 men were missing; a total loss of 609 out of 1281—nearly 48 per cent. of the strength. 80

On the night of July 18th the 3rd Battalion, 58th Infantry, bivouacked on the southern slope of Hill 172. 80

During the day of the 18th the battalions of the 59th Infantry had been moved up close to the battle line. At 11:00 A. M. the 1st Battalion, under the command of Major J. C. Williams, was placed in a supporting position about 800 yards east of Hautevesnes. Upon orders from Lieut. Col. Kiffer it moved forward at 9:00 P. M. to relieve one French battalion in the front line, between the Bois de l'Orme and the village of Courchamps. The relief was completed at 11:00 P. M. Liaison was established with a battalion of the 133rd French Infantry on the right and the French liaison battalion on the left. The battalion immediately dug in. 83

The 2nd Battalion of the 59th, Major H. A. Musham commanding, was placed under the orders of Lieut. Col. Dussauge and by him moved, about 6:00 P. M., to the northeastern slope of Hill 172, about $1\frac{1}{2}$ kilometers east of Chevillon, where it
83 dug in for the night.

Thus, at midnight of July 18th, the elements of the 58th and 59th Regiments and the 12th Machine Gun Battalion, were disposed as follows from right to left: 1st Battalion, 58th, and Company "C," 12th Machine Gun Battalion, in the Ravine de la Folie, both in support; 1st Battalion, 59th, and Company "D," 12th Machine Gun Battalion, between Courchamps and the Bois de l'Orme, in front line; 3rd Battalion and Machine Gun Company, 59th, in front line, west of the Bois de l'Orme and separated from the 1st Battalion by the 41st Battalion, Chasseurs à Pied; 2nd Battalion, 59th, and Company "B," 12th Machine Gun Battalion, in front line, on the left of the 3rd Battalion; 3rd Battalion, 58th, and Company "A," 12th Machine Gun Battalion, both in support, on the southern slope of Hill 172. The 2nd Battalion, 58th, and the attached units were preparing to march back to Chézy.

The first day of the great offensive had seen the Allies not only completely surprise the Germans but also reap a great success by their dash and power. In this success the American troops had fought with an audacity and recklessness that had shocked the Germans into a realization of what the entry of America into the war meant. The men of the 4th Division who had spent their first day in action, not as a division but as battalion units under French command, had shown a contemp for danger and a resolution that had impressed the French as little else could have done.

It was not only the troops in the front line but also those which, like the 47th and 59th, served as reserve battalions or had other important duties in the rear, who suffered casualties on that day. The engineers organizing the captured territory; the signal battalion maintaining communication under heavy artillery fire and assisting as stretcher bearers; the medical service attending and evacuating the wounded; the ammunition and supply trains operating under the orders of the French; all worked with a devotion to their tasks that reflected nothing but credit on the Division and its commander.

Nor was the fighting without its humor. In the sunken road in front of the Bois de l'Orme a French soldier was found shining his shoes with American polish. When asked why he was doing this, he said, "I have found this polish. It is good polish, and I may not get another chance to use it." A little farther on an American soldier had dug a pit in the side of a bank and was calmly seated with his back to the enemy, shaving from his tin cup.

Early next morning the advance was continued. The Germans had recovered from the first shock of surprise and, quickly reorganizing their retreating troops, they threw up new defenses and were in a better position to offer resistance than on the first day.

On the front of the 164th French Division the 58th Infantry had seen much hard fighting and had lost many men. It had been severely tested. The 59th Infantry was now to undergo a similar test.

For the operations of July 19th the Commanding General, 164th Division, had designated as a first objective the unimproved north and south road between Sommelans and Monthiers. Beyond this line battalions were to advance as far as a line through Halloudray Farm and Bonnes, without regard to the progression of units on their right or left. 84

At 4:20 A. M., July 19th, the men of the 59th Infantry were awakened by the toe of somebody's prodding boot. The order had just been received that the general advance would begin at 4:25. No one had time to estimate the situation or to rub tired eyelids. Seizing their rifles, and only partially awake, they sought their places in the half-light of the gray dawn. Many forgot their bandoliers of ammunition, which they had flung off for the night. Down came the French barrage at 4:25. To the soldiers everything seemed in wild disorder. But they went forward behind the barrage. Soon black bursts in their ranks and flying pieces of steel told them that the enemy was sending over a counter-barrage, and then momentarily hugged the ground for protection. Their advance carried them up and down gently sloping hill-sides, through fields of wheat, waist-high. The whistle of small arms bullets was everywhere. The rat-tat-tat of the machine guns could be heard above the roar of the artillery. Blue cross gas hung heavy in the valleys. But no enemy could be seen. Not a German was in sight. The men began to fall, first singly, then in groups. From all sides came the cries of the wounded. Forward they lunged, firing as they went, usually without aim. All they wanted was to see their opponents, to get at close quarters with them. They paid the price. Soon the ammunition gave out. But they did not stop. They were looking for the enemy, to kill him with the bayonet. The hostile machine gun positions were reached and passed. They were empty except for a few dead Germans and some abandoned guns. Then the rifle and machine gun fire from the enemy doubled in volume. The intensity of German artillery fire increased. The advance was temporarily stopped.

The casualties among the officers and men had been heavy. One French officer remarked: "Had I not seen it with my own eyes, I would never have believed that green troops would advance under such fire."

The line of advance of the 1st Battalion, 59th, lay due eastward between Courchamps and the woods of la Remise. It found its progress disputed with a determination that cost many lives. On the crest, about 200 yards east of the Courchamps-Priez road, the battalion was definitely stopped by German machine gun and artillery fire. The battalion commander was wounded, and seven other officers were either killed or wounded. Every movement drew intense fire. Disheartened by the loss of their officers, the men fell back to the Courchamps-Priez road, where they dug in. Here the battalion remained until 5:30 A. M., when it was relieved by a French battalion of the 152nd Infantry Regiment and moved back to its original position between Courchamps and the Bois de l'Orme. Here it was reorganized and, late that evening, returned to its fox-holes on the Courchamps-Priez road as a support for the French battalion which was then dug in on the crest 200 yards to the east. 83

At 4:25 A. M. the 3rd Battalion, 59th Infantry, under cover of a strong barrage, attacked in conjunction with the French battalion on its right. Its route was due east, through the Bois de l'Orme, on la Grenouillère Farm. The Bois de l'Orme was occupied by only a few scattered German machine guns. These were soon wiped out, and the battalion, in good order, advanced through the enemy barrage across the Courchamps-Priez road to the high ground about 300 yards east of that road. Here a withering machine gun fire was encountered which took a heavy toll of lives. Unable to advance farther or to overcome the hostile fire with the weapons at their command, the men dug in. German artillery then reinforced the machine guns. High explosives and gas were showered on the Americans. The battalion commander was badly wounded, and fifteen other officers were either killed or wounded. For thirty minutes the battalion held its position under this hail of missiles and then retreated to the high ground between the Bois de l'Orme and the Courchamps-Priez road, where it again dug in. Throughout the rest of the day it remained here, under heavy artillery fire. Upon order from Lieut. Col. Dussauge the battalion was placed at the disposition of the Commanding Officer, 59th Battalion of Chasseurs à Pied, at 10:00 P. M., and at 11 o'clock that night was moved forward and dug in along a line facing slightly northeast, with its left resting on the Courchamps-Priez road at a point where the latter is joined by the road from Orme Signal. It was in support of the French battalion. 83

The advance of the 2nd Battalion, 59th, which was on the left of the 3rd Battalion, led first into the valley in which the Bois de Cassel is situated. On the sides of this valley the wheat was almost breast-high. The Germans had placed a long line of machine guns at the bottom of the very gradual slope. These were completely hidden by the tall wheat. Not an enemy could be seen. The Americans were permitted to come half-way down the slope; then all the machine guns opened at a preconcerted signal. They fired low, shooting through the wheat. The slaughter was terrible. Hardly a soldier was hit in the head; nearly all the dead had chest or abdominal wounds. But the men pushed forward in the face of this fire with magnificent gallantry, overran the machine guns, killing the gunners at their posts. On they went over the crest, crossed the Orme Signal-Montmenjon road and entered the next valley. From the Bois de Cobourg and the high ground south of it German machine gunners rained bullets upon them. Our soldiers could endure no more. They fell back behind the crest they had just crossed, to rest and reorganize. About noon they moved forward again and occupied the Bois de Cobourg and the Courchamps-Priez road. At 4:00 P. M., in conjunction with the French battalion on its left, the 2nd Battalion once more advanced and took the Bois de Leipsig. Here it entrenched and remained for the
83 rest of the day.

In the meantime the 1st Battalion, 58th Infantry, had supported the attack of a French battalion of the 133rd Infantry Regiment, launched from the eastern edge of Courchamps. This attack was stopped on the crest 800 meters east of Courchamps by fire from the Bois Pétret and la Remise. The battalion dug in about 700 meters east of Courchamps and stayed there. The 3rd Battalion, 58th, remained in essentially the same position it had occupied
83 on the 18th.

During the night of July 19th food and ammunition was

brought up, and the battalions were reorganized and made ready for the struggle on the following day.

On the morning of the 20th the attack began at 4:30. The 1st Battalion, 59th Infantry, remained throughout the day in support of the French battalion of the 152nd Infantry Regiment, and occupied trenches along the Courchamps-Priez
road. 83

The 3rd Battalion, 59th, acting under the orders of the French commander of the 59th Battalion of Chasseurs, attacked, with a French battalion on its right, in the direction of Hill 184. La Grenouillère Farm and the woods of la Remise were taken after a hard fight. Throughout the day this position was held by the battalion. The French, in the meantime, pushed forward. At 6 P. M. the battalion, under orders, occupied a position northwest of the town of Bonnes on the Bonnes-Sommelans road. Here it was, by mistake, fired into from the rear by machine guns and 37 mm. guns from five French tanks. At 9:30 P. M., acting under orders from the French, the battalion moved back to Hill 184 and reoccupied the trenches there for the night. 83

The 2nd Battalion, 59th, had been placed in support of the French and on the afternoon of the 20th moved forward to the western edge of the village of Sommelans, where it remained during the day.

The 1st Battalion, 58th Infantry, in support of the 133rd French Infantry Regiment, attacked at 3:30 A. M. after a 30-minute artillery preparation, in the direction of Pétret Farm and the Bois Pêtret, from which points a heavy machine gun and 37 mm. gun fire had stopped its advance on the 19th. The attack progressed without effective resistance, as far as Pêtret Farm, when the Germans counter-attacked strongly from the Bois Pêtret. They took the French line in flank and forced the French and Americans back to the hill-crest 800 meters east of Courchamps. The German machine gun fire was extremely effective, coming from both flanks. During the day our artil-

lery and 37 mm. gun fire bombarded the German machine gun positions. At 4:00 P. M. our troops again advanced and occupied, with little opposition, Pêtret Farm, Bois Pêtret and the crest on which is the Monthiers-Sommelans road. Along this road the 1st Battalion took up an outpost position for
83 the night.

The 2nd Battalion, 58th, remained in Chêzy until 4:00 P. M., July 20th, when it marched to Hill 172 and occupied trenches there. The 3rd Battalion did not change position.

During the night of July 20th-21st orders were received relieving the units of the 4th Division, except the 10th Machine Gun Battalion, from duty in the front line and directing their assembly in the vicinity of Brumetz, in the valley of the Clignon River. At 2:00 A. M. July 21st the march began, and that evening the troops were back under the wing of their own
85 division.

The losses of the 59th Infantry were 9 officers and 172 enlisted men killed and 29 officers and 831 enlisted men wounded;
86 a total of 1041 casualties.

The 10th Machine Gun Battalion remained in action until July 23rd, serving as a supporting unit for the French. Its advance carried it as far forward as the Bois du Roi, just west of the Château-Thierry-Soissons road. The Germans, unable to withstand the driving power of the Allied troops began, on July 21st, to withdraw from the Marne salient. When the 10th was relieved the Bois du Châtelet had already been entered and the main highway, from Château-Thierry to Soissons, had been crossed. Fourteen kilometers of French territory had been recaptured and the menace to Paris once more removed.

The success of the Division had cost 15 officers and 382 enlisted men killed and 63 officers and 1806 enlisted men wounded. One officer and 66 enlisted men were captured by the enemy or were missing; a total loss of 2333.

One soldier, in particular, could look back with satisfaction to his part in the fight. He was freckled of face and his hair, if not exactly red, was inclined to be sandy. He was a quiet individual and did not howl when he found his name on the K. P. list. No one thought much about him, but when the wounded were lying on the battlefield, he was one of the first to volunteer to go and get water two kilometers distant. On the morning of July 19th, when the 59th, his regiment, attacked, every officer in the company with the exception of one was either killed or wounded during the first hour. The second lieutenant who was left in charge was at his wit's end. The game was new to him and his men were scattered all over the country. He had managed to round up two platoons in pretty good shape and get them in position to jump off for their second objective, for they had taken the first within the first hour. Sweating, fuming, running up and down the line, he begged and prayed for assistance from someone. It may be that his prayer was heard. At least it was answered, for over the ridge of a little hill to his left and slightly behind him he saw a line of men coming. Fifty there were in all and, if the truth must be told, there were some non-commissioned officers among them. At their head was a freckled-faced private. Up and down the line he went bossing them, yelling at them, ordering them this way and that, a seasoned veteran, a natural leader. He had marshalled them from shell holes, from secure places behind trees and in the corners of the trenches. When he came up to where the lieutenant was standing he was grinning from ear to ear and his sandy hair was standing pretty much on end. "Lieutenant," he said, "I got fifty of 'em here; where'll I put 'em?" For two days he bossed the men in that platoon in exactly the same manner as though he were their commanding officer and they accepted it without question. That is the simple tale of a private's rise to sergeant. In the eyes of the company he was no more than a freckled-faced, sandy-haired boy that didn't amount to much. But he was a leader of men. He had made a name for himself.

The men, as they marched back, reflected on their experiences and mourned their lost comrades. They had received their baptism of fire; they had demonstrated their courage; they had proved their fortitude. They had the satisfaction of knowing that "green" as they were, unaccustomed to fire as they had been when they started on their mission, they could say that they had upheld the honor of the Division. They had won confidence in themselves.

Their officers appreciated better than did the men that this experience had shown that both officers and men had still much to learn. These men had demonstrated their reckless bravery, their supreme courage, their careless disregard of danger. They had overrun hostile machine guns regardless of losses. But, for the results achieved, their losses had been too great. They had still to learn how to play the game.

The French officers and soldiers viewed the conduct of the Americans with mingled admiration and regret; to them it was glorious, it was heroic, but it was not war.

The question as to whether, in achieving a certain goal, a brief engagement, with heavy losses, is more advantageous than a protracted engagement with small daily losses is one that has been much discussed; whether to lose 1000 men in one day and attain a certain result or to lose 100 men per day for ten days in attaining the same result. Its answer must be found in the temper and resources of the nations engaged. The French, for several years, had been forced to conserve, by every possible means, their man power until help could arrive. They had, in consequence, acquired a conservatism in their method of fighting that greatly diminished their losses while, on the other hand, it equally diminished their gains. Such a course was not in consonance with the temper of the American nation.

Certain it is that the advent in the war of the American troops brought about a complete change in the tactics of the Allies. The daring of the Americans infused new hope, new vigor, new courage. The cautious, carefully-prepared offensives, with limited objectives, began, during the latter period of the war, to give way and were replaced by operations more daring in concept, more brilliant in execution, more costly in man-power, more decisive in results.

CHAPTER VII

SERGY

ONCE out of the line, the whole Division was reassembled
87 and regrouped as a reserve for the Sixth French Army. The 7th Brigade, by an order from the II French Corps, reverted to the control of the Division Commander on July
88 22nd. By the evening of July 23rd the Division was bivouacked in the area Marizy-St. Mard, Bonnes, Hautevesnes, Brumetz, St. Quentin, Marizy-Ste. Genevieve and Division Headquarters established at Bourneville.

Amid the shattered villages and on the ground they had so recently helped to conquer, the men, happy in the enjoyment of hot food and what appeared to be rest after the recent strenuous days, wrote letters home telling of their thrilling experiences and begging relatives and sweethearts not to worry about their safety; that they would come home all right some day. Arriving in the rest area, mostly in woods, they dug in and pitched their shelter tents over the holes. That was the last seen of them for about a day and a night. Not since they had left the reserve line on July 16th had they had an opportunity to get any sleep except that which they could snatch in bits. Their rest was undisturbed, save for the bombing by German aeroplanes and the ever-present boob who automatically yelled "Gas" in echo of a horn on some French car that happened to pass. Awaking, they raided the kitchens. The cooks battled with the men's appetites as the men had battled with the enemy. Wagonloads of "chow" were prepared and eaten. It was a case of meals at all hours. For once the lid was off.

While in the rest area the men cleaned up, received some

new clothing and equipment and replenished their supply of ammunition. Several hundred replacements came in and the Division again began a course of training. Salvaging parties were sent out over the battlefields. Under the supervision of the chaplains the survivors buried the dead. The bodies of 302 dead of the 4th Division were collected and buried in soldierly rows in two cemeteries, one just outside Chevillon, the other on the Courchamps-Priez road opposite the Bois de Cobourg. The Division Chaplain personally identified the dead and was able to place the correct marks over all the graves, except in a few cases where the men's identification tags had been shot away. A neat inclosure with an ornamental archway was constructed by the engineers and each cemetery surrounded by a fence of barbed wire.

The people back home were meanwhile reading about the successes of the Allies, of the withdrawal of the Germans from their recently won territory and were mingling tears with smiles, anxiety for their loved ones with the joy of victory. With Paris no longer threatened and no "Big Bertha" to disturb the equanimity of the boulevards the French capital had good reasons to feel both joy and gratitude. The entry of the Americans into battle had marked the retreat of the enemy from the gates of Paris. The victories at Cantigny, at Soissons and on the Marne aroused the French people to great enthusiasm for their ally from across the sea. In Paris, in particular, much hero-worship was lavished on the Americans who enjoyed a popularity and were regarded with a gratitude that seldom falls to the soldiers of one nation from the people of another. Everywhere it was "Vivent les Américaines" and every orchestra in Paris considered it its duty to play "Over There" as many times during one evening as the duration of the performance would permit.

The 4th Division, however, like most of its fighting brethren in the American Army, was far from these scenes. The little villages at the front, like all the actual war zone, had long since been evacuated and there were no inhabitants left to greet their saviors. This was a desolated area where even the dead had been thrown out of their graves by shell-fire.

On July 24th the 7th Brigade was again detached from
the Division and placed under the orders of the VII French
Army Corps. The 47th Infantry, which had remained in re-
serve throughout the recent action, was sent forward on the
night July 24th-25th to perform what is known familiarly as
89 "mopping up" the Bois du Châtelet. To "mop up" is to
exterminate the remaining machine guns and snipers in a local-
ity after the first line of infantry has passed. It is trying work
necessitating quickness of both eye and ear and skill with
the rifle. The remainder of the 7th Brigade was re-attached
to the II French Army Corps on the same day and marched to
the vicinity of La Croix, where it took up a position between
90 La Croix and Rocourt.

The 58th Infantry, on July 25th, was placed at the disposal
of the XI French Army Corps and, on the night of July 25th-
26th, marched to a point just west of Oulchy-la-Ville where it
went into second line position and remained until the follow-
ing night when it marched back to Brumetz to return to the
91 divisional fold.

On July 26th Brigadier General Ewing E. Booth joined the Division and was assigned to the command of the 8th Brigade.

At 3:00 A. M., July 28th, orders were received placing the
Division under the command of the I Corps, U. S., as corps
92 reserve. The Division (less the 7th Brigade which was now
concentrated in the Bois du Châtelet) marched at 6:00 A. M.
via Priez and Sommelans to the Bois de Bonnes, where it
bivouacked for the night.

Tramping along through the darkness,
Splashing our way through the rain
With chafing packs slung on our backs,
Bound for the trenches again.

Flashes of light in the distance,
Splotches of red in the sky,
The sound of shell creating hell
In a convoy creeping by.

Our line moves on like a shadow,
Pushing its way through the wreck,
Each man in his place, rain in his face,
And streaming cold down his neck.

Silent and grave, moving forward,
Each having thoughts all his own,
As we tramped the path of the War Lord's wrath
Where the fires of Hell are blown.

The Headquarters of the Division was established at noon, July 28th, at Artois Farm, a short distance south of the town of Beuvardes. At 6:00 P. M. orders were received directing the Division Commander to place two battalions of infantry at Artois Farm at 8:00 P. M., subject to the disposition of the Commanding General, 42nd Division. This Division was at 93
the moment having a hard fight on the banks of the Ourcq River in an endeavor to force a way to the Vesle. The 47th Infantry was ordered by General Cameron to furnish the two battalions.

At midnight, July 28th-29th, orders were received from the I Corps to concentrate the Division in the Bois du Châtelet. 94
The movement was completed by noon July 29th. The 7th Brigade was returned to the control of the Division Commander on July 29th and on the afternoon of that day was marched to the Bois de Beuvardelle, west of Artois Farm, where it bivouacked for the night. 95

On July 30th the entire 7th Brigade was placed at the disposition of the Commanding General, 42nd Divison, for emergency use. The 39th Infantry, 4th Engineers and 11th 96
Machine Gun Battalion were ordered to construct a defensive position along the northern and eastern edges of the Forêt

97 de Fêre. On July 31st the 58th Infantry and the 12th
Machine Gun Battalion were directed to relieve the 39th In-
fantry and the 11th Machine Gun Battalion of the task as-
signed the latter units in the Forêt de Fêre, in order to leave
98 them free for call from the 42nd Division. The relief took
place about 10:00 P. M., August 1st. The remainder of the
8th Brigade and the auxiliary troops of the Division were
moved from the Bois du Châtelet to the Bois de Beuvardelle
during the day of July 21st.

The Germans, alarmed by the speed with which the Americans and French were pressing on their heels as they withdrew from the Marne salient, decided to make a temporary stand on the grassy slopes and amid the woods in the valley of the Ourcq. The Ourcq River rises about seven kilometers southeast of Sergy and flows generally northwest past the southern edge of the town. From a point one kilometer south of Sergy to a point three kilometers northwest its channel is bordered by a rather heavy growth of young trees. The northern bank rises gradually but steadily to such a height as to command the entire region south of the stream. Four kilometers to the northwest of Sergy is Fére-en-Tardenois. From there the German position ran southeast through Sergy to Hill 212.

It was a position particularly favorable for the enemy. His experience and skill in machine gun tactics, his ability to choose the best positions for securing the most effective fire and to utilize every fold in the ground, conferred on him advantages over the American troops which he was determined to exploit to the utmost while his troops and supplies were being safely withdrawn from what remained of the salient. But, as each day passed, the reckless determination of the Allies to force a way through and defeat his scheme of withdrawal, gave rise to an increasing German anxiety. Heavy losses were inflicted on the Americans. But the German troops, too, were being slowly decimated by the severity of the fighting.

When the 4th Guard Division (Prussian), having spent a month resting at Metz, arrived to stem the American wave, its members determined that they, at least, would acknowledge no defeat from these young and inexperienced soldiers. They 99
fought with particular bravery. They counter-attacked with a dash that indicated their determination to drive the 42nd Division back to the Marne.

While this battle was raging along the Ourcq, while the 42nd Division, weakened by severe losses, was in need of help, the 1st and 3rd Battalions of the 47th Infantry, which had not heretofore seen front-line service, were called upon to enter the action.

It was on the evening of July 28th that the two battalions marched to Artois Farm and were reported to the Commanding General, 42nd Division. Major Gulielmus V. Heidt was in command of the force. Major Troy H. Middleton commanded the 1st and Major James P. Cole the 3rd Battalion. The battalions were permitted to rest at Artois Farm, and then shortly before daybreak, took up the march toward the Ourcq via le Four à Verre. Major Heidt and his adjutant preceded the command and reported to Brigadier General R. A. Brown, commanding the 84th Infantry Brigade, 42nd Division, at la Croix Blanche Farm, for orders.

Under General Brown's instructions, the two battalions of the 47th went into action independently of each other. They were attached to separate regiments. The 3rd Battalion, 47th, was placed on the left flank of the right regiment (the 168th Infantry) and the 1st Battalion, 47th, was placed on the right flank of the left regiment (the 167th Infantry). So it happened that the two battalions of the 47th were adjacent to each other during the action although attached to different regiments of the same brigade. 100

There are many towns and villages in France whose names will go down in American history. Cantigny, the first to be captured by American troops; Château-Thierry, scene of the

Brigadier General Ewing E. Booth
Commanding 8th Infantry Brigade

Unloading wounded at field hospital at Château de la Forêt, in the Forêt de Fére. August 10, 1918

Infantry and supplies moving forward to the Vesle River. Sergy, August 3, 1918

remarkable stand of a regiment of the 3rd Division; and many others where Americans shed their blood and proved their fighting prowess. No place, however, saw a more desperate combat than Sergy. It stands to-day a shattered monument to American heroism.

Sergy, before the war, was an unpretentious village in the valley of the Ourcq. The river itself is only a few feet wide and in places can be easily taken at a leap. Many American officers, when they first crossed it, reported that they had passed a creek but were unable to find the river. From the sides of the Ourcq rose the green slope of its banks on which, about 200 yards from the stream, were the outlying buildings of Sergy. These were of stone or had stone foundations, with yellow plaster sides built on wooden frames and topped with red tiles. A stone wall surrounded each dwelling house and its barn.

In 1914 the village had been overrun by the Germans. After the First Battle of the Marne Sergy had lived again to welcome French troops. Four years passed. In May, 1918, the German drive from the Chemin des Dames once more shrouded the little village in an agony of war. Once again the Germans retreated and again Sergy came into the roar and confusion of the battle zone.

Late in the evening of July 27th the 42nd Division had crossed the Ourcq River. The Germans had defended the passage of the river with desperation. From the heights, from the woods and villages a devastating fire had been poured into the Americans. There were attacks and counter-attacks, but
101 the 42nd secured the reverse slope north of the Ourcq. At Sergy, however, was seen the bitterest struggle of the advance. The 168th Infantry penetrated into the village early on the 28th but the Germans, favorably located on neighboring heights and in woods, made the cost of retention so heavy that the Americans were forced to withdraw. The 168th renewed the attack. The Germans, who wished to hold the

village themselves, counter-attacked and drove the 168th out and down the slopes of the Ourcq. The battle for Sergy swung back and forth throughout the day. At night the 168th held it but early in the morning of the 29th was driven out by the Germans. It was then that the 3rd Battalion came on the scene.

The two battalions of the 47th, marching through the Forêt de Fére, were met by Major Heidt at the crossroads west of la Croix Blanche Farm at about 7:00 A. M. Pursuant to his orders the 3rd Battalion marched through the northeast corner of the woods toward Favière Farm. From this point the advance on Sergy was begun in combat formation. Companies "I" and "L" were in the lead, with Companies "M" and "K" in support. The battalion advanced through the tall wheat that covered the slopes to the thin belt of woods along the Ourcq, south and west of Sergy.

Meanwhile the 1st Battalion had marched from the crossroads due north for a distance of 1½ kilometers to the northern edge of the Forêt de Fêre. Under the cover of the trees it then moved east to the Rû (Ravine) de la Taverne, where the battalion was halted while the company commanders went forward to reconnoiter the route of approach to the Ourcq. The battalion had instructions to relieve one batta- 102
lion of the 167th Infantry, holding a position on the north bank of the Ourcq a short distance southeast of Meurcy Farm. The battalion followed down the Rû de la Taverne under cover of the trees along its banks, which afforded such protection for the troops that there were but two casualties before the battalion took up its position two hundred yards north of the Ourcq. The relief was effected, and from that moment the 1st Battalion of the 47th became known as the 3rd Battalion of the 167th Infantry. In the position Companies "D" and "B" were in front line, Companies "C" and "A" in support. The line rested on the unimproved river road running northwest to Fêre-en-Tardenois. The marked maps provided by the 167th

Infantry showed the boundaries of its sector as follows: east boundary, Sergy (exclusive), Nesles (exclusive), les Bons Hommes Farm, Dôle; west boundary, from a point 500 meters east of Meurcy Farm parallel to the east boundary.

The immediate objective of the 1st Battalion was the Forêt de Nesles and more particularly the Château de Nesles which lay two kilometers away. The battalions on either flank were unable to advance in the face of a murderous machine gun fire and the heavy artillery bombardment of the Germans. The 1st Battalion moved forward at 9:00 A. M. and gained the heights to the north of the road from which it had started. With the Germans entrenched on the hillside, it took fierce hand-to-hand fighting on the part of Companies "B" and "D" to dislodge them. But the Americans were not to be denied and the advance elements of the battalion rested the first night, July 29th, just south of the unimproved road running southwest from Nesles. Here it dug in, with its line extending along the summit of the crest, and strengthened its position during the night in anticipation of further severe fighting on the morrow. To meet the fire from the flanks Company "D" had been swung partly around to the northwest. On the other flank Company "B" had done the same to the northeast, Company "A" coming in between the two in the meantime. Losses had been heavy. The evacuation of the wounded was difficult and the water question serious.

On the morning of July 30th the situation remained unchanged, the battalion being under heavy fire from both enemy machine guns and artillery. The Germans had control of the air and their planes were very active.

Late in the afternoon the Germans counter-attacked, after a fierce preparation by machine guns and artillery. After severe fighting the attack was repulsed without gain for the enemy. On July 31st the battalion advanced in liaison with troops of the 167th Infantry on the left and pushed north to the unimproved road running southwest from Nesles. The remainder of the day saw no change in the alignment of the battalion. That night it was relieved by the 167th Infantry, and marched back to the Forêt de Fére.

The advance of the 3rd Battalion from Faviére Farm toward Sergy began about 9:00 A. M. and was soon met by a hot enemy fire. Liaison was established with the 168th Infantry on the right. Its men were storming Hill 212 southeast of Sergy; they could be seen going up in combat groups and being driven back by machine gun and artillery fire. 100

Upon approaching the Ourcq the 3rd Battalion encountered a hostile barrage that took a heavy toll of life. A determined attempt was made to enter the village but when the command reached the narrow gauge track just south of the river, the fire became so intense that it was forced to halt and seek the protection afforded by the railroad embankment.

Roughly speaking, the German front line before Sergy at this time was an arc of a great circle with the town just inside the edge. As the action developed each of the two battalions of the 47th Infantry drove a wedge into the arc in a general northerly direction, that of the 3rd Battalion being the more blunt of the two and gradually including the entire town within its limits.

At the railroad track the first line was strengthened and then all moved forward, Company "L" taking the left of the village and the ground to the west, Company "I" the right of the village and the wheat-covered slopes of Hill 212 to the east. The troops advanced until the hostile machine gun and artillery fire became so murderous as to make further progression impossible. But the line held. Company "L" had succeeded in advancing through three fourths of the village, but had suffered heavy casualties. Company "I" had advanced approximately half-way through the village and the wheatfields on the east, when an enemy aeroplane swooped low, dropped a few flares, and disappeared after inflicting machine gun wounds on several of the men. Immediately a terrific artillery fire was placed by the Germans upon the village and the wheat-fields. Our whole line fell back to the Ourcq River, where it held for a short time and was then established along the narrow gauge railroad just south. This was about noon.

Major Heidt, who accompanied the 3rd Battalion, was wounded just before the battalion entered the woods bordering the Ourcq. Major Cole, the Battalion Commander, was hit on entering the village of Sergy, and Captain Louis T. Roberts, Company "I," who succeeded him, was severely injured by fragments of a high explosive shell almost immediately afterward. Captain Ross Snyder, Company "M," then assumed command of the battalion.

Some of the bolder spirits immediately started to re-enter the village to look after the wounded. The enemy in the meantime had advanced through half the village with a strong party, at which the 47th took pot-shots. The Americans began to creep forward by ones and twos, their fire gradually developed in intensity and, late in the afternoon, the Germans disappeared from view and the survivors of the 3rd Battalion re-entered the village and organized it for defense. Strong points were established, covering the streets and the open spaces. Some eighty wounded were gathered in an improvised first-aid station at the southern edge of the village. Ammunition was running very low and the men had eaten most of their extra rations. Food and ammunition were collected from the dead. The greatest shortage was in Chauchat ammunition; this was carefully conserved.

That night the situation was reported to the Commanding Officer, 168th Infantry, and medical assistance and some ammunition were procured. Before morning the number of wounded at the first-aid station had increased to more than a hundred.

Shortly before daylight of July 30th information was received from the battalion of the 168th Infantry on the right that the artillery of the 42nd Division would place a barrage north of the village and that the 3rd Battalion was to press forward at nine o'clock, behind the barrage, to the ridge one kilometer north of Sergy, in conjunction with the 168th on the right.

At nine o'clock the battalion moved forward through the village and beyond its northern edge. To the right no movement of the 168th could be observed. No barrage came but the men were met by a severe machine gun and artillery fire which was followed by a German counter-attack that drove the battalion back through the village to the river. The Chauchat ammunition had all been expended. Some of the men were using German rifles and ammunition, also German "potato-masher" hand grenades. What was left of the battalion promptly reorganized, counter-attacked, drove out the Germans and reoccupied the village about noon. It was then learned that Captain Snyder had been killed and seven lieutenants wounded. First Lieut. Gustav Braun now assumed command of the battalion. The village was filled with dead and wounded, both Americans and Germans. Many men with two to four wounds continued in the fight; many men, still fighting, had mustard burns.

A machine gun company from the 42nd Division moved up late that afternoon to assist the 3rd Battalion in holding the village. Before its guns could be placed, the company suffered the loss of a considerable number of its men and several of its machine guns. The remaining guns moved into the strong points established by the 3rd Battalion. On the left of the village Company "K" was pinned to the Ourcq by enemy machine gun fire. There was marked German aerial activity, the enemy using about twenty planes which dropped grenades, machine-gunned the Americans and directed German artillery fire. This artillery fire continued intermittently all day.

During the night Major Harrison Webster, senior surgeon of the 47th Infantry, who was later killed in the Battle of the Meuse-Argonne, came to the river at Sergy with a motor ambulance and, regardless of the German artillery fire, started the evacuation of the wounded.

The next morning, July 31st, the enemy put down a heavy artillery barrage on Sergy. This was followed by an infantry attack which was broken up by our machine gun and rifle fire at the northern edge of the village, before the Germans entered it. They sustained heavy losses. The enemy, however, kept up his artillery fire throughout the day. About five o'clock in the afternoon word came from the Commanding Officer, 168th Infantry, that Company "K" of that regiment plus two other platoons would be sent to relieve the battalion that night. About six o'clock an intense artillery fire, which raked every corner of the village with marvelous accuracy, filling the valley with gas and high explosive, showed that another hostile infantry attack was impending. It came, and with the assistance of the machine gun company from the 42nd Division, one captured enemy machine gun and our rifle fire, was broken up as it crossed the open space north of the village. The Germans fell back. Everything now became quiet. The sky was lit by hostile flares. Patrols were later sent forward nearly a kilometer and found the area in front deserted.

About midnight what was left of the 3rd Battalion was relieved by Company "K" of the 168th Infantry, and moved back to the Bois de Beuvardelle, west of Artois Farm. Here it discovered that Lieut. Col. Robert H. Peck was in command of the regiment, having relieved Colonel Roudiez who had been ordered to duty with the Quartermaster Corps. Company commanders of the 2nd Battalion came out to meet the men who had just been relieved while men of the 2nd Battalion did the kitchen police for the exhausted troops in order to give the latter a chance to rest. Cooks and wagoners prepared meals, making doughnuts for the occasion. That night the regiment was ordered forward again but the 3rd Battalion was permitted to remain in the Bois de Beuvardelle during the night with orders to catch up on the following day.

Such is the story of the capture of Sergy.

The work of the 1st and 3rd Battalions, 47th Infantry, in winning and holding the positions on the heights north of the Ourcq was performed at heavy cost. When relieved, they were found to be badly cut up. The two battalions had lost 27 officers and 462 men killed, wounded, or gassed; 6 enlisted men were officially reported as missing in action. However, they had met and inflicted sanguinary losses on the 4th Guard Division, which in later actions never forgot the terrible mauling it received from the Americans on the Ourcq. There were very few prisoners taken on either side.

On August 1st the 39th Infantry was in support of the 42nd Division in the Forêt de Fére. At eight o'clock that night it passed through a memorable experience. The regiment was taking up new positions in the woods preparatory to the advance on the following day. The 1st Battalion was in column of twos, the platoons ready to move to their new positions, when a German bombing plane glided quietly over them. Flying almost on a line with the column the aviator dropped a string of bombs so rapidly that the separate explosions could hardly be distinguished. The result was a scene of death and horror worse than battle. Every company in the battalion was hit, the total casualties being 27 killed and
94 wounded. 103

Apparently the success of the aviator was entirely accidental. It is believed that he was either searching for large dumps of ammunition which had been left behind by the Germans when they were forced to retreat or for a battery of American artillery which had inflicted considerable damage on the Germans during the day. For more than an hour he hovered over the woods dropping bombs wherever he had reason to believe that German dumps or American artillery might be concealed. As a result of this the 2nd Battalion, 58th Infantry, also suffered 4 casualties.

Meanwhile the 4th Artillery Brigade had left Camp de Souge and was hastening to join the Division. On July 30th it began to detrain at Château-Thierry.

CHAPTER VIII

FIGHTING ON THE VESLE

THE 4th Division was now to fight, for the first time, as a
complete unit. On August 1st orders were received from
the I Corps directing the 4th Division to relieve the 42nd Divi-
sion the night of August 2nd–3rd. The 51st Artillery Bri- 104
gade (26th Division), which had been attached to the 42nd
Division, and the 67th Artillery Brigade (42nd Division) were
to remain attached to the 4th Division.

On the night of August 1st the Germans began their re-
tirement to the Vesle, leaving behind numerous machine gun
detachments to dispute the advance of the Americans. 105
This withdrawal was discovered on the morning of August
2nd and the 42nd Division inaugurated a pursuit which, how-
ever, soon encountered active machine gun opposition.

On the night of August 2nd–3rd the relief of the 42nd
Division was accomplished, according to schedule, along a
general southeast and northwest line from the southern point
of the Bois de la Pisotte along the Les Bons Hommes Farm–
Mareuil-en-Dôle road to the northeastern tip of the Bois de
la Porte d'Arcy. Division Headquarters was at the Château 106
de la Forêt.

With the two infantry brigades side by side, the 8th on the right and the 7th on the left, the pursuit of the Germans was begun at four o'clock on the morning of August 3rd and vigorously pressed. In the 8th Brigade the 59th Infantry, with two battalions in front line and one in support, was in the lead, followed by the 58th Infantry. In the 7th Brigade the 39th Infantry, with two battalions in front line and one in support, was in the lead, followed by the 47th Infantry.

Each infantry battalion had a machine gun company attached to it. Strong advance guards were thrown out in front and the Division moved forward rapidly through the thickly wooded country, on its way to the Vesle. The rain was falling in torrents, the roads were bogs of mud. The opposition encountered consisted almost entirely of spasmodic artillery fire that shelled towns and roads—especially crossroads—and machine gun rear guards which were left behind by the Germans, in isolated units, to hold up the advance. Some of these remained with their guns until they were killed but the majority, after using all their ammunition, either managed to escape or threw up their arms when they saw the Americans.

At 9:30 A. M. the leading regiment of the right brigade had reached a northwest and southeast line, a little north of Chéry-Chartreuve, where it encountered heavy artillery fire from across the Vesle and determined machine gun opposition
107 from the front. The regiment stopped and dug in. The 39th, on the left, reached the northern part of the Bois de Dôle
108 at 2:30 P. M. Here it came under heavy machine gun fire and dug in for the night along a general line marked by the Chéry-Chartreuve–Mont Notre Dame road.

In rear of the advance elements, the artillery and transport encountered difficulties that seemed almost insurmountable. The road from Sergy to Mareuil-en-Dôle, leading through the Forêt de Nesles, was knee-deep in mud. Only an unimproved road at best, it had been badly cut up by the retreat of the Germans. The heavy rains during the first few days of August had rendered it impassable for vehicles. The artillery horses could not pull the guns through the mud. The engineers, the artillery and the infantry hastily filled the deepest holes with stones and stumps of trees, and corduroyed the road with saplings, branches and camouflage material. During the
109 night of August 3rd the transport began again to move.

During the late afternoon of August 3rd orders were issued from Division Headquarters that the brigades would form

strong advance guards, and with these push across the Vesle and form a bridgehead on its northern slope in advance of the line Vauxcéré-Blanzy-les-Fismes; the main body was not to cross the river until further orders. One company of engineers was attached to each brigade. This advance was begun, 110
about 10:00 P. M., in a pouring rain. The 39th Infantry moved by the unimproved road leading from Montbani Farm toward St. Thibaut. When the column neared the Vesle valley it encountered the artillery fire which the Germans were placing on all approaches to the Vesle River. The command was moved back a short distance and bivouacked off the road wherever the men could find a place. They settled down to await the coming of dawn, too tired even to notice the German shells that fell among them. The inky darkness of the night and the torrential rain made reconnaissance impossible. 111
The 59th Infantry failed to gain ground to the front. Two battalions, in an attempt to march across country to the river, became lost. They finally halted not far north of Chéry-Chartreuve and awaited daylight. Just before dawn the 112
58th Infantry and Companies "A" and "B," 12th Machine Gun Battalion, relieved the 59th Infantry and attached machine gun units in front line, and the latter became the reserve. The 47th Infantry and the 11th Machine Gun 113
Battalion (less two companies) halted that night in the Bois de Dôle, a little east of Mottin Farm.

Early on the morning of August 4th the 51st and 67th Artillery Brigades reached and established themselves in rear of the high east and west ridge about one kilometer north of Chéry-Chartreuve.

Shortly after daybreak the advance toward the Vesle was again taken up. The 39th Infantry, finding the Montbani Farm–St. Thibaut road still shelled by artillery, moved eastward across country, under cover of the woods, to the Chéry-Chartreuve–St. Thibaut road and approached the river by that route. About 8:00 A. M. St. Thibaut was en-

tered by the advance detachment with but little opposition from the village; but from the low ground across the river and the high slopes on its northern banks a deadly machine gun
111 and artillery fire was poured on the Americans. St. Thibaut, on the south bank of the Vesle, was under the direct observation and fire of the enemy who occupied an excellent position on the formidable heights across the river. The inevitable result was that the slightest movement of troops in the vicinity of St. Thibaut brought down a deluge of machine gun fire and of shells, particularly gas.

Looking out across the Vesle, the officers studied the positions that could be occupied by the enemy and began to visualize the magnitude of the task of crossing the river, should the Germans offer any decided opposition. In every respect the enemy held the advantage. The Vesle is not a formidable river, as rivers in America go. It is not, in fact, much wider than the Ourcq and the Ourcq was always called a creek by the soldiers. But there were marshes on each side of the Vesle's banks, and slowly stretching away on the northern side were heights that reached to the gorge of the Aisne. From the high ground on the American side it was possible to view the whole of the German positions in turn and, from the precarious security of the observation posts constructed in the trees to see the tall spire of the cathedral of the famous city of Laon, a mocking finger on the horizon.

Patrols were sent out by the 39th Infantry at 8:35 A. M. on August 4th to cross the Vesle and discover the strength of the enemy around Bazoches. Six patrols succeeded in crossing and returned with the information that the Germans had placed lines of barbed wire, staked down, in the bed of the stream and on the north bank; that they had been fired upon from Bazoches, which was apparently strongly held with machine guns, and that the Germans had machine gun nests scattered all through the valley of the Vesle, from the low ground in front of the river to the heights north of it; that

from observation posts on these heights, which gave a clear view of the river and of the entire valley, German artillery could be brought to bear on the Americans at any point and at any moment. These patrols were the first Allied troops to cross the Vesle. The Germans, as if to add emphasis to the 114
report that they were prepared to defend their formidable positions, poured shell after shell into St. Thibaut and on the narrow road leading to it through the forest. This bombardment was kept up all day long and the flying fragments and gas caused many casualties.

In the meantime the 58th Infantry which, after marching all night through the mud and rain had reached Chéry-Chartreuve shortly before dawn, had moved out on the Chéry-Chartreuve–St. Thibaut road about 9:00 A. M. Near the Farm 115
des Filles the aid station and rear elements of the 39th Infantry were encountered and it was learned that the advance of the latter regiment had been definitely stopped. The 3rd Battalion, 58th, was then sent up a draw leading southeast from the Farm des Filles to cross over to Villesavoye and ascertain if the country between Villesavoye and St. Thibaut, as far as the river, was cleared of the enemy. The remaining battalions counter-marched almost to Chéry-Chartreuve and then cut across country through Les Pres Farm to the main road near Resson Farm, with the intention of moving on Villesavoye via Mont St. Martin. The main road was reached about one o'clock in the afternoon. Here General W. G. Haan, commanding the 32nd Division, met the column. He declared that progress on the road to Mont St. Martin was impossible because of enemy shell-fire and advised the two battalions to seek shelter temporarily in the Bois de Cochelet and the small woods east of Les Pres Farm. Regimental Headquarters was established at Les Pres Farm.

In the meantime the Commanding General, 8th Brigade, who, in the progress of a personal reconnaissance had visited the rear elements of the 39th Infantry on the Chéry-Char-

treuve–St. Thibaut road, had cut across country, on foot, through the Bois de Cochelet and reached the 1st and 2nd Battalions, 58th Infantry, shortly after they had been placed in the woods above mentioned. He immediately issued orders that the advance on the river be resumed at once. The 3rd Battalion, which had started up the draw from the Farm des Filles, was now discovered entering the Bois de Mont St. Martin; instead of moving northeast from the draw, on
116 Villesavoye, it had continued southeast up the draw. The advance on the river began about 3:00 P. M. The 2nd Battalion, moving from Resson Farm, advanced east of Mont St. Martin; the 3rd Battalion at 4:00 P. M., west of Mont St. Martin; the 1st Battalion followed the 3rd at 600 meters. As soon as the troops left the shelter of the woods they were met by an intense artillery fire. The 2nd Battalion pushed forward, moving east of the town of Villesavoye, and reached the river with its most advanced elements at 4:30 P. M. On account of the destructive enemy fire the movement from the woods northeast of Villesavoye to the narrow gauge railroad, and from there to the river bank, was made by individual soldiers. By 8:00 P. M. three companies were along the riverbank, dug in, and one in support along the railroad track. Machine guns and one-pounders were placed on the high ground east of Villesavoye to deliver overhead fire. The Stokes mortars were in the support line. The losses of the
117 battalion had been heavy.

The 3rd Battalion advanced west of Mont St. Martin, over perfectly open ground, due north toward the river. Shortly after it left the shelter of the Bois de Mont St. Martin it ran into a German barrage which, providentially, was so placed as to fall almost entirely between combat lines. As a result, although the fire was heavy, the battalion suffered few casualties. Moving across the open plateau, the battalion descended the steep hill south of the Villesavoye–St. Thibaut road and established itself in the woods that cover the slope.

Patrols were sent forward to scout the valley of the river as far as the slopes of its northern banks. They returned and reported the enemy in strength in the valley of the Vesle and on the heights. The battalion remained in this position for the night. Liaison was established with the 39th Infantry on the left. 115

The 1st Battalion had followed the 3rd at 600 meters. In advancing across the open ground its Commander, Major Samuel H. Houston, was killed. Captain Rutherford H. Spessard assumed command of the battalion, which halted for the night in the ravine running southeast from St. Thibaut. 115

The 59th Infantry that night was in reserve position; two battalions were in the Bois de Cochelet and the woods just west of the Chéry-Chartreuve–St. Thibaut road, and one battalion dug in on the reverse slope south of the Bois de Cochelet. The 47th Infantry and one battalion of the 39th were west of the 59th, along the Chéry-Chartreuve–Mont Notre Dame road.

At midnight of August 4th-5th the 51st Field Artillery Brigade (26th Division) was relieved from duty with the 4th
Division and directed to rejoin its own division. Intermit- 118
tent enemy artillery shelling continued throughout the night.

During the day of August 4th the engineers had brought their bridge material as far forward as Chéry-Chartreuve. All things necessary for bridging the river had been secured. During the late afternoon of the 4th, and all that night, engineer officers reconnoitered the river valley selecting points for placing bridges that would enable the infantry to cross.

On August 5th a determined effort was made to cross the Vesle. From midnight until 5:00 A. M. a concentrated artillery fire was placed on the town of Bazoches and on the positions which observation had shown to be occupied by Germans. The hostile artillery positions could not be located with any degree of certainty. The German aviators had control of the air and effectually prevented our planes from securing the information so essential to successful operations.

Testing and tagging telephone cables at 4th Division Headquarters exchange at Mareuil-en-Dôle, August 11, 1918

Looking north from the Bois de Cochelet. The 58th Infantry advanced in combat formation on August 4th, over the plateau shown in the middle distance, under a heavy German barrage and with remarkably few casualties

The heights north of the Vesle

The infantry advance began at five o'clock in the morn-
119 ing. All along the line it met with strong resistance from
machine guns placed in the valley along the north bank of
the river and from artillery fire on the heights. The lat-
ter was perfectly controlled from the German artillery obser-
vation posts cunningly hidden near the crest of the hill.

The 2nd Battalion, 58th Infantry, succeeded in pushing
two strong patrols across the river. These suffered severely
in crossing, and on reaching the north bank were counter-
120 attacked and driven back across the river. The Germans
did not follow. Late in the afternoon several patrols finally
established themselves on the north bank and stayed there.

The 1st Battalion, 58th, moved forward at 4:30 A. M. and,
pursuant to orders, passed through the 3rd Battalion and ad-
vanced to the narrow-gauge railroad track south of the river,
the left of the front line resting on the track about half a kil-
ometer northeast of St. Thibaut. The battalion support dug
in along the St. Thibaut–Villesavoye road. In this ad-
vance the battalion suffered severe losses from artillery, min-
enwerfer, and machine gun fire directed on the open ground
115 between the bottom of the bluff and the river. Patrols
were at once sent forward to cross the river, only to be killed
almost to a man. After repeated attempts four small patrols
finally secured a foothold on the opposite bank where they
121 maintained themselves during the afternoon.

The 39th Infantry, facing the town of Bazoches, had by far the most difficult task. Skilfully placed machine guns, scattered all over the town in favorable localities, enabled the enemy to control every foot of the river. Minenwerfers added their nerve-racking fire to that of the machine guns. The German artillery fire was controlled as easily as a man would handle a hose. The forward movement began at 5:00 A. M. As soon as the troops appeared, the Germans concentrated their artillery on the area between St. Thibaut and the river. The advance was to be made with two companies in first line

and two in support. The right companies were unable to advance. One company moved to the left and succeeded in reach- 122
ing the river bank between the St. Thibaut–Bazoches bridge and La Maladrerie Farm. The men began to advance individually, most of them over the fragments of the demolished bridge. Thirty-eight had succeeded in crossing when the artillery concentration which had been placed on Bazoches was lifted. Immediately the German machine gunners came out of their holes, and the fierceness of their fire, together with that of the minenwerfers, made it impossible to get any more men across. The 38 men who had crossed established themselves along the low rise of ground just north of the river. One other company occupied the railroad track between St. Thibaut and the river. Several determined attempts were made during the day to increase the strength of the force on the north bank but each time the deadly nature of the machine gun and minenwerfer fire rendered the attempts abortive. 123

At 5:10 P. M. the Chief of Staff transmitted the following orders to Generals Poore and Booth: 124

1. In order to avoid any more serious losses, the General has decided to wipe the towns of Bazoches and Haute Maison and the possible emplacements east, north and south of these towns, off the map.

This artillery preparation will be made to-morrow and last until the late afternoon. At H hour the troops are to go across (not in mass, but in small, rapidly-moving patrols). The destructive fire on the western half of the sector north of the river will consume about 4 hours and will be delivered after adjustment of about 3 hours. The troops will go across under cover of a smoke barrage and preceded by a rolling barrage along the whole sector front. Details will come out in orders and any small matters will be arranged to-morrow forenoon at a conference of the brigade commanders.

2. Therefore you are directed to withdraw, after dark this evening, all patrols and combat groups north of the Vesle. Only small security patrols need be put out along the southern bank of the river. The troops should be held on the alert ready to repel any possible counter-attack from across the Vesle. Our artillery

will be directed to adjust for the night on the river. Its fire can be called for, if necessary, by rockets.

3. H hour will be given you at conference. In order to have troops ready to-morrow they should only be drawn back, south of the river, as far as is absolutely necessary.

During the afternoon preparations were made by the engineers for placing a number of bridges across the river, one an artillery bridge composed of heavy oak stringers, and the others, eight in number, foot bridges to enable the infantry
125 to cross rapidly. During the night of August 5th-6th these bridges were hidden near the points where they were to be used.

On the night of August 5th-6th the 4th Artillery Brigade made its first appearance in action, occupying filial positions
118 with corresponding elements of the 67th Artillery Brigade. The 4th Brigade had received a thorough course of training at Camp de Souge. With the exception of Colonel Henry W. Butner, who had replaced Colonel Newbold of the 16th, and Lieut. Col. A. F. Commiskey, who had assumed command of the 77th, vice Colonel Preston promoted to Brigadier General, the personnel of the organization commanders was unchanged. Toward the last of July orders had come to move and at midnight of July 28th-29th the 4th Brigade had entrained, leaving in the early morning for an unknown destination. Two days on the train passed slowly. At the close of August 1st the train entered a town that had been ill-treated with bombs and shells. There was a sudden shock of surprise as the soldiers read on the sign the magic name of Château-Thierry. In a moment every head was out of the window. An excited cheering broke forth as realization came that this was the place where they got off. But little time was required to unload, harness up and get ready to pull out.

Late that evening, amid the boom and flash of anti-aircraft guns and the crackle of machine guns directed at hostile bombing planes, the artillery commenced its march toward the front, crossing over the Marne on the one remaining bridge and passing through the portion of the town once held by the Germans. For five nights the march continued, each bringing closer the lightning-like flashes on the horizon and the rumble of the big guns. Each day was spent in going into position as reserves in case the enemy should counter-attack. There was little time to sleep. It seemed as though the Germans would never halt for them to catch up. On the night of August 5th-6th, in rain and mud, they marched through Fére-en-Tardenois, where the Brigade got its first taste of being under fire, and arrived finally at a position in an open field to the west of Chéry-Chartreuve. By daybreak the guns were entrenched and camouflaged. At 12:30 P. M., August 6th, the first shot sped on its way toward the enemy. 126

On the morning of August 6th the artillery preparation commenced. Arrangements had been made, during the night, with the 62nd French Division, on the left, for a French advance in conjunction with that of the 4th Division. All day 127
long the whine and roar of the shells passing over the river made it almost impossible to hear the spoken word anywhere among the waiting infantry. The men knew what was coming and viewed with satisfaction the clouds of smoke that arose on the north bank while they themselves were taking refuge from the return fire of the Germans.

H hour was fixed at 4:30 P. M. At that time the artillery barrage would leave the river and progress. Beginning at H hour a smoke barrage was also to be put down on Bazoches and la Haute Maison for one hour. 123

On the right, the plan of the 58th Infantry contemplated that the 2nd Battalion would hold the south bank of the river from a point one kilometer east of St. Thibaut to a point south of Château du Diable, the 3rd Battalion in support of the 2nd. The 1st Battalion was to make the attack, passing through the 2nd Battalion. 129

Following closely behind the artillery barrage, the 1st

Battalion, with two companies in front line and two in support followed by a machine gun company, crossed the river. Stumbling over the shell-sown field, muddy with falling dirt, clothes torn by shell-fragments, trying not to see their fallen comrades as they passed, the men gained the railroad, where a number of prisoners and eight machine guns were captured. About a hundred German infantry retreated hurriedly from the railroad cut toward the Château du Diable as the Americans advanced. The embankment rocked with the impact of large shells. Machine-gun bullets ricochetted off the rails, striking fire as they passed. Pushing on, through these, the companies gained the Rouen–Rheims road where they hurriedly
130 dug in on the south side beneath the shattered poplar trees. The support companies and the machine gun company were established in the railroad cut about 500 yards south of the road. The enemy abandoned the trenches immediately north of the road, but on both flanks hostile machine guns were very much alive. The battalion had lost heavily, but it had reached its objective. It was alone. There were no friendly troops on its left and none on its right.

During this attack the regimental 37 mm. guns and Stokes mortars and all machine gun companies were actively engaged and, from well-located positions, gave powerful support to the advancing units.

On the left of the Division sector the troops did not progress. The 1st Battalion, 39th Infantry, which had been in reserve with the 47th Infantry near the Farm des Dames, had been designated as the attacking battalion. The Battalion Commander apparently miscalculated the time necessary for
131 his command to reach the river. He was one hour late. In consequence, the barrage had passed on when he arrived, and the Germans had emerged from their bomb-proofs and dugouts in Bazoches and la Haute Maison. When the battalion attempted to cross the river just east of St. Thibaut it was met with a fire so severe, from rifles, machine guns and minenwer-

fers, as well as artillery, as to make impossible the crossing of a force of any size. A considerable number of men filtered across through the fire and established themselves on the north bank. Most of these, however, were driven back. 132

During the night of August 5th-6th, and during the bombardment, the engineers perfected the bridges that were to enable the infantry to cross. The artillery bridge was finished and needed only to be placed. At 4:20 P. M. the engineers started for the river to place their foot-bridges As fast as they were laid down they were destroyed by enemy artillery. Again and again, sweating, panting, the attempt was made. Trees were finally chopped down and lashed together with wire to form temporary bridges. On two such improvised structures the 1st Battalion of the 58th crossed the river. In the area of the 39th even such makeshifts could not be maintained. The enemy artillery fire concentrated on the river was too heavy. Upon all sides geysers of earth sprang up and vanished. A direct hit on a bridge section tossed the two carriers high in the air, twisting and turning in horrible limp shapes until they struck the ground.

About 7:00 P. M. an attack directed against the 58th Infantry on the Rouen–Rheims road by a German detachment of about 300 men coming from the hill north of la Haute Maison, was broken up and the detachment practically exterminated by an accurate 30,000-round machine gun concentration delivered by the two companies of the 10th Machine Gun Battalion from the high nose of the Montagne de Fére just
southwest of St. Thibaut. During the evening two other 133
strong German attacks were broken up by artillery fire. 134
These German attacks and the prospect of other similar ones caused the Brigade Commander to order two companies of the 2nd Battalion, 58th, to cross the river and protect the
flanks of the 1st Battalion. At 3:15 A. M. these two com- 135
panies were relieved by two companies of the 59th and re-
joined their battalion. Night fell but the darkness that 136

the troops had hoped for did not accompany it; instead, the night became a white glare of starshells, so many, it seemed, that the slightest movement could not remain undiscovered.

Along toward morning the 39th Infantry and Companies "A" and "C," 11th Machine Gun Battalion, exhausted by heavy losses, gas, marching and fighting, were relieved by the 47th Infantry and Companies "B" and "D" of the 11th Machine Gun Battalion. The 39th was moved back to a reserve
137 position near the Farm des Dames for a well-earned rest.

It was now up to the 47th Infantry to effect a crossing. The 62nd French Division, on the left, had established strong patrols along the railroad track north of the river with its right about one half kilometer west of the St. Thibaut–Bazoches road. The 58th Infantry, on the right, had reached the Rouen–Rheims highway. It is worthy of note that the entire front was dominated by the high ground north and northeast of la Haute Maison. This made the holding of the north side of the river by the Americans almost impossible unless possession could be secured of Bazoches and la Haute Maison.

Orders were issued to the 47th to attack, with the heights behind Bazoches as the objective. The 2nd Battalion was designated to lead the attack, with the 3rd in support and
138 the 1st in reserve.

At 3:45 on the morning of August 7th the 2nd Battalion began to move through St. Thibaut. By noon one company had attained the south bank of the river. Shortly after, the enemy aeroplanes made a reconnaissance of the American front line and soon the shells began to fly. Heavy artillery fire was put down on the river and on the American troops. In the face of the enemy machine gun and artillery fire the battalion continued to advance, fighting over every inch of the ground and by five o'clock in the afternoon approximately 350 men of the 47th had crossed the Vesle, some on trees which had been felled across the water, some on a foot-bridge, which

was destroyed by direct hits from the enemy only to be rebuilt by the persistent efforts of the engineers, and still others by swimming and wading. Most of Company "H" and parts 139
of Companies "G" and "F" crossed the river and established themselves along the railroad track with the left about one-half kilometer east of the château of Bazoches. All the while machine guns placed on the tops and in the cellars and windows of buildings in Bazoches were pouring their deadly fire into the men. In crossing the river, too, many of the gas-masks had become worthless owing to their being soaked. As the enemy threw over between three and four thousand shells during the day, one third of which it is estimated were gas shells, it may be realized what this effort meant. Liaison was established with the support companies of the 1st Battalion, 58th Infantry, on the right. Two platoons were sent forward to the unimproved road half-way between the railroad and the Rouen–Rheims highway. The battalion had suffered heavily as the cost of this success, but as at Sergy, the men had once more demonstrated of what splendid material they were made.

The 1st Battalion, 58th Infantry, on the right, was struggling desperately to maintain its position on the highway. At 4:00 A. M. an attack launched against the left flank was repulsed, with slight losses to the Americans from German hand-grenades.

At six o'clock on the morning of the 7th the Germans launched an attack in force against the right of the battalion and partially penetrated the protective barrage of the American artillery. However, the attack was stopped by automatic and rifle fire before it reached the road. About 10:30 A. M. 140
a German trench just north of the Rouen–Rheims road and opposite the left of the line was raided and ten machine guns and five prisoners were captured. Ten of the enemy were killed and five wounded. Others were shot as they retreated. 141
About noon two companies of the 1st Battalion, 59th Infantry, and Company "D" of the 12th Machine Gun Battalion re-

lieved the units in support, and that night the front line com-
panies of the 58th were relieved by the other two companies
115 of the same battalion of the 59th. Liaison was established,
by the support companies, along the line of the railroad north
of the river, with the 112th Infantry Regiment (28th Divi-
141 sion), on the right, at 6:30 P. M. This regiment had just
come up. That night the entire 58th Infantry was relieved
by the 59th.

In Villesavoye there stands the ruin of an old church. On
the corner of this church at one time there was built a tower.
On August 4th, the town was captured and occupied by the
58th Infantry. It was necessary that the enemy's move-
ments across the Vesle be observed. The one position from
which these observations could be made was the tower of the
old church, and it was by far the most outstanding target for
the Germans. First Lieutenant Peter W. Ebbert and Second
Lieutenant Lowell H. Riley, 58th Infantry, volunteered to es-
tablish an observation station in this tower. For two days
they occupied it and sent in very valuable information. Dur-
ing the whole of the time they were subjected to heavy artil-
lery fire. On the evening of August 7th an enemy shell crashed
142 into the tower and both officers were instantly killed.

There appeared to be no sign of weakening in the German resistance. There they were, on the heights, inflicting a monotonous list of casualties, attacking only when the Americans made some slight progress, and defending themselves with an outpost line of machine guns and expert snipers. The main line of resistance was well back on the heights and all of the ground in front covered by very accurate artillery fire. Meanwhile the French were endeavoring to dislodge them from all their positions on the Vesle by an attack to the north of Soissons in an easterly direction.

On August 8th enemy artillery was especially active. Nearly three thousand shells of various kinds were dropped on St. Thibaut and its vicinity and on the troops of the 47th Infantry that day. The battalion north of the river spent the day digging in and reconnoitering for favorable points from which to attack Bazoches. One company was moved over to the left to maintain liaison with the French. 143

The 59th Infantry sent out strong reconnoitering patrols to determine the enemy's main points of resistance. During the afternoon a heavy artillery barrage fell on the companies along the Rouen–Rheims road. This lasted for almost three 144
hours and caused many casualties. Machine gun and sniper fire, coming from the woods surrounding the Château du Diable and enfilading the American line, caused the Battalion Commander to reform the right flank of the right company so that at the end of the day this company faced northeast in the direction of the château, its right resting on the railroad. On this day Major Max B. Garber assumed command of the 59th Infantry.

On August 9th the 47th Infantry made another determined effort to take Bazoches. This attack was to be made in conjunction with troops of the 62nd French Division on the left, the French moving against Bazoches from the southwest, the Americans from the southeast. Sending patrols forward, protected by a heavy barrage, the 47th Infantry advanced against the town. Strong opposition was encountered, but throughout the afternoon machine gun nests and snipers were carefully sought out and eliminated. Progress was slow but steady. On the left the French had failed to advance; only one platoon began the attack and it withdrew upon encountering heavy fire. The liaison company of the 47th, west of Bazoches, moved forward, however, and was almost exterminated by machine gun fire. Nevertheless the troops persisted. At 7:30 P. M. they were half-way through Bazoches, and the right of the line was on the Rouen–Rheims highway. At this juncture five German aeroplanes came over and, flying within 100 yards of the ground, dropped bomb after bomb on Americans and Germans indiscriminately. Flesh and blood could stand no

more. The men fell back from the town and, crossing the river, took up a position along the line of the narrow-gauge railroad. Some of the Germans followed almost to the river
145 but did not attempt to cross. Over a hundred dead bodies were left on the open ground south of Bazoches. These dead men could not be recovered without sacrificing live men to do it.

The 59th Infantry, on the right, was finding the woods surrounding the Château du Diable a hotbed of trouble. During the day repeated attempts were made, with artillery assistance, to dislodge the enemy but without success. That evening, owing to the exposed position occupied, all front line troops on the Rouen–Rheims road were withdrawn to the line
146 of the railroad north of the river. That night, also, the 1st Battalion was relieved by the 2nd.

During the day five German machine gunners dressed in American uniforms had worked their way from the Château du Diable around the flank of a platoon near the Rouen–Rheims road. The platoon was commanded by one of the oldest sergeants in the regiment. The Germans called out in English that they were "coming up in support." "They have no business coming from that direction," said the sergeant to his men; "Let 'em have it." The Germans were killed. The sergeant himself was later killed in action.

During August 10th there was no change on the left. The day was notable for increased enemy aërial activity. At one time eight enemy planes came over the front line of the 47th and dropped between forty and fifty bombs. On the right a portion of the woods surrounding the Château du Diable was captured after a bitter struggle. The Germans still occupied the remainder. That night the 67th Artillery Brigade was relieved from duty with the 4th Division and marched back to
147 join the 42nd Division.

On the late evening of the 10th, the 59th Infantry heard a German wagon train moving on the Rouen–Rheims highway, northeast of Bazoches. The information was promptly transmitted to the artillery, which fired 50 shots. That the shells found their target was proved by a great flame and a tremendous explosion which occurred a few minutes after fire was opened. A German ammunition train had been destroyed.

The 11th was characterized by heavy artillery firing on both sides, which lasted nearly all day. Strong patrols were sent to the Rouen–Rheims road and into the woods around the Château du Diable. The latter, while gaining some ground, failed to drive out the enemy completely. That day the 47th Infantry held the line of the southern bank of the river by troops placed along the narrow-gauge railroad, and the 59th held the line of the railroad on the northern bank between the river and the Rouen–Rheims highway. On the night of the
11th-12th the Division was relieved by the 77th Division. 148
The 4th Artillery Brigade remained in line until the nights of
August 15th-16th and 16th-17th, when it was also relieved. 149

Throughout all these adventurous days the rival artillery took measure of each other. The Germans viciously shelled every town and locality in front and rear. Their planes had complete control of the air and kept them accurately informed of the location of the American artillery. The only safety for the latter lay in frequent shifts of position. St. Thibaut, Villesavoye, Mareuil-en-Dôle and Chéry-Chartreuve came in for their share of artillery attention.

To the officers and men of the 67th and 51st Artillery Brigades, who, under trying conditions, served so gallantly with the 4th Division, all honor is due. Ever loyal, cheerful and uncomplaining, they earned and received the cordial respect and good will of men they so ably supported.

The 4th Artillery Brigade, too, had been weighed by the infantry and not found wanting. The batteries vied with one another in seeking out targets and destroying them. Officers and men were keen to "make good." They demonstrated their personal courage in a hundred different ways. One morning the men of one of the batteries were spread in the thin underbrush surrounding the position, partaking of their favorite dish—beans. A German shell came whistling through the air and landed squarely on the muzzle of a gun. A gunner corporal and private were killed, five men wounded, and the lieutenant's right arm torn off near the shoulder: At the dressing station near by he remarked: "The Boches haven't got me —I write with my left hand."

The test imposed on the Division in its first appearance as a full combat unit had been of the most severe character. The fighting with the French, in July, while bitter, had been brief. The combat around Sergy, while exceedingly desperate, had been waged by but two battalions. On the Vesle, however, were born and developed those powers of courage, fortitude and endurance that were later to meet their supreme test in the battle of the Meuse-Argonne. Here in the river valley, winning positions under heights dominated by the enemy, and maintaining those positions in the face of the whole repertoire of devices which the Germans had gathered during four years of the great game of war, the men learned how to "carry on." Amid the woods and hills, shelled and gassed all of the time and unable to move without encountering severe opposition, they maintained a cheerfulness of outlook that enabled them to endure. It was the spirit of the Ivy Division.

The Aisne-Marne offensive, launched by the Allies, officially ended on August 6th. That was the date on which the advance, all along the line, was definitely stopped. Actions subsequent to August 6th were considered as operations in sector warfare. The 4th Division may claim the distinction of being the first Allied division to cross the Vesle in this offensive and establish and maintain itself on the northern side.

Since they first went into line on July 18th, the divisional troops, in the counter-offensive, had advanced to a total depth of 19 kilometers. No accurate record was kept of material and prisoners captured, for most of the captures were made with the French and turned over to them. But the list was large, particularly during the 18th and 19th of July. The total losses of the Division during the operations from July 18th to August 17th were 53 officers and 1,357 enlisted men killed and died of wounds; 1 officer and 37 enlisted men died of other causes; and 179 officers and 5,296 enlisted men wounded and gassed; a total of 6,923 casualties.

All arms proved their worth in the fighting on the Vesle. It was not only the infantry and machine guns and artillery, but the medical services, the engineers, the Signal Battalion, and the trains, who bore the strain of the bitter fight with distinction to themselves and credit to the Division.

The fighting imposed a hard test on the medical detachments. Dressing stations were established immediately behind the front line. Ambulances came up under shell-fire to Villesavoye and St. Thibaut and by excellent organization and sturdy courage the fields were kept clear of such wounded as did not fall into the hands of the enemy.

The work of the Signal Battalion was of special value. To maintain communication, with roads blocked by transport, with shells constantly severing telephone lines, was its task. The battalion performed it to the complete satisfaction of the Division.

During the night of August 11th-12th the two reserve regiments and the remainder of the divisional troops (except the artillery) marched to the Forêt de Nesles. The two front line regiments and the attached units, which were not relieved until nearly daybreak, marched to the Bois de Dôle and biv-
ouacked there. Command of the sector passed to the 148
77th Division at noon, August 12th. 150

CHAPTER IX

MENDING A DIVISION

AFTER the severe fighting on the Vesle it became necessary to rest and build up the 4th Division. Leaving the Bois de Dôle and the Forêt de Nesles, the Division, less the Artillery Brigade and the Ammunition Train, was marched on the night of August 12th-13th to the Forêt de Fére where it rested until the night of August 14th. While bivouacked in the forest the Division was bombed on the night of August 13th-14th by German aeroplanes as the 39th Infantry had been on a previous occasion in the same locality. The 10th Machine Gun Battalion lost eight men as well as several motorcycles and a motor car. There were few other casualties. While the rest was welcome to the Division the men, once back in the forest through which they had passed on their way to the Vesle, could not but recall the many comrades they had left behind on the river's bank, and although the morale of the Division remained high, something of the melancholy of war was reflected in the survivors as they moved about waiting for orders. While in the Forêt de Fére some new equipment was distributed and a batch of replacements came in.

During this period the Division was visited by Mr. Winston Churchill who was accompanied by an English General. It was the first time the British statesman had ever seen a "pup tent" and he asked many questions about its use and origin. Although, speaking generally, there was not much difference between American and European equipment the two things that never failed to excite the curiosity and admiration of the Allies were our shelter tents and the American pack.

The Division was moved from the Forêt de Fére to the Reynel training area, one of the best and most agreeable in France. Reynel is situated about thirty kilometers northeast of Chaumont, where the American General Headquarters had been located since September, 1917. The movement was made partly by marching and partly by train.

The march was made on four approximately parallel routes. The troops were divided into five columns (two 151
being on the route farthest east) and distributed in such manner as to place at the five entraining points the exact number of officers and men that would fill the trains scheduled to depart from those points. The two columns on the route farthest east, crossing the Marne at Jaulgonne, followed the pleasant valley of the Surmelin River along its eastern bank to Condé-en-Brie, then up the western bank of the Dhuis River, through Artonges to Montmirail. Here one column halted and the other continued to Esternay. The route just west of this, after crossing the Marne at Chartéves, also followed the Surmelin River, on its western bank, to St. Eugéne and then led through Courboin, Viffort, Fontenelle to Joiselle. The third route, crossing the Marne near Gland on a "ship" bridge, followed the southern bank of the river to Chézy-sur-Marne, thence through La-Chapelle-sur-Chézy and Viels Maison to La Ferté-Gaucher. The fourth route led through Château-Thierry, thence along the north bank of the Marne, crossed the river at Nogent l'Artaud and passed through Hondevillers to St. Siméon. The march began after dark on August 14th. On the 15th the troops rested. On the nights of August 16th and 17th the march was continued. On the 18th and 19th the columns were entrained at Montmirail, St. Siméon, La Ferté-Gaucher, Joiselle and Esternay. At each of these points a company of engineers provided the necessary facilities for entraining. All marches, except from the last bivouac to the trains, were made at night.

View of Bazoches taken from the north. St. Thibaut in the distance. From the tree-covered nose on the sky-line at the right the 10th Machine Gun Battalion smashed a German counter-attack launched, on August 6th, from a point a little east of where the photographer is standing, against troops of the 58th Infantry on the Rouen–Rheims road. Taken September 9, 1918, after the Germans had retreated to the Aisne

Major General John L. Hines
Commanding 4th Division

Before the march was begun strict orders were issued that whenever an aeroplane was heard approaching the columns at night, the men were to be halted off the road and stand in absolute silence as, when the moon was up, a dark column moving or standing on the white chalk roads could be easily seen from the air. The wisdom of these orders was demonstrated within an hour after the columns had started on their march south after dark on August 14th. Just as the 8th Brigade, going in the direction of Condé-en-Brie, cleared the Forêt de Fére several enemy planes flew over the column. The German aviators dropped flares searching for a target. Although lights fell on both sides of the column, which had halted according to orders, the airmen failed to spot it and flew away.

During the day the men slept and smoked and loafed luxuriously. In the fields they could see women, children and old men at work; the younger men had all been taken away for the war. France owes a debt to her splendid peasant women that she can never repay. They tilled the fields; they gathered the crops; they made it possible for the men to "carry on." No one ever heard them complain.

Above all things, the soldiers enjoyed the silence. There were no big guns here. There were no snipers and machine guns to keep them keyed up to concert pitch. With a sigh of relief they relaxed. There were streams in which they could bathe and this luxury was utilized to the full. It was the first opportunity that had been afforded them, for nearly a month, to clean up at leisure. This is how one soldier described the feeling:

"After spending thirty days on the battle line or living in woods, where one sneaks about in a crouched position like a hunted beast; where dirt, death, vermin and decay are all about you; where even fresh air and sunshine are denied you; then, on a warm day, to stand, in clean clothes, on the banks of a stream where cooling waters have soothed your aching feet and refreshed your weary frame; to feel the tingle of a new life through your body made healthy and strong by your hardships; literally to drink in the fresh air and sunshine of the vast outdoors; that is an experience known only to a soldier of the great World War. You stand, a man transfigured. Then, if ever, it is that "you pack up your troubles in your old kit bag and smile, smile, smile."

And so the march south continued. The scenes of war were behind them. The country through which they were passing was a peaceful agricultural section something like the more densely populated parts of the Middle West. One thing was noticeably missing here, as everywhere in France. There was not a shock of corn to be seen. In general the roads were better than those in America. They were fine, straight highways, rock-surfaced and lined with stately trees that arched overhead. Occasionally the solitary grave of an aviator showed that even these verdant fields had been visited by the iron hand of war.

Division Headquarters was temporarily established at Montmirail, from which point the largest number of men were entrained. Here train after train of American troops of other divisions went by, some bound for training areas, others for some part of the front. All motor transport was sent overland. It was the general concentration on the Lorraine front of what was to be the First American Army.

The 4th Artillery Brigade remained in action on the Vesle, attached to the 77th Division until such time as the artillery of that Division could be brought up. A diligent searching of the sector was maintained. The casualty list grew steadily. One heavy shell which struck directly along the line of guns in "C" Battery, 13th Field Artillery, during the firing of a mission killed 5 men, wounded 27, and put 2 guns temporarily out of action. The fire of the undamaged guns was not interrupted for a moment.

On the nights of August 16th–17th and 17th–18th the brigade was relieved by the Artillery of the 77th Division and marched south through Fére-en-Tardenois and Château-Thierry, entraining on the 20th and 21st at the same points as did the infantry.

The arrival in the Reynel area, on August 19th and 20th, of the Ivy Division, fresh from the fighting on the Vesle aroused great interest among the American soldiers who had not yet seen action. Many a tale was told of the perils and hardships of war and of the amusing side of battle and it must be confessed that, to the inquiring novices, the men of the Ivy Division, proud of their experiences, poured out advice and wisdom with something of the exalted air of a successful father telling of the world to a wide-eyed son. Many an American soldier was "gassed" with "hot air" long before he got to the Argonne.

Division Headquarters was established at Reynel; the Headquarters of the 7th Brigade at Rimaucourt; of the 8th Brigade at Prez-sous-Lafauche, of the Artillery Brigade, at Signéville. The railhead was at Rimaucourt.

To clean up was every man's first desire the moment a billet had been found; to get deloused became an obsession. As one soldier described it: "We were comfortably billeted in private homes, public buildings, barns and barracks. As soon as we got ourselves located we declared war on the great international army—an army whose numerical strength was twenty times greater than that of the Allies and Central Powers. The personnel of this vast army of all nations are highly efficient individual fighters, seasoned veterans who have seen service on all fronts. They are decorated with Croix de Guerres, Iron Crosses, V.C.'s and D.S.C.'s and wear from one to eight service chevrons. For boldness and daring they are unexcelled. They are always on the offensive. They are equally at home whether attacking from the front or maneuvering for a flank position. When it comes to infiltration, they are in a class by themselves. They are the most adept troops in the world at gaining and keeping contact with the enemy. It is the only arm in the world rationed by the enemy. Only the most violent methods can be successfully employed against this vast horde, which threatens the peace of the world. We boiled them; we steamed them; we ran hot irons over them; we fought them single handed in the open; at night we attacked their ration parties by candle light; we soaked them in gasoline. Three days of incessant battle brought its reward. The cooties in our midst were at last temporarily exterminated."

The training area, located by the Americans in one of the choicest parts of France, was greeted with enthusiastic approval by the entire Division. In the vicinity of many of the towns Adrian barracks had been constructed which provided dry, comfortable, quarters. Bathing facilities were excellent and the weather was delightfully clear and warm. The peaceful surroundings, the regular work and hours, the bracing climate, the environment of pretty clean villages and green uplands worked wonders in producing a new mental and physical attitude among officers and men.

But a division does not exist solely for fighting and resting. It has to be rebuilt after every action. It has to be mended. The repairing of a division that has been battered during a campaign and rebuilding it for the next embraces many activities. Requisitions for officers and men necessary to bring each of the divisional organizations up to full strength are sent to the Adjutant's office, there consolidated and forwarded to Army Headquarters. The replacements upon arrival at the railhead are escorted to the divisional replacement depot and there assigned to organizations as the needs of the latter may demand. Regiments require carpenters, blacksmiths, painters, cobblers, clerks, musicians—in fact almost all trades and crafts are needed. From the descriptive list of each replacement soldier is ascertained his former occupation. The needs of the organizations for specialists are first filled, as far as can be, from the new men. The remainder of the replacements, who have no specialized occupation, are then distributed among the organizations in accordance with the needs of these units. Army Headquarters is charged with the responsibility of supplying to the divisions all officers and men that are called for.

Similarly supplies and equipment of all sorts are requisitioned for by the various organizations; clothing, arms, ammunition, signal, engineer, ordnance and medical supplies are asked for by the units concerned. Soldiers have a habit of losing their equipment in action, except, of course, their weapons. These requisitions all pour into the office of G-1 under whose direction they are consolidated and forwarded to Army Headquarters to be filled. After the replacements have arrived and been distributed, after new equipment has been provided, the division must be welded into a homogeneous whole, the newcomers must be trained up to the standard of the rest of the men, the defects noticed in action must be remedied and physical fitness maintained. Indeed much hard work is done by a resting division.

Replacements in officers and men were sent to the Division without loss of time. Most of the men thus received were already quite well trained in all but actual front-line experience having been drawn from certain divisions which, upon arrival in France, had been designated as replacement divisions. Supplies of all kinds also poured in and were promptly distributed. Of only one class of supplies was there a shortage—spare parts for motor vehicles—and this shortage had existed since the arrival of the American divisions in France and continued to exist until they sailed for home. No one in America can appreciate the troubles of the Divisional M.T.O. (Motor Transport Officer). Despite his utmost efforts he could not get from Army Headquarters what was needed. He bought, he borrowed, he lied, he begged, he stole. He sacrificed his hopes of salvation every day of his life. When he was not being reviled by generals, colonels and truck drivers in the Division he was being cursed at Army Headquarters for an infernal nuisance. The one comfort in the otherwise cheerless existence of the 4th Division M.T.O. was the satisfaction he derived from knowing that the motor transport was kept moving. With the active aid of G-1 he sent out a live wire—an intelligent young officer who spoke French—in a motor car, who scoured France for spare parts. He got them from the British, from the French, from shopkeepers, from depots. His journeys carried him from Calais to St. Nazaire, from Brest to Dijon. What he did no one knows. How he got the parts has never been told. But he got them and the Division motor transport functioned.

Training went rapidly ahead in the Ivy Division. It now possessed a new interest as the lessons learned from the war could be adapted with excellent effect. The men were keen, the replacements sought to emulate their veteran comrades and a soldierly happiness, that brought gladness to the hearts of the commanders and pride to the line officers, diffused itself throughout the Division.

Much real, constructive training was done in the Reynel area. Many mistakes had been made in the first two engagements. Many lessons had been learned. Officers applied in their training the knowledge and experience gained under fire. Field problems reproduced fighting conditions. Great pains were taken to train the men in the attacking of strong points, especially the reduction of machine gun nests. There was constant drill in approach formations under various kinds of fire. As rifle ranges already existed in the area target practice was resumed at once. There were schools for officers and non-commissioned officers.

Special attention was devoted to liaison, both between the units of the Division and between troops on the ground and in the air. The necessity for this work had been impressed upon all officers and men by their experience in the fighting on the Vesle River. Great disadvantages had resulted from difficulties in maintaining communication between adjoining units and between units and their next higher command. Too many runners had been killed. So visual signalling was emphasized. Besides the communications between the infantry units, there was the all-important question of liaison between the infantry and the artillery. Artillery cannot win a battle. Its sole function is to help the infantry and it can do this effectively only if the knowledge of what is required and the means of communicating this knowledge are as perfect as possible. The very special interest that attaches to good liaison between the artillery and the infantry was thoroughly instilled into the minds of all. By means of problems both branches were impressed with the importance of the work of the artillery liaison officers who accompany the infantry and telephone back to the supporting artillery the immediate requirements of the infantry commander; with the value of artillery observation posts located as closely as possible to the front for quickly directing destructive fire on enemy concentrations or batteries; with the necessity of having all means of communication functioning at the same time. During the last week in August a liaison maneuver was held in which the entire Division participated.

In the training particular emphasis was placed upon the essentials of control and discipline of the smaller units and upon target practice, the officers realizing to the full the prime importance of the rifle and of perfection in fundamentals. Ranges already constructed were in constant use during the day. Under the instructions of the more experienced riflemen the new men made excellent progress.

The outstanding features of these weeks were the rapidity and completeness of re-organization, the thoroughness with which the replacements were assimilated and equipped and the untiring energy and good will of all concerned. A month in a rest area was usually allowed to put the finishing touches to the training; but at this time events were moving fast and when orders were received, on August 31st, to move, the Division found itself in good shape for active service.

The artillery, as well, had its re-organizing tasks. Lost equipment having been replaced, the materiel was cleaned and oiled and worked on by men who had now acquired a real affection for their guns. The horses were doctored and grazed and given light exercise in an attempt to overcome the effects of the exhausting work they had been called on to perform on the Vesle. It was here that Lieut. Col. Rucker rejoined the 16th Field Artillery and assumed command, Colonel Butner having been transferred to the 1st Division on August 17th. Command of the 77th Field Artillery also changed hands when Colonel Creed F. Cox was assigned to command that regiment, Lieut. Col. Commiskey, the former commander, remaining with it as its lieutenant colonel.

Here, too, Colonel F. M. Wise, U. S. Marine Corps, who had seen much active service with the Marine Brigade, 2nd Division, reported for duty and was assigned to the 59th Infantry taking over the command from Lieut. Colonel Max B. Garber. Colonel George H. McMaster, Infantry, also joined and was assigned to command the 58th Infantry.

During the time spent in this area the Division was training and working constantly during the daylight hours. The atmosphere was one of expectancy. Every one knew that an offensive on a large scale was in course of preparation although all were more or less ignorant as to just when and where the blow would be struck and one guess was as good as another.

With the return to their old units of some of the men who had been wounded in the action on the Ourcq, the wound stripes began to appear and even came, such is the nature of a soldier, to excite a curious envy among the unwounded.

Meanwhile there had been important changes in the divisional command. On August 16th Major General Cameron was transferred from the Division to command the V Corps. It was with very real regret that the officers and men learned of his departure although they were the first to congratulate him on his promotion. But it is the fate of a division of a growing army to lose its generals for the higher commands. The record of the division itself, usually bought with blood, is the best commendation for its general. General Cameron had commanded the Division since its organization. With his departure officers and men alike felt they had lost a real friend.

Brigadier General Benjamin A. Poore, commanding the 7th Brigade, took temporary command of the Division while Colonel Frank C. Bolles acted as Brigade Commander. On the 27th of August, however, the Ivy Division came under the command of one of the ablest and most popular American generals in France, Major General John L. Hines. General Hines was born on May 21, 1868, and graduated from the Military Academy, class of 1891. He came to the 4th Division armed with invaluable knowledge and experience gained as a successful regimental and brigade commander in combat. He knew the weapons that were placed at his disposal—arms and men. He knew arms from the standpoint of the man who has used them. He could estimate the temper and spirit of men with unerring accuracy. His military judgment was sound and he had the faculty of placing his finger at once on the weak spot of a plan or project. He had the viewpoint of one who sees big things and the driving power to push them to a successful conclusion. He expected a lot and got it. He was just, loyal, indifferent to hardship, full of human sympathy—a man generous in giving credit, impatient of failure, who marched always with his eyes on the goal.

The Division now formed part of the First American Army under the personal command of General Pershing and, on
153 August 27th, was assigned to the V Corps. Rumors of a new offensive ran their course through the villages where the

men were resting and training and when, on September 1st, all units of the Ivy Division were moved north by truck to an area about Vavincourt, a little north of Bar-le-Duc, the rumor
seemed to gain confirmation. One of the most striking in- 154
cidents that happened in the Vavincourt area was the manner in which the majority of the officers at Division Headquarters received their first, and what proved to be accurate, information of where the next blow would be struck. A French woman who kept a village wine shop told the French liaison officer at Division Headquarters that the American Army was going to reduce the St. Mihiel salient and that the 4th Division would participate in the attack. She even knew details of the proposed operation, which showed that her connection with the French Army Intelligence Service was perhaps very real. Back of the front in many of the villages the French had established "souricières" or "mouse traps," generally in small wine shops or eating places frequented by officers and men passing to and fro behind the front. Here the winning smile of the fair daughters of France often drew confidences from those whose lips should have been sealed, or, the good wine aiding, led men to speak thoughts which had been better locked in their bosoms. Many a man has found promotion slow or been removed to distant parts of the front because somehow his superiors had been informed that, good soldier as he was, he lacked discretion; many a doubter, many a man who lacked confidence in ultimate victory has found himself either suddenly sent forward with attacking troops or else relegated to semi-civilian work at the rear; above all, many a spy has been shot as the result of the clever work of patriotic French women and men who served drink and food in the "souricière."

At Vavincourt more training and less rest was the order of the day. Although still a considerable distance from the lines, every precaution was taken to guard against enemy aërial observation. While troops were drilling, sentries were constantly on the watch to signal the approach of enemy planes. The men were forbidden to circulate on the roads except in case of necessity and were ordered to keep off the streets in the towns and to stay under cover as much as possible. Camp sites were carefully concealed in the woods and great care taken to secure only dry wood for burning so that the smoke from rolling kitchens might be reduced to a minimum. Training progressed satisfactorily and rapidly. On September 5th another successful liaison maneuver was held. On September 7th the 39th Infantry staged an exhibition demonstration of an attack on a strong point by a battalion, using Stokes mortars, 37 mm. guns, machine guns, and Chauchats, with live ammunition. The exercise was highly instructive. The preparations for the offensive of St. Mihiel were being completed.

CHAPTER X

THE ST. MIHIEL OFFENSIVE

THE reduction of the St. Mihiel Salient will go down in military history as a well conceived, carefully planned, and efficiently executed operation. It opened a new chapter in the history of the American Army in France. Ever since the day when General Pershing had offered Marshal Foch all the American troops and materiel to use as he saw fit these units, whatever their experience and training, had been placed unreservedly at the disposal of the Allied High Command with a willingness and freedom from conditions that proved of extraordinary advantage to the Allies. But, although American divisions were freely given to serve with both French and British Corps, that was by no means the plan on which General Pershing had been working ever since his arrival in Europe. The American idea was to end the war in the shortest possible time by the only way in which war can be ended—decisive victory. General Pershing took a broad view of the problem that confronted him. When he saw that every available man was needed by the French until England could send over the last reserves she was holding at home, he was ready to fling everything he had into the furnace. Losses did not appal him. He knew that the whole man power of the United States was behind the army and he realized that a short war, even if costly, is better in the long run than a prolonged war of attrition.

But although he had offered all his forces to Foch, in the emergency that had arisen, he was determined that America, as one of the great nations engaged in the war, should play its own part. He was determined that there should be an Am-

erican Army, just as there was a French Army and a British Army; that, from the large number of soldiers already in France and the larger number still to come, he would form an army which would fight as a separate and independent force on the western front. General Pershing appreciated that differences of language and temperament would operate to prevent that complete understanding and sympathy without which full co-operation is impossible. Even while victorious and advancing side by side with French units, it was difficult at times for American brigades and divisions to fight in harmony with the French. A reverse, under such conditions, would have been disastrous. Squabbles as to responsibility would have ruined international friendship and injured American morale. In this matter the American commander had the British experience to guide him. While admitting General Pershing's viewpoint, the French High Command very strongly expressed the belief that it would be better if—during the year 1918 at least—he would consent to attach his troops to the French and British armies and distribute them as individual units not exceeding a division in strength. In an effort to convince him of the wisdom of this course an important conference was held at Bombon in July, 1918, at which, in addition to Marshal Foch and General Pershing, there were present General Pétain, Sir Douglas Haig and several members of the inter-allied General Staff. Another was held on August 30th and a third on September 2nd. General Pershing remained ob-
155 durate and resisted all the pleadings of the Allied leaders. He said that, if at any time an emergency arose which demanded the placing of American divisions under French or British Command for short periods, he was quite willing to accede to the Allies' desires; but he insisted that he had been sent to Europe to form an American Army which was to operate independently, although, of course, under the supreme direction of Marshal Foch.

General Pershing, having interpreted the sentiments of

the American people and of their army to the Allied High Command, started on the work of forming the First Army. His divisions had been massed in an area extending roughly southwest from Pont-à-Mousson on the Moselle River. The St. Mihiel salient lay right before them. It had been formed in September, 1914, when, in a counter-attack after the Battle of the Marne, the German forces had broken into the Woëvre defenses and captured the two important positions of the Camp des Romains, just south of St. Mihiel, and of Mont Sec, 13 kilometers east of it. This placed the main railroad from Paris to Nancy directly under their fire, prevented the use of the Canal de l' Est along the Meuse River and greatly hampered communication between the Capital and the Lorraine front. The reduction of this salient was the first task assigned by Marshal Foch to General Pershing who gathered together fifteen American divisions for the attack. Of these nine were in the assaulting line and six in reserve. Four French divisions were added by Marshal Foch to this force. 156

It was to be the first independent operation ever carried out by the American Expeditionary Forces and, at the same time, it was the first occasion in history that large bodies of troops belonging to a foreign nation—in this instance French—were to fight under a purely American command. Every preparation was made for the attack. The artillery was massed in the greatest possible strength, the aeroplanes were gathered together in flocks—French, American, and British—and the infantry was concentrated for a swift, sure blow.

The general plan was to make simultaneous attacks against the flanks of the salient. The objective was fixed as the line Vigneulles–Thiaucourt–Regniéville. On the southern 157
face of the salient two American Corps, the I, on the right, with four divisions and the IV, on the left, with three divisions, were to advance generally northwest. On the western face of the salient the V Corps, with three divisions, was to push generally southeast. The IV and V Corps, advancing toward

each other, were to meet and form the neck of the bottle that should inclose the mass of Germans in the salient. In the centre, between the IV and V Corps, was the II French Colonial Corps consisting of three small French divisions. These were to follow up the retirement of the enemy from the tip of
156 the salient.

The V Corps, on the western face, consisted of three divisions arranged in order from north to south: 4th Division, 15th Colonial Infantry Division (French), 26th Division. Of these the 26th alone was to make a deep advance, directed toward Vigneulles. The French division was to make a short progression to the edge of the heights in order to cover the left of the 26th. The 4th Division was not to advance without orders from the V Corps.

On the night of September 5th-6th and 6th-7th, troops of the 59th Infantry and the 12th Machine Gun Battalion began the relief of French troops in what was known as the Tou-
158 lon sector, southeast of Verdun. This sector, very quiet at the time, formed the extreme northwestern edge of the salient. It had been held continuously since 1914 and some places, like Fresnes-en-Woëvre and Manheulles and, farther south, Les Eparges, bore bitter memories for all the many French divisions which had fought there throughout the long months of trench warfare. Both the opposing lines were heavily entrenched with the utmost thoroughness. Nothing was lacking. Every new device, every new intricacy of trench warfare had been introduced in the sector which the 59th took over when it relieved its French comrades. The terrain in No Man's Land bore all the scars of former suffering. Shell holes were so thick that it seemed as if a projectile must have fallen every ten feet. Here and there they overlapped one another. The roads in many places had completely disappeared. Hills devoid of every form of vegetation; woods that knew only the hanging, blackened branches of what had once been trees; all the surrounding country white with the

overturning of the chalk subsoil; this was the scene of devastation which greeted the eyes of the men of the Ivy Division as they looked from their commanding position over the flat plain of the Woëvre.

The 59th Infantry under the command of Colonel Wise, together with the 12th Machine Gun Battalion, which had been moved by bus on the night of September 5th-6th to the woods in rear of the front line completed the relief of the French garrison about 12:30 A. M. on September 7th, the line of occupation running from north of Watronville in the north to just opposite Trésauvaux in the south. The regiment occupied 9 kilometers of front. North of Haudiomont, the line crossed the famous Paris–Metz road which the men had last seen at Château-Thierry. The 3rd and 2nd Battalions held the right and left flanks, respectively, with the 1st Battalion between them. Each battalion had three companies in front line and one in reserve. Behind them the rest of the 8th Brigade, which had been moved up by bus on the night of September 6th-7th, was in support north of the Sommedieue–Haudiomont road. The 7th Brigade and 10th Machine Gun 159
Battalion were in reserve, having been marched up on the nights of September 9th-10th and 10th-11th to positions in concealment in the woods east of Haudainville. The en- 160
gineers as usual were busy. They were repairing roads for the 26th Division and were later to perform the same exacting task for their own division. Division Headquarters was established at Haudainville.

During the day of September 6th the officers of the 8th Brigade studied the prearranged plans for the defense of the division sector, and carefully reconnoitered the area in which they were to operate and the ground in their more distant front. The men of the 59th were soon viewing with amazed curiosity the intricate system of trenches, the deep, electric-lighted, concrete dug-outs, the up-to-date means of communication, and striving to familiarize themselves with the way in and the way out of such a maze. So these, at last, were the trenches for the warfare in which they had been so assiduously trained in the United States and which, up to that time, although they had suffered many casualties, they had never seen. The men had read of how careful they must be in a front-line trench; how if they stuck so much as a finger above the parapet the German snipers would immediately pop it off. They had read of the many weird adventures of patrols roaming about at night in the wire of No Man's Land or forced to lie for hours in shell holes half full of water. They did not know at what moment a barrage would drop in their trenches and for a few days they pussy-footed around like real Indians. Many a man on night duty jumped as some prowling rat caved in the side of the trench. Those rats were big but the darker the night the bigger they got. They carried off hard tack, bully beef, mess kits, packs, rifle grenades; in fact, some of them must have been collecting an arsenal, for everything lost was blamed on "the rats."

Meanwhile, officers were inspecting the well-placed observation posts replete with telephone lines and powerful glasses. From where they stood they could see stretching before them a broad, gently rolling plain which rose gradually to a hill on which, 35 kilometers away, was the fortified city of Metz. The visibility was excellent. From where they stood they could see, 250 feet below, band after band of wire entanglements which zigzagged across the front and ran parallel to it for miles and miles. Manheulles, with a few shattered walls still left standing to show where the town had once been, lay two kilometers across the line, a mass of powdered stone. It was the same with Fresnes-en-Woëvre and Bonzée-en-Woëvre. Off in the distance, far behind the war zone, the smoke of factories and the plants in the Briey iron basin could be discerned and, so excellent was the observation, the railroad trains could be seen moving across the industrial region. St. Mihiel was the last salient left on the western front. To reduce it and place the region beyond under fire, to interrupt and finally stop the German exploitation of the Briev iron mines, was the task entrusted to the First American Army. 157

The 4th Division was directed not to attack without express orders from the V American Army Corps, commanded by Major General George H. Cameron, to which it was assigned. At the extreme edges of the St. Mihiel salient, however, there were to be raids and a general harassing of the enemy.

On September 11th information reached the Division Headquarters from the V Corps that "D" day would be September 12th; that the artillery preparation would commence at 1:00 A. M.; that "H" hour for the I and IV Corps would be 5 A. M.; that the infantry attack of the V Corps would commence at 8 A. M. The troops were immediately informed. 161
The 59th was not to attack in the morning but was to stand by, holding its front.

The night was very quiet, nervously, expectantly quiet. At exactly one o'clock on the morning of September 12th the greatest artillery concentration ever brought together on the western front roared forth in unison. The drive was on. In a few seconds after the first discharge ton after ton of explosives was being hurled into the front line of the enemy and the immediate rear. Long range naval guns merged their heavy discharges with the sharp reports of the "75's" as they sought out objectives many miles behind the German lines. The whole sky, reflecting the flashes of the thousands of guns, looked, as one officer said, "like a great white way." It stretched along the entire front as if miles of munitions factories, placed side by side, had been fired simultaneously. The flashes merged into one another while the gas and smoke created a mist that preserved a fantastic outline in the light of the cannon. As day dawned the men of the 4th Division, looking out at the amazing scene, noted that the barrage became more and more intense and at 7:30 A. M. it seemed as if

the artillerymen themselves were frantically endeavoring to outspeed one another in getting the greatest number of shells into the German lines in the shortest possible time. What the men had seen before in the way of artillery fire was totally eclipsed by what they now saw. They at last realized that this was the real mechanism of war, the sublime destructive power of modern shell fire. To see this was well worth coming all the miles they had travelled. They were right. That morning bombardment fired away over ten million dollars worth of
162 shells but the result was worth it.

During the early afternoon the first patrol left the 8th Brigade trenches for the German lines. At 4:10 P. M. another followed to reconnoiter the town of Manheulles and to reach a point just beyond on the Paris–Metz road. Heavy machine gun fire greeted them and they returned. All that night and the next day patrols were kept out endeavoring to determine the exact movements of the enemy. On September 14th, at 10 A. M., the reserve company of each battalion left the trenches to attack. Advancing behind a barrage they reached their objectives and "D" Company, entering Manheulles, captured three of the many Germans who could be seen fleeing
163 from the rear of the town. But the great surprise of the soldiers was to find hot meals awaiting them in Manheulles. In their haste to escape, the Germans had left their kitchens, on which the noon meal was being prepared. Good roast beef, fried potatoes, sauerkraut, coffee, loaf sugar, bread and beer were the delicacies that they had arrayed for themselves.

Meanwhile Fresnes-en-Woëvre was being attacked. When the Americans first approached the town they were heavily fired upon. They withdrew to cover but at 2:00 P. M. the town
163 was captured and remained in their possession. Patrols sent out against the Ronvaux wood and Hill 259, on the left, met with artillery and machine gun opposition which, in itself, was sufficient confirmation of the Germans' determination to hold that sector. At Riaville a similar fate befell another

patrol although it approached within 500 yards of the village which lay far beyond the new line. 164

In the meantime the 7th Brigade had not been idle. During the night of September 12th-13th, under orders from V Corps, it marched to the junction of the Grand Tranchée de
Calonne and the Mouilly–Les Eparges road. Here it was 165
joined on the morning of the 13th by the 10th Machine Gun
Battalion. This force was held as a reserve for the V 166
Corps. It did not, however, participate in the action. On the night of September 13th-14th these troops marched back to their former billets. 167

The following summary of the operations is taken from the report of General Pershing:

> At dawn on September 12th, after four hours of violent artillery fire of preparation and accompanied by small tanks, the Infantry of the First and Fourth Corps advanced. The Infantry of the Fifth Corps commenced its advance at 8 A. M. The operation was carried out with entire precision. Just after daylight on September 13th, elements of the First and Twenty-sixth Divisions made a junction near Hattonchatel and Vigneulles, 18 kilometers northeast of St. Mihiel. The rapidity with which our divisions advanced overwhelmed the enemy and all objectives were reached by the afternoon of September 13th. The enemy had apparently started to withdraw some of his troops from the trip of the salient on the eve of our attack, but had been unable to carry it through. We captured nearly 16,000 prisoners, 443 guns and large stores of materiel and supplies. The energy and swiftness with which the operation was carried out enabled us to smother opposition to such an extent that we suffered less than 7,000 casualties during the actual period of the advance. 168

The American offensive had proved a remarkable success. The St. Mihiel salient was closed. The line when finally established was about five kilometers in advance of that designated in the original plan.

During the night of September 14th-15th part of the 59th Infantry was relieved by the French and the following morning the relief was completed. The 8th Brigade was with-

Replacement troops for 4th Division in railroad yards at Rimaucourt, August 25, 1918

3rd Battalion, 39th Infantry, entering French trucks to be taken to the Vavincourt area preparatory to entering the St. Mihiel Drive

Desolation in the Bois des Eparges after the American artillery fire

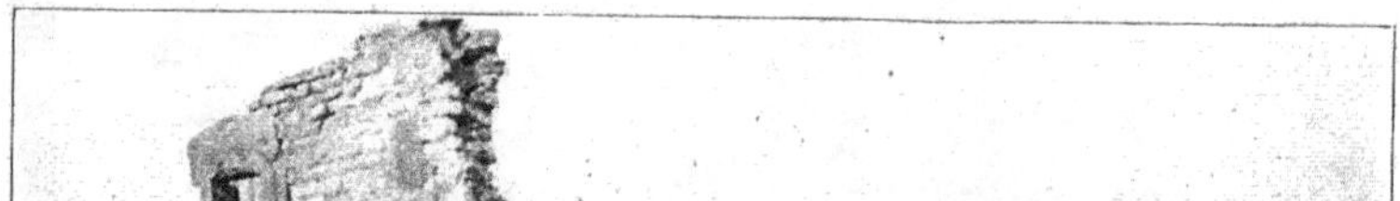

Ruins of homes made into machine gun nests by the enemy.   Manheulles

drawn from the front line and placed in bivouac in the woods
169 between Haudainville and Sommedieue. One task had been performed. The 4th Division was getting ready for another and a greater one.

Far more trying to the men than hostile shell fire and bombing was the rain that fell continually from September 8th to 13th. In order to avoid these fall rains, which usually began about the middle of September, General Pershing had hastened the operation by every means possible. Bivouacking in sodden woods, plodding at night along muddy roads in utter darkness with clothing soaked and no prospect of dry blankets or a cheering fire, the Ivy men felt, more than once, that when it came to bad weather during combat their luck was certainly against them.

The 4th Artillery Brigade did not fight with the Division during the St. Mihiel operation but all of its units were in action throughout the entire period. The 4th Trench Mortar Battery at last found an opportunity to cast its clumsy projectiles into the enemy lines. Like the 16th Field Artillery, it was brigaded with the 15th French Colonial Infantry Division on the immediate right of the Ivy Division's sector. The 13th and 77th Field Artillery were attached to the 26th American Division on the right of the French Colonials. Thorough reconnaissances were made of the areas and, on the night of September 7th-8th, the regiments moved forward and
170 the guns were put in position.

The more difficult of the two sectors in which the units of the 4th Artillery Brigade operated was that of the 15th French Colonial Infantry Division. Colonel Docteur, the French artillery commander, had prepared a barrage for the combined French and American artillery of that division which, by general agreement, was pronounced a masterpiece. All the experience of four years of warfare was compressed into the chart issued to the officers and the Americans considered it quite a privilege to fire a barrage under such conditions. The

16th Field Artillery fought in the southern half of the French divisional sector and the condition of the terrain after the battle was, in the minds of the men, the best testimony to their accuracy in firing. The number of Croix de Guerre awarded the regiment by the admiring French confirmed this
impression. The fire of the 13th and 77th Regiments, serv- 171
ing with the 26th Division, was no less accurate and effective.

The difficulty of moving the guns over the muddy roads and especially the difficulty of bringing up ammunition were far more formidable than all the stray shells from German batteries that aimlessly sought their targets on the morning of the attack. The men had to work harder than ever before in their lives to move the guns and caissons. Much of the ammunition, in fact, had to be carried up by hand. At night, men with two shells on their shoulders would, in the impenetrable blackness, walk two hundred yards down a steep muddy hill, across the face of which ran a trench. This it was impossible to see; so a man stationed at the trench would cry "Jump" and the carrier with his shells jumped, another man on the other side catching him so as to keep his balance. The weight of the shells coming down on the shoulders after the jump can be imagined. Six of those trips and a man had to be rested. It was only the training and physique of the soldier that enabled him to endure such a strain.

The advance of the 26th Division was so rapid that the light artillery commenced to move forward immediately after the first wave had broken the German defensive system. Many of the batteries had advanced to new positions over roads made almost impassable by German trenches. There was consequently some difficulty in reaching them with the with-
drawal orders which had been issued on September 13th. 170
On the 15th and 16th, however, the regiments arrived in their
bivouac area in the Forêt de Souilly. 170

During the entire period of this operation no unit in the Division had harder or more trying labor than did the Ammunition Train. All work was done at night, all the nights were inky dark, every trip was made in pouring rain, and all the roads were bad. It was necessary here to detail two drivers to each truck, each driver taking his turn. Even then they could hardly stand up under the strain of supplying, for this tremendous artillery concentration, not only the units of the 4th Division but those of the 15th French Colonial Division as well. The trucks of the Ammunition Train delivered their loads, not at dumps, but to the individual guns wherever this was possible.

During the 15th, 16th, 17th, and 18th of September the troops received special training in liaison preparatory to the Meuse-Argonne Offensive. To the disgust of the men, gas masks were required to be worn two hours continuously each day. The latter instruction later vindicated itself. In the Vesle the Division had suffered many gas casualties; in the Meuse-Argonne, under concentrations still heavier, these were materially reduced.

On the night of September 19th-20th the Division (less the Artillery Brigade) moved to the Bois de Landrecourt and the Bois de Lemmes, just northeast of Lemmes. The march was made in three columns, the troops crossing the Meuse at
172 Haudainville, Dugny, and Dieue. Daybreak of the 20th found them in their new localities. The artillery adjoined them on the east. Division Headquarters was established at Lemmes. The utmost secrecy of movement was enjoined on all. The "big push" was impending.

CHAPTER XI

THE MEUSE-ARGONNE OFFENSIVE—1ST PHASE

THE Battle of the Meuse-Argonne will always stand out as the greatest in American history and as the supreme achievement of the American Expeditionary Forces in France. In order to realize its overwhelming importance it is necessary to understand its relationship to what is sometimes called the great Battle of France, which began on July 18th and ended the war on November 11th and of which the Meuse-Argonne was an essential part. In this Battle of France are included all the operations in which the 4th Division participated; the Aisne-Marne Offensive, known generally as the Second Battle of the Marne; the St. Mihiel Offensive, usually called the St. Mihiel Drive; the Meuse-Argonne Offensive, better known as the Battle of the Meuse-Argonne.

It will be recalled that, after long years of trench warfare, the Germans had smashed through the British front in March, 1918, nearly reaching Amiens and making the deep Montdidier salient. In May they had attacked again, this time breaking through the French front on the Chemin des Dames, threatening Paris and creating the Marne salient.

The Battle of France began on July 18th with General Mangin's counter-offensive south of Soissons which drove in the German flank and developed into the Second Battle of the Marne in which the 4th Division played its distinctive part. After the Marne salient had been reduced, the British and French forces wiped out the Montdidier salient. The St. Mihiel salient, the only one remaining on the front, was reduced by the First American Army under General Pershing and again in this phase of the great battle the 4th Division participated.

The reduction of these three salients left the Germans on the western front behind defenses of great strength. After St. Mihiel fight the front was like a huge door shut in the face of Marshal Foch. Its hinges were on the Meuse, east of the Argonne forest, and the door was closed by an immense boulder of defenses around Lille. The whole plan of the main and final phase of the Battle of France was, after the salients had been reduced, to push back the door by throwing forces against it all along the line and simultaneously to take it off its hinges by smashing through on the Meuse-Argonne front. This was the task assigned by Marshal Foch to General
173 Pershing who, at his own request, had been given this sector in which the now thoroughly organized American Army could fight as a whole.

In giving General Pershing the task of breaking through these formidable defenses Marshal Foch said to the American Commander-in-Chief: "*La Meuse-Argonne est un morceau dur à avaler; il y a là des obstacles considérables. Mais c'est bien, vos hommes ont un mordant endiablée. Ils auront raison de tout çela. Allez-y!*" ("The Meuse-Argonne is a hard nut to crack; there are great obstacles to be overcome. But it is all right, your men have the devil's own punch. They will succeed despite everything. Go to it.")

Nowhere on the western front were the German defenses stronger than on the Meuse-Argonne. Apart from the natural strength of the terrain, the defensive lines were closer together here than anywhere else on the front. The reason for this was that the entire German system of defense in France was like a series of huge steel cables strung between two pillars, the northwestern pillar being the fortified area of Lille, and the southeastern the Metz-Thionville system. These several cables sagged in the centre, just about Laon, so that between cables at this point there was a distance of some 20 kilometers. But on the Meuse-Argonne front, close to the Metz-Thionville pillar, all the cables were so grouped together that they covered

a total depth of not more than 20 kilometers. It was against 174
this mass of defensive lines, the Hagen Stellung, the Volker Stellung, the Kriemhilde Stellung (which constituted part of the Hindenburg line) and the Freya Stellung, not yet fully organized, that the American Army launched its attack.

All these lines followed the heights, fortified not only by the natural strength of the positions on which they were built, but generally concreted and wired. They were the last word in defensive warfare as experience since 1914 had taught it. Standing on the crest of Montfauçon, on the roof of the steel and concrete observation post from which the Crown Prince watched his troops during the Verdun attack in 1916, a splendid view of the battlefield is obtained. Toward the north there is a series of rolling hills covered here and there with patches of woods, each hill and each wood well adapted to sustained defense; on the left lies the difficult massif of the Argonne forest; on the east are the heights of the Meuse, dominated from the eastern bank of the river; and to the southeast appear the hills made famous in the Battle of Verdun, Hill 304 and le Mort Homme.

It was known that the Meuse-Argonne area would be more solidly held and more desperately defended than any other part of the front for if it were broken through, the retreat of the German armies to the Rhine would be cut off and they would be forced to surrender. The Carignan–Sedan–Meziéres railroad was absolutely essential to the enemy. With this cut, before they had completed their retreat, their ruin was certain.

The German High Command, in the last days of September, 1918, had come to realize that all hope of victory had been
shattered; that the counter-drive of Marshal Foch and the 175
renewed energy of the Allied troops, the collapse of Bulgaria
on September 22nd and the destruction of the Turkish Armies 176
on the Syrian front, had reversed the whole situation. The 177
Germans believed that the Americans would attack east of the Meuse, in the direction of Metz, as an exploitation of the St.

Mihiel success. This impression was confirmed by some very clever and ingenious wireless messages, sent out from American G. H. Q. by a fictitious army commander in charge of a phantom American army to the east of Verdun. All of these were caught by German wireless stations and decoded and resulted in a sudden increase of anxious information-seeking raids by the Germans on that front. Hence, General von Marwitz, commanding the Fifth German Army, issued an exhortation to his men to "fulfill their duty to the utmost as the Fifth Army may have to bear the brunt of the fighting of the coming weeks, on which the security of the Fatherland may depend. The fate of a large portion of the western front, perhaps of the nation, depends on the firm holding of the Verdun front." While expecting the attack to be launched from positions east of the Meuse instead of west of the river, he correctly divined the Allies' intention and emphasized the importance of frustrating them.

The Americans, however, were confronted with a task such as few armies in the history of any war have been called upon to face. The Meuse-Argonne sector, though the scene of desperate fighting during the French attempt to capture the Argonne Forest in 1915 and the battles of Verdun in 1916, had for a long time lived in that quietude which the French, with a wise regard for the necessity of preserving what they could of shattered France, had no intention of disturbing. The consequence was that no light railroads, so necessary for an advance had been built from the rear to the front lines. The road communication still depended on the only direct artery, the Clermont-Varennes highway, and the lateral road from Verdun, through Dombasle to Ste. Menehould. It would have been impossible to find, on the whole western front, a sector less adequately provided with avenues of communica-
178 tion. Furthermore, the Americans had never seen the sector before. Very little of the ground was known to them. The difficulties of communication would have to be surmounted

after the offensive had started. To this sector was transported the First American Army.

The staff work, at Headquarters First Army, involved in moving, in a period of twelve nights, over roads insufficient in number, the entire personnel and transport of 15 divisions can not be grasped by a person unfamiliar with such tasks. Of these 15 divisions 7 were involved in the St. Mihiel drive, 3 were in sector in the Vosges, 3 in the neighborhood of Soissons, 1 in a training area and 1 near Bar-le-Duc. To these must be added scores of batteries of French artillery, balloon units, American Army and Corps artillery and other auxiliary troops. The routes of all were bound to cross more or less. The ammunition necessary for the tremendous artillery concentration that was to precede the attack, had to be brought up on truck trains whose routes threaded in and out through the mass of moving men and transport. For each a road must be determined that would least interfere with others. Every night the entire area south of the front line and west of the Meuse was a wriggling, seething, ant-like mass of men, animals, wagons and trucks. They covered every road, trail and footpath. All this was in utter darkness; not a light was shown; and night after night it rained.

No more creditable piece of technical work was ever performed by the American Army. The troops were not moved into front line trenches until the night that preceded the attack. At "H" hour every unit was in place, the ammunition and guns had arrived, the troops were ready to begin the fight.

Three army corps were assigned to the sector of the First Army. This extended from the Meuse River on the right to the western edge of the Argonne Forest, on the left, where it joined that of General Gouraud's Fourth French Army. In an air-line, the front held by the Americans covered twenty miles from east to west. The actual length of the line was much greater, owing to its winding character. The I Army Corps, under General Hunter Liggett, was on the left, the V

Major General John L. Hines, commanding 4th Division, and Staff, at Haudainville, Sept. 15, 1918. Reading left to right: 1st Row: Lt. Col. E. W. Savage, G-2; Lt. Col. E. Canfield, Jr., G-1; Lt. Col. C. A. Bach, C of S; General Hines; Col. A. H. Acher, Div. Engr.; Lt. Col. T. L. Sherburne, Div. S. O.; Lt. Col. R. L. Carswell, Div. Surg.; Lt. Col. Arthur Hixon, Comdg. Trains; 2nd Row: Lt. Col. H. A. Parker, G-3; Capt. H. J. Savage, Div. Adjt.; Maj. H. H. Fletcher, D.M.G.O.; Capt. M. Hommet, French Mission; Maj. L. N. Hine, D.O.O., Capt. Ralph Smith, ADC; Lt. Chas. O. Lawrence, ADC.; Lt. W. M. Carson, ADC.; Top Row; Maj. F. Gilbreath, D.Q.M.; Lt. Col. C. C. Teare, D.J.A.; Chaplain R. R. Rankin, Div. Chaplain

American trenches south of the Ruisseau de Forges and on the northern slope of Hill 304, north of Esnes. There is not a square yard of ground in this area that has not been struck by a hostile shell. On the morning of September 26, 1918, the advance of the 4th Division began from the north slope of this hill

Army Corps, under General George H. Cameron, in the center, and the III Army Corps, under General Robert L. Bullard, on the right, its eastern flank resting on the Meuse River. Across the river the XVII French Army Corps and, east and southeast of Verdun, the II French Colonial Corps were placed
179 under General Pershing's command.

It was to the III Army Corps that the Ivy Division was sent. It had as companions the 33rd Division, newly arrived from the Somme where it had been fighting with the British and the 80th Division untried, as yet, in battle. The 33rd was on the right of the Corps, with its right resting on the Meuse; the 80th was in the center; the 4th Division was on the
180 left of the Corps. On the left of the 4th was the 79th Division as the right division of the V Corps; this division also had not heretofore participated in a battle.

The general plan of attack was that the Americans, in concert with the French on their left, were to press northward toward the Meziéres-Sedan-Carignan railroad, leaving the Argonne Forest itself to be "pinched out" by the process of driving troops up each side and meeting around Grandpré at its northernmost extremity; the 77th Division, in the forest,
181 following up as the two wings went forward. Five German divisions and part of a sixth were in line opposite the Ameri-
182 cans the night before the attack. The activities behind the American front had not passed unobserved by the Germans and they had been able to make some preparations for the attack when it finally began.

From the night of September 21st-22nd until the night of the attack the troops of the Division were moving up, by successive stages, to the front, endeavoring in every way to conceal their movements from prying Germans in the sky. One unit would advance from its bivouac, which would then immediately be occupied by another unit. All the marches were short to enable the infantry to become settled and the artillerymen to locate and camouflage their guns before morning. The 7th Brigade was moved up first as it was to make the initial attack. It was followed by the Artillery Brigade and the latter by the 8th Brigade.

On September 25th, the advanced P. C. (headquarters) of the Ivy Division moved from Lemmes to two dugouts on a steep hill overlooking Montzéville. On the night of September 25th the troops of the 7th Brigade marched to their positions in the trenches north of and around Hill 304, north of Esnes and west of Dead Man's Hill, famous in the former battles for Verdun. The 8th Brigade similarly took up initial positions in the trenches north and west of Vignéville, in reserve.

Within the sector of the 4th Division there was only one road leading up to the front. It ran from Montzéville to Esnes, and over this all the troops had to pass. The night before the drive commenced was dark and misty. The scarcity of roads and the density of the transport contributed, together with the unfamiliar surroundings, to make it a problem demanding much patience and skill to reach the positions at the ordered hour; in fact, before the barrage opened, the transport question had already assumed overwhelming importance.

All along this road and south of Montzéville, miles of convoys, ammunition trains, guns and troops competed with one another in their anxiety to reach their destination and made the grim night resound with their varying noises. The crunch of feet, the rattle of equipment, the snorting of impatient tanks, the grind of wheels, the roar of camions were all intermixed with the curses and cries of warning of the officers and non-commissioned officers as they shouted down the line "Keep to the right there" and "Hold it."

The mule drivers admonished their animals with a charming variety of phrases, some of which were French. Through the inky darkness never a light shone. The hum of an aeroplane engine overhead and the trail of tracer bullets, the spit of anti-aircraft batteries, told the men that the Germans were looking for them, but they had become used to this type of warfare and were much more immersed in the difficulties of moving forward than in worrying when a bomb would fall among them. Then the roads ceased, the camions turned off into the by-ways, the discordant noises of the struggling transport became fainter and fainter and only the machine gun carts and the artillery carriages were left with the infantry to block the highways. They were nearing the front line. The villages became more and more shattered, the way underneath full of shell holes and the fields covered with barriers of wire. Occasionally a stray rocket would pierce the heavens; a shell would screech through the air and land with a dull thump, while a gun would suddenly send a missile on its way to the Germans—for silence breeds suspicion and, to blind an enemy, the normal is the most necessary.

The men of the 7th Brigade struggled along the road from Montzéville to the heaps of stone that mark the site of Esnes and then, with cautious footsteps, swung slightly to the right and passed over the slopes of Hill 304, still keeping to the road. Here the trench systems swallowed their figures. They were on one of the most famous of battlefields. Behind them in the far distance was the citadel of Verdun and all around were the scenes where the drama of the fiercest of attacks and the most stubborn of defenses had swung its course day after day and where thousands upon thousands had been wiped out of existence.

They found the trenches an organized maze where French signs pointed in all directions and where acceptable dugouts afforded some protection. From the forward slope of Hill 304 they looked down on the valley of the Forges Brook and across it to the German trenches on the opposite slope. Malancourt lay in the valley off to the left; Béthincourt, in the valley, to the right. Desolation, upturned earth and barbed wire was all that could be seen. The men knew that they had to pass the valley and ascend the hill opposite and they realized that, if the Germans made a stand, there would be as difficult a task ahead of them as they had ever dreamed.

Farther to the rear the division, brigade and regimental staffs were wrestling with the problems of directing the movements of the men and providing them with supplies and ammunition. Telephones announced that the Signal Battalion had connected up, that dressing stations had been established. Farther back hospitals were being made ready for the inevitable inmates of the morrow. America was in arms and busily in arms. There were few slackers on the front that night.

At last order began to appear out of the seeming chaos. The infantry was in place and the men snatched a little sleep. The artillery impatiently waited the coming of the hour to begin. But the transport had still to struggle with congestion and poor facilities and was slipping into greater and greater difficulties when the new attack was heralded by the opening of the heavy artillery.

As the fight was almost certain to be a long one, the Division plan contemplated the use of but one brigade in the attacking line and one in reserve. When the forward 183
brigade became exhausted the reserve was to "leap-frog" and become the attacking brigade. The exhausted troops would then have an opportunity to rest before being employed again in the attack.

The Division sector was funnel-shaped. From a width of two and a half kilometers between Malancourt and Béthincourt it increased gradually to a width of five kilometers between Brieulles and Cunel. The left boundary passed east of Malancourt, west of Cuisy and Septsarges, east of Nantillois and Cunel, and west of Aincreville and Villers-devant-Dun. The right boundary extended from a point on the Forges Brook one kilometer west of Béthincourt, through Hills 262 and 280, and thence north to the Meuse just east of Brieulles. The 7th Brigade was designated to make the initial pro-

183 gression. The two regiments of the brigade were formed side
by side, each with one battalion in the assaulting line and
one in support. Two battalions were in brigade reserve. Upon
arrival at the Corps Objective the right regiment was to
place two battalions in the assaulting line, its battalion in
brigade reserve then being released to become its support
183 battalion. The 47th Infantry was on the right, the 39th on
the left.

Besides the three regiments of the 4th Artillery Brig-
ade, the Brigade Commander, General Babbitt, was given the
10th Field Artillery (75's) under Colonel Lloyd and the 18th
Field Artillery (155's) under Colonel Fuger, both from the
3rd Artillery Brigade, the 250th R.A.C. (a nine-battery motor-
ized French regiment of 75's) under Colonel De Chaunac, and
the 2nd Battalion of the 308th R. A. C. (French 155's) under
183 Major Bavarez. This made a total of 108 pieces of 75 mm.
and 60 of 155 mm. caliber organized into three main groupings,
the barrage grouping under Colonel Cox, the light harrassing
grouping under Colonel Lloyd and the heavy harrassing
grouping under Colonel Smith. Each of these groupings was
in turn divided into groups. Certain special missions were
assigned to specific units. One battalion of the 16th Field
Artillery was designated to accompany the infantry, one bat-
tery with each leading infantry battalion and the third battery
183 with the support battalions. The leading batteries were or-
dered to commence their forward movement as soon after "H"
hour as possible. Each gun was furnished with two days'
supply of ammunition, which was sufficient to carry it through
the fire prescribed for the original assault. In addition all
caissons were filled in anticipation of a forward movement.

At 2:30 on the morning of September 26th the infantry, who were crouched in all kinds of positions at the bottom of the trenches and who had gone to sleep, were suddenly startled by a tremendous burst of vibration and noise. It was the opening of the artillery bombardment. In a few seconds the sky which had been black became red. The air was buffeted about. The earth rocked with the tornado. The trenches seemed to tremble in answer to the artillery's call.

The blackness of night became transformed into the grayness of dawn and the men began to find the wait wearisome. They fidgeted and were restless. "Surely no one could be surviving in the German lines. Why not go over?" they argued in thought. Four o'clock passed, then five. Half an hour, and then for it. At 5:20 A. M. the artillery began combing No Man's Land very thoroughly, seeking out the wire and machine guns and preparing the way for the advance of the infantry. Officers anxiously watched the hands of their illuminated wrist watches. Minute after minute slipped slowly by. At last it was five-thirty. "They're off!" officers in observation posts exclaimed. The drive had commenced.

The entire valley containing the Ruisseau de Forges, from the hilltops on the south to the hilltops on the north, was shrouded in thick mist. It was impossible to see more than fifty feet ahead. The leading elements of the 47th and 39th Regiments were lined up on taped lines placed by the engineers. During the night preceding the attack the engineers had cut wide lanes through the barbed wire entanglements in front of the American lines. Detachments of engineers equipped with heavy wire cutters also preceded the infantry to cut the German wire. At "H" hour the men went forward under the command of their officers. Each front line company commander was accompanied by an engineer officer or sergeant who carried a compass having an illuminated dial. The advance across No Man's Land and through the German area was made entirely by compass bearing. The fog had 184
swallowed all land marks. Later, when the sun had dissipated the fog and the troops had passed through the wire of the Volker Stellung, the engineers joined the infantry and served as infantry throughout the remainder of the day. The work performed by the engineers cannot be too highly commended.

As the men helped one another scramble out of the deep trenches and started for the creek at the foot of the hill, the 10th Machine Gun Battalion and two companies of the 11th Machine Gun Battalion added to the roar of battle by placing a concentrated barrage on the enemy front line just
185 north of the brook. The Germans were thus kept below ground until the artillery barrage could reach them and prevented from firing through the barrage on the Americans. The fog was of great assistance in covering the advance, but it caused many difficulties in maintaining liaison.

Slowly the khaki-clad figures poured down the hill side, picking their way over shell holes and passing through the lanes in the barbed wire that the engineers had already cut. When they reached what they thought would be a creek nothing but a treacherous swamp could be seen, with patches of water showing here and there through the mud. The engineers, with the front line of the infantry, quickly bridged the swamp by means of ladders which they carried with them and the men were soon on the other side, slowly ascending the hill where the German trenches lay. The artillery fire had shattered the enemy trenches and wire into the most fantastic of shapes and, as the Germans were not replying to the bombardment with any heavy fire, progress was rapid.

Into the German trenches the men of the 39th and 47th Infantry went, searching for their occupants. The appearance of the Americans was the signal for small files of German soldiers to emerge from their dugouts all ready to surrender and spend the rest of their warring days in captivity, for they carried bundles and pipes in approved travelling fashion. A solitary doughboy was sufficient to take them back to the M. P.'s who marched them off to the cages.

Small mines had been placed on the parapets by the enemy but they exploded without apparently doing any great harm. Down one hillside and up another brought the two regiments on a line with Cuisy. South of Cuisy the Germans had a formidable line of trenches, the intermediate defensive position. Here the barrage was to rest for thirty minutes to allow the support battalion to "leapfrog" the assaulting battalion and carry on the fight. It was 9:30 A. M. when this position was reached. The 1st Battalion of the 39th Infantry, in support, which was to pass through the lines of the 3rd Battalion, had lost its way in the fog and had moved over to the left in the direction of Montfauçon. This town is on a high hill, and the hill was strongly defended with barbed wire entanglements and machine guns. It was on the eastern slope of this hill that the battalion found itself. Two platoons moved forward through a communicating trench, flanked the hill and captured more than a hundred prisoners. More Germans now appeared, machine guns became active, and in the progress of a series of engagements the 1st Battalion lost heavily. It was not until three o'clock in the afternoon that the battalion broke off this indecisive fighting and proceeded toward Septsarges to join its regiment. 186

In the meantime the 3rd Battalion, after considerable opposition, had entered Cuisy and cleaned it up, capturing a large number of prisoners, 12 pieces of artillery of 77 mm. caliber, and many machine guns. The advance was continued through Septsarges. West of the town strong machine gun resistance developed which cost many lives before it was overcome. The battalion then advanced to the Corps Objective and dug in. One company and a machine gun platoon faced
west to protect the exposed flank. The 79th Division was 186
still held up south of Montfauçon.

On the right the 47th Infantry advanced with comparatively little opposition. Prisoners were captured apparently dazed by the terrific artillery fire to which they had been subjected. At 9:30 A. M., when the barrage rested for 30 minutes at the hostile intermediate position, the 3rd Battalion "leapfrogged" the 2nd Battalion and the advance was continued, with but slightly increased machine gun resistance, to

German trenches north of the Ruisseau de Forges about 1 km. northeast of Malancourt. Observe the destruction wrought by the American artillery fire, and the wire entanglement in the upper part of the picture, still intact

A few of the 2,500 German prisoners captured by the 7th Brigade. Montzéville, Sept. 26, 1918

View ot Esnes showing traffic on the road from Montzéville at a time when it was not as heavy as usual. Sept. 29, 1918

the Corps Objective. Up to this time the 7th Brigade had captured about 1700 prisoners and more than 40 guns of 77 and 150 mm. caliber, together with numerous machine guns and minenwerfers. The enemy was slowly withdrawing, leaving only small machine gun and sniping outposts behind to delay the advance, and firing with such artillery as he found it impossible to withdraw. The 7th Brigade had reached the Corps Objective at 12:30 P. M., on a line which ran approximately east
187 and west through Nantillois. The 80th Division, on the
right of the 4th, was also on this objective and was almost
abreast of the 7th Brigade, but the 79th Division, on the left,
188 was held up by Montfauçon. Corps orders had directed
that the advance of the Division should be continued beyond
the Corps Objective as soon as the right or centre division of
180 the V Corps had reached the Corps Objective. As neither
the right nor the centre division of the V Corps had reached this
line, the advance of the 4th Division was halted and was not
again resumed until late in the afternoon, when orders from
III Corps directed the division commanders to press the ad-
189 vance regardless of the progress of the V Corps. At 5:30 P. M. the advance was resumed without artillery support, but such strong resistance developed on the part of the Germans that but little progress was made. Owing to the fast approaching darkness, the advance was halted, and the troops bivouacked for the night approximately on the line of the Corps Objective.

The increased German resistance initiated in the late afternoon can only be ascribed to the fact that they had discovered the long halt made by the Division, had taken heart and brought back machine guns, snipers and troops sufficient to make further progress difficult.

This experience illustrates clearly the danger, to which every tactician is exposed, of being dominated by precedent. For almost four years all attacks had been launched with limited objectives—well-defined lines, marked on maps, at a certain distance from the jumping-off line—beyond which troops should not advance until orders were received from higher headquarters or other stated conditions had been fulfilled. It was an inheritance from the trench phase of warfare, where it had been found necessary. The American Army, under direction of the French and following French precedent, could not escape the contagion of the limited objective. The result was that when, pursuant to orders, the troops halted on the Corps Objective they stopped the pursuit of an enemy that was in retirement; halted long enough for the enemy to become aware of the halt, muster up his courage, not only cease his retirement but bring back his machine guns and troops and thereafter put up a resistance that took a heavy toll of American lives. It is believed by those officers who were at the front and who are competent to speak, that, had the advance not been halted for five hours on the Corps Objective, the 4th Division could have reached, on the night of September 26th, the Kriemhilde Stellung and possibly the Army Objective, along the northern edge of the Bois de Forêt—an achievement which later required fifteen days and cost 6000 casualties.

The fallacy of the limited objective in open warfare and against a shaken enemy was later admitted by the Commander-in-Chief of the Allied forces. Under date of October 26th, Marshal Foch in a letter to General Pershing sent him the following memorandum and requested that it be transmitted to the units under his orders:

Operation plans of the 4th French Army as well as those of the 1st American Army, prescribe for those armies a method of attack by limited objectives.

Such a plan, which was very much used during the war of position, can still be useful in certain special cases, when only a restricted result is contemplated for the time being. It is not possible of general application, for the reason that it can produce only limited results.

By limiting in advance the progress of the troops to lines traced on a map, and consequently by preventing these troops from taking advantage of the opportunities which always present themselves after a successful start, the high command by its restrictive orders compromises the final result—and renders it in every case more costly.

Important results such as we are striving for in the present period of the war and the presence of an enemy whose strength is decreasing daily, are to be hoped for only by a progress as rapid and deep as possible.

Troops thrown into the attack have only to know their *direction of attack.* In this direction they go as far as they can without any thought of alignment, attacking and maneuvering the enemy who resists, the most advanced units working to help those who are momentarily stopped.

In this manner they operate, not toward lines indicated ahead of time according to the terrain, but against the enemy, with whom they never lose contact once they have gained it.

(Sgd). F. Foch.

However, long before that, on September 28th, in his Field Order No. 27, General Pershing had issued instructions as follows:

3. The 3rd, 5th and 1st Corps will advance within their zones of action as specified in Field Orders No. 20, *without regard to objectives.*

The 8th Brigade, at 8:45 A. M., was moved from the reserve
trenches west and north of Vignéville to the front line trenches
on Hill 304, to hold that position in case of a determined German
190 counter-attack. At 4:00 P. M. it was again moved forward,
this time to occupy the trenches of the German intermediate
191 position, south of Cuisy. Here, under the desultory shelling of the enemy, the troops spent the night. Toward morning a drizzling rain began to fall. In this intermediate position there were but few dugouts and these had been almost totally destroyed by the American artillery fire. As the trenches were cramped and uncomfortable the men abandoned them, preferring to take their chances in the open with weather and German shells.

The Germans, though retiring, were by no means disorganized. Casualties, especially in the 39th Infantry, had become numerous. A long line of wounded sought the dressing stations behind the lines. Evidence was rapidly accumulating that the enemy was recovering from the shock of the first blow. The morrow promised stiff fighting. What had been happening at the rear while the infantry was driving forward was of great importance. Artillery was needed, there were calls for ammunition, and the men were already in search of food.

The necessity for immediately rebuilding the roads behind the advancing infantry and thus making a way for the guns and ammunition was realized by General Hines before the attack commenced. On the left of the Division sector the maps showed a road running from Esnes to Malancourt and continuing up the hill to Cuisy and Montfauçon. Four years of incessant bombardment had entirely obliterated this road between Esnes and Malancourt, but it still existed in a fairly well preserved condition beyond the spot where the latter town had been. Between Esnes and Malancourt the surface was a mere mass of overlapping shell holes clogged with barbed wire. Every square yard of ground in this area had been struck by a shell. This road lay partly in the area of the III Corps and partly in the area of the V Corps, but it had been assigned
by the III Corps, for what it was worth, to the 4th Division. 192
The engineers were instructed to rebuild it at once.

At 9:30 on the night of the 25th, the 4th Engineers commenced their great task. Three miles of road had to be built over an area which was nothing but a mass of shell holes. In order to avoid arousing the suspicions of the Germans it was necessary for the men to work in silence and complete the task at the earliest possible time on the following morning. As the sound of road building is audible at a certain distance the work was at first begun on only the short stretch between Esnes and a point slightly beyond where the Avocourt road joins the Esnes-Malancourt road. The work was then halted and was not again resumed until the beginning of the artillery bombardment. Under cover of the noise of the artillery fire the work progressed with feverish haste. There was no rock to be had except the stone from ruined buildings, and filled sandbags had to be carried and flung into the holes. An occasional shell would take its toll of the men but the work went forward hurriedly and even frantically. Soon Haucourt was reached, and here, amid the swamps, work was even more difficult. The sand bags were getting farther and farther away and the only resource left was to pull down the remains of the shattered stone houses and fling them into the road. In a few days the ruins of the church at Malancourt had disappeared, and the stones of the graveyard were no more. They had all found a firm resting place in the road beneath the tramp of marching soldiers, the unceasing tread of horses and the grinding of wheels. At 1:35 on the afternoon of September 26th—the day of the offensive—the way was open.

Forty thousand sand bags had been used. The artillery could move forward, and the ammunition trains could follow. It had been ordered by General Hines that these should have absolute precedence, food coming next. But, unfortunately, the 79th Division, crowding on the road, did not organize its transport with similar forethought, and soon the road became hopelessly jammed.

At two o'clock on the morning of September 26th the order was sent to the battalion of the 16th Artillery, which was to accompany the infantry, to harness its horses and prepare to advance. The guns were equipped with ramps to facilitate the moving of the carriages over the tortuous and treacherous No Man's Land. There was a long wait for the road to Malancourt to open but at last the batteries moved forward. Progress, however, was exceedingly slow. It was already getting on toward dusk when the 16th Field Artillery reached Malancourt. Orders were received to proceed to Cuisy at once. It was found that the direct road to Cuisy could not be used because of enemy snipers and machine gunners around and in Montfauçon. A reconnaissance party was then sent out to scout the road to Béthincourt, which ran due east from Malancourt, and discover whether Cuisy could be reached by the Béthincourt–Cuisy road. Soon the artillery was heading in the direction of the Meuse River. Progress was exceedingly difficult. When a ditch or shell-hole was reached through which it was impossible for the horses to pull the guns, officers and men pulled them, and pushed them through on their shoulders. All through the night the batteries struggled on. In the morning they were stretched between Béthincourt and Cuisy, caught in traffic congestion. 193

The regimental commander of the 77th Field Artillery, having observed the complete blocking of the Esnes–Malancourt road, undertook, on his own responsibility, to send one battalion via the Esnes–Béthincourt road and encountered no difficulty with the traffic of the 80th or 33rd Divisions until the town of Béthincourt was reached. Here a tremendous congestion had developed. However, the battalion managed to make its way through Bethincourt and reached Cuisy about sundown on the 26th. It went into position on the Cuisy–Montfauçon road, about 500 meters west of Cuisy, to cover the exposed left of the Division. The other battalion bivouacked just north of Malancourt on the night of the 26th. Early on the 27th it moved, via Béthincourt, to a position on the Cuisy–Béthincourt road, 1500 meters east of Cuisy. This position was reached about 5 P. M. 194

It was 9:00 P. M., September 26th, when the 2nd and 3rd Battalions of the 13th Artillery started to move forward over the Esnes–Malancourt road, taking with them the horses of the 1st Battalion which was left at Esnes. From south to north the regiment encountered first a seemingly interminable column of machine gun carts, rolling kitchens and ammunition wagons, and then an even more trying succession of stalled trucks of all characters, that reminded one of ships that had gone aground in a narrow channel, and, with bows stuck fast on either side, had swung their sterns completely across the fairway. Light transportation was ruthlessly ditched. Trucks were started ahead under their own power or fitted with a tackle and pulled over and off the trail by straining men. Heavy howitzers, weighing, with their limbers, nearly five tons, which because of faulty driving in the darkness or crumbling of ground under their weight, had slid bodily into enormous shell craters, were lifted and pulled out by a hundred men, who with lever and rope reinforced the doubled teams. All night long the struggle with the road and its congestion continued. At daybreak the two battalions reached Malancourt after eight hours of ceaseless labor of the most exhausting kind.

Orders received during the night from the Commanding General, 4th Field Artillery Brigade, had directed Colonel Smith to advance his battalions to positions near the junction of the Malancourt–Montfauçon and Malancourt–Cuisy roads. Reconnoitering parties started out at once while the command snatched a brief rest. By 10:00 A. M., September 27th the 2nd Battalion was in position about one kilometer south of the road junction. The 3rd Battalion was placed in position near
195 the road junction about 1:00 P. M.

Once settled for the night, the men of the 4th Division began the old cry, "When do we eat?" Some fortunate ones, on investigating a dugout that had been hurriedly evacuated by the enemy, found, to their happy surprise, a quantity of German bully beef and biscuits which they ate with the relish of an unexpected feed. But without blankets to sleep in they soon found the night very cold. Search parties were sent out from the various units and soon a small collection of German overcoats, uniforms, sweaters and blankets were found which afforded some, if very insufficient, protection from the weather and which were often alive with "cooties."

Division Headquarters was established at Cuisy at mid-
196 night, September 26th-27th. Some German dugouts facing

the enemy were secured for this purpose but the accommodation was amusingly scanty. When General Hines, with General Babbitt, arrived there he found three Brigadier Generals —for Brigadier General John S. Winn was attached to the 7th Brigade as an observer—the Chief of Staff, some other officers and a flock of orderlies, all in one small dugout. They were sleeping on a cement floor built over a spring, from which the water was oozing. General Hines himself sought out a pile of three-cornered wood and there he slept until morning, in the drizzling rain. General Babbitt slept outside on a duckboard in the drizzling rain.

At the time the P. C. was established here it was advantageously located for communication, being placed between the 7th Brigade in front and the 8th in rear. The determining factor in this selection, however, lay in the fact that by placing it well forward it would not be necessary to move it again very soon. In this case the headquarters, when established, was only three kilometers from the front line. There is a distinct advantage in maintaining a headquarters in one spot as long as possible; it permits the establishment of a number of wire lines, a fact which is almost impracticable when daily moves are made. The requirements of open warfare also demand that a headquarters shall be as close to its troops as possible. The leader must make and transmit rapidly, his decisions; he must sometimes inspire his men by personal example.

The difficulties of preserving communication between the various headquarters were great. The Signal Battalion worked valiantly in maintaining wire lines and suffered severe casualties. Owing to the frequent destruction of telephone wires many messages were delivered by runners who had to dodge in and out of the shell holes to escape the enemy fire. Radio was employed. From the front lines pigeons were used as a means of communication. Baskets of these birds were dispatched to the infantry by the Corps and whenever com-

The Esnes–Malancourt road

Scene on the road from Esnes to Malancourt, Sept. 27, 1918. The 4th Engineers, in 16 hours, built five kilometers of this road across what had for four years been No Man's Land

Cuisy, Sept. 28, 1918. The men in the truck in the foreground are Americans wounded, but not seriously. On the litters are Germans, seriously wounded. In the background, in front of the church are F and G Companies, 58th Infantry, going forward to Septsarges.

munication could not be established in any other way a message was attached to the pigeon which flew back to the loft at Corps Headquarters. The messages were then telephoned to the divisions. Some of these birds arrived at the lofts wounded.

During the day the German aviators were exceedingly active. Not only did they bomb and machine-gun the American infantry and artillery but they daringly attacked the balloon line. One German aviator set fire to two balloons in as many minutes. Suddenly, from out of the sky, before even the anti-aircraft batteries had seen him, he dived down on the first balloon. There was a sputter of machine gun fire, now joined by the anti-craft fire from the ground but the aviator, to the thousands of eyes watching from below, had apparently missed his target for, as he left, the balloon appeared to be still unharmed. Suddenly a figure beneath a white parachute was seen falling to earth—the observer had jumped. And then a streak of flame burst from the silvery back of the balloon and soon it was descending to the ground, a mass of fire. Meanwhile the observer in the second balloon had jumped while the German dived toward him. There was another streak of flame and then another mass of burning wreckage. A third, however, escaped, the men on the ground pulling the balloon to earth before the German reached it. He was soon on his way back to his own lines.

197 On September 27th the attack was resumed at 6:30 A. M., with the support of one battalion of the 77th Artillery, less one battery. Two batteries of the 16th Artillery joined in support between 9:00 and 10:00 A. M., one battalion of the 13th about 10:00 A.M., and the other about 1:00 P. M. The direction of attack for the two regiments now began to diverge slightly. The 47th inclined toward the town of Brieulles-sur-Meuse with two battalions in the front line and one in support; to the latter was attached Co. "A," 10th Machine Gun Battalion. The 39th pushed on in the original direction with one

battalion in the front line and one in support; its remaining battalion was in brigade reserve.

When the troops left their fox-holes that morning they soon found that the German resistance had greatly increased compared with that of the previous day. Where it had been possible, twenty-four hours before, to walk upright, the men now had to crouch or crawl, and both the artillery and the machine guns of the enemy combined to contest the Division's advance. The 39th Infantry, with three companies of the 11th Machine Gun Battalion, advanced across the open ground between the Bois de Brieulles and the town of Nantillois. From the moment the troops started forward they met strong opposition. On the top of Hill 266 the machine guns were placed in the cut dug for the track of the narrow-gauge railroad that runs from Hill 295 through the heart of the Bois de Brieulles. It was here that Lieut. Col. William E. Holliday was killed. The infantry, in an attempt to advance down the northern slope of this hill, received such a withering fire of shrapnel, high explosive and gas, from the southern borders of the Bois des Ogons and the Bois de Fays and machine gun fire from Nantillois, that the men broke and retreated southeast into the western edge of the Bois de Septsarges and behind the cover of Hill 295. This was about 9:45 A. M. Through 198
the efforts of the Regimental Commander and his officers and of Generals Poore and Winn and the officers at Brigade Headquarters, the battalion was re-organized and took up a position running northeast from Hill 295 along the railroad, toward Hill 266. The machine guns and about 50 infantrymen remained on Hill 266 throughout this entire period. The regiment bivouacked in this position on the night of the 27th, the support battalion digging in on the reverse slope of Hill 295. The losses had been severe.

The advance of the left of the 47th Infantry was held up by the German resistance on the front of the 39th but the right of the regiment continued to move forward into the Bois

de Brieulles despite severe machine gun and sniper fire. The line, at the close of the day, extended from Hill 295 northeastward along the edge of the Bois de Septsarges, through the
199 Bois de Brieulles to the edge of the woods on Hill 280. Here it connected with the left of the 80th Division.

The 79th Division had succeeded in capturing Montfauçon
200 shortly after noon of the 27th. Its line that night was about a kilometer north of Montfauçon, its eastern end reaching the Septsarges-Nantillois road one kilometer south of Hill 295.

The determined resistance of the Germans on the second day of the drive demanded that clear lines of transport should exist to the rear for the bringing up of food and ammunition. The weather grew hourly worse. The rain poured. The men were wet and miserable, and the roads were in an almost impassable condition. The struggle of the artillery to keep pace with the infantry during those two days almost equalled that of the men of the front line fighting the Germans. Fortunately an extra battery of German 77's had been added to the division artillery. When the 39th Infantry captured Cuisy and drove out the Germans twelve 77 mm. guns were captured. Eight of these had been destroyed by American artillery fire or rendered unserviceable by the Germans but the advance of the 7th Brigade had been so rapid that the enemy had no time to complete the work of destruction and four of the guns fell into our hands in good condition. They were organized into a battery by officers and men of the 13th Field Artillery and later became known as "Q" Battery. German wire and telephones were salvaged and an O. P. established. A goniometer was borrowed from the 77th F. A. Details who served these guns took the greatest interest in them; horses were supplied and the countryside scoured for German ammunition. On the morning of the
201 27th these guns were turned against their late owners. On October 3rd the battery was attcahed to and manned by the 77th Field Artillery and more than 3,000 rounds of German

shells were fired before it was finally discarded and sent to the rear with other captured materiel.

About sundown of the 27th the entire 16th Field Artillery, after a most trying march over shell-torn roads and across trench-covered fields, arrived at the eastern slope of Hill 302,
north of Cuisy, and went into position there. 202

In order to furnish protection for the left flank of the Division, the 58th Infantry was moved from the trenches south of Cuisy to the vicinity of Septsarges during the early afternoon of the 27th. The 318th Infantry (80th Division)
was, by a Corps order, attached to the 4th Division. It was 204
placed under the orders of the Commanding General, 8th Brigade, and rendered valuable service until October 2nd when, with the exception of one battalion, it was relieved; the last battalion was relieved on October 5th.

The attack was resumed at 7 o'clock on the morning of
September 28th, this time with complete artillery support. 205
The 47th Infantry, fighting its way through the tangled undergrowth of the Bois de Brieulles and eliminating enemy machine guns in rapid succession, reached a line close to the northern edge of the woods. A German ordnance storehouse containing considerable quantities of rifles, grenades, ammunition and machine guns was captured and the latter at once turned against the enemy.

The 39th Infantry, following behind a barrage, moved forward from its position, over Hill 266, toward the southeastern corner of the Bois des Ogons. Passing through a curtain of machine gun fire from these woods and from the Bois de Fays and artillery fire from the north and from east of the Meuse, the 39th continued to press forward into the southern
edge of the Bois des Ogons where it was held up, about 2:00 206
P. M., by heavy machine gun fire from the woods in front and the clumps of trees on the left. Here Colonel Bolles was wounded. Despite the hostile fire, the regiment dug in on the crest one half kilometer south of the Bois des Ogons and

tenaciously clung to this position until the 8th Brigade passed
207 through its lines at 5 A. M. next morning. It was partly in the area of the 4th Division and partly in the area of the 79th. Colonel J. K. Parsons, who had been attached to the regiment as an observer, was now directed by General Poore to take command.

The 79th Division, after taking Nantillois, had succeeded in penetrating the Bois des Ogons only to be driven out by the enemy. Their line at 6:00 P. M. extended east and west about
207 one half kilometer north of Nantillois.

During the day two German aeroplanes drove down an American plane, which was working with the 4th Division, setting the machine in flames. The pilot, within sight of the men of the 16th Artillery, steered his aeroplane to earth and the observer, hanging over the side, jumped clear. The pilot was fatally burned but he saved his comrade.

The units of the Artillery Brigade, which had been placed on a general east and west line south of the Bois de Septsarges, gave most satisfactory support to the infantry. Unfortunately the supply of ammunition was insufficient. The roads at the rear of the whole American sector were in a state of great congestion. In some places never a wheel turned for hours and to get ammunition up to the batteries was a stupendous feat. Infantry was sent out to get the traffic moving and all kinds of devices were employed to bridge over the difficulty. The spirit of the Division, however, was the force of the hour. Every man worked with a will. Small arms ammunition was brought up by men as they carried messages to and from the rear. Food was searched for. Each man felt his personal responsibility in the success of the Division.

On the morning of September 29th the 8th Brigade, reinforced by the 318th Infantry, 80th Division, passed through the 7th Brigade, at 5 A. M., and attacked. The attack was preceded by artillery preparation and the troops followed a
208 barrage. The Bois de Brieulles was cleared of the enemy

but the increased machine gun resistance from the Bois de Fays and the Bois des Ogons and the exceedingly heavy artillery fire from east of the Meuse prevented a permanent establishment of the line in the Bois de Fays. During this action two accompanying guns, belonging to Battery "A," 16th Field Artillery, were demolished by direct hits while occupying a position in the infantry front line. The 8th Brigade now took over the line, the 7th going into bivouac in the southern part of the Bois de Brieulles and in the Bois de Septsarges.

Company "M," 58th Infantry, reinforced by a platoon of machine guns, acted as a combat liaison group between the 4th and 79th Divisions. Early in the forenoon of September 29th it established itself on the southeastern slope of Hill 274, one kilometer north of Nantillois. Its left extended 800
meters into the area of the 79th Division. Half a kilometer 209
in its front, at the southern edge of the Bois des Ogons, were Companies "C" and "D" which had attacked that morning with the rest of the 58th Infantry. During the day Company "C" fought its way through the Bois des Ogons, from south to north, until within sight of Madeleine Farm. Not seeing any Americans on his right or left and noticing that German troops held the farm, the company commander then withdrew to his former position. Later in the day Companies "C" and "D" repelled two determined attacks from the Bois des Ogons.
These two companies were relieved on the night of October 210
1st-2nd by Company "I," 58th Infantry, which continued to hold the position until the attack of October 4th. Company "M" maintained its position against repeated enemy attacks and effectively protected the exposed flank until the attack of October 4th, when it also moved forward with the 8th Brigade.

On the night of September 29th the front line of the Division extended along the northern and northwestern edges of the Bois de Brieulles to the Moulin de l'Etanche, then west to the position held by Companies "C" and "D" at the southern edge of the Bois des Ogons.

Very little progress was made during the next four days. The Germans were so well established and the woods and heights so difficult to clear of the enemy that only by the slow process of suppressing machine gun after machine gun in the woods could any progress be made. The infantry had to wait until the artillery ammunition necessary for another drive could be brought up. The conditions at the front were a terrible strain on the men. Wet, cold, and with little hot food, they had inadequate protection from the enemy and none from the weather. A constant stream of bullets came from the Germans during the day. The enemy artillery never ceased its activity; its fire came not only from the front but, in increasing volume, from the right flank, east of the Meuse. Constant patrolling was kept up to acquire a greater knowledge of the terrain and learn the disposition and movements of the enemy.

On October 3rd the line had not changed. The first phase of the battle ended. The increasing resistance of the Germans had slowed down the advance and the congestion of the roads had forced a halt. Full preparation had to be made for a further attack. The American Army had not progressed as rapidly as it had expected but the task was harder than had been anticipated. Difficulties almost insurmountable had been overcome: the Army had driven the Germans back to their reserve defenses and now faced the Hindenburg line. The condition of the roads gave cause for anxiety. But the spirit was good and vigorous preparations went forward for the second phase of the battle.

CHAPTER XII

THE MEUSE-ARGONNE OFFENSIVE—2ND PHASE

THE first phase of the Battle of the Meuse-Argonne lasted from the initial attack on September 26th to the resumption of the general attack on October 4th. The American Army had broken through the first two lines of the enemy's defenses, which had been lightly held, and was now right up against the main German position—the Kriemhilde Stellung. Although, on September 28th, the Corps Objectives had been passed, only slight progress had been made on the five days following.

The slowing down of the advance after September 28th and its cessation in front of the German main position were due to three factors. First, and least important, was the resistance of the enemy, which had gradually stiffened; next in importance was the inability of many of our divisions to exploit their success, due to lack of training and experience. The controlling factor, however, was the lack of artillery and infantry ammunition with which to prosecute a further advance. This lack of ammunition was the direct result of the conditions then obtaining on the one and only road in the Division sector. The Esnes–Malancourt–Cuisy road carried all the traffic of the 4th Division; but that part of it constructed by the 4th Engineers from Esnes to Malancourt and the part beyond, as far as the fork of the Malancourt–Cuisy and Malancourt–Montfauçon roads, carried, in addition, all the traffic of the 79th Division and later of the 80th and 3rd Divisions.

The congestion on this road can hardly be described. Beginning at 1:30 P. M. on September 26th there was not a minute of the twenty-four hours when the road was not cov-

"Reading his shirt"

Looking west from Cuisy. Germans shelling the Cuisy–Montfaucon road. Oct. 7, 1918. The bursts of three shells can be seen in this photograph

German regimental P. C. which was hastily abandoned by the Germans on Sept. 26th, opened at midnight of that date by the 4th Division, and used as a Division P. C. thereafter until Oct. 19, 1918. General Hines is shown standing in the centre of the picture.

ered for every yard of its length with everything it could carry. Whenever a truck broke down, an animal dropped dead, a wagon broke a wheel—and these were not infrequent incidents—the entire traffic halted. When a shell struck the road it caused a jam which could not be moved again until the road had been repaired. Consequently, when traffic did move, it moved very slowly. Frequently it did not move at all. During the whole period this road was constantly under repair by two entire regiments of engineers; the 4th Engineers, the engineers of the 79th Division and the Corps Engineers worked on it. Rock had to be brought from Esnes, Malancourt and Haucourt because the soil along the roadside was a clay which the heavy rains had reduced to muck.

Every conceivable effort was made to remedy the road condition. Damaged trucks, supplies which were not absolutely necessary, were ruthlessly ditched. All sorts of ways were devised to improve the situation. Corps Headquarters struggled with it. It made the road a one-way road, ordering all return traffic to pass by way of Béthincourt. Traffic control posts were established at various points, especially at the Esnes and Malancourt forks. By tireless efforts the engineers managed to widen the road throughout its whole length and, at a later stage of the operations, about the middle of October, the road had been so widened that it could be used for two-way traffic.

On September 26, 27th and 28th the hospitalization of the wounded was a most serious problem. There can be no question that, on the first two days of the battle, the road congestion prevented prompt evacuation and that some men died in the ambulances for lack of sufficient attention, which could not possibly be given to them because the ambulances were caught in the traffic jam and blocked. To remedy this condition two field hospitals were opened at Cuisy on September 27th. Here, although surrounded by friendly artillery and under frequent hostile shelling, which caused a number of

casualties, the more seriously wounded were operated on or received care for which they would otherwise have had to wait
for several days. The less seriously wounded were littered 211
or sent on returning empty trucks to Esnes, there to be placed in ambulances and transported to the rear. Later, after the Cuisy–Béthincourt road had been made a one-way road, for south-bound traffic only, a concentration of field hospitals was established at Béthincourt for the 4th and 80th Divisions. The movement of wounded to the field hospitals and the evacu-
ation of the latter was thereby greatly facilitated. 211

On September 30th the 79th Division, which had been fighting on the left of the 4th, was relieved by the 3rd Division
along the general line of the Cierges–Nantillois road. On 212
the night of September 28th-29th the 80th Division, which had been fighting on the right of the 4th, was withdrawn from the front line for a few days, its area being taken over by
the 33rd Division. Then, on October 4th, the 80th Divi- 213
sion, as part of the III Corps, was sent in again; this time on the left of the 4th Division between it and the 3rd, taking over, a little north of Nantillois, a sector from the 3rd Division.
The 80th thus became the left division of the III Corps. 214

By extreme effort on part of the ammunition and supply trains enough materiel had been accumulated on October 3rd to warrant a general advance. The effort was to be made by the entire First Army. For the 4th Division the line of the
Army Objective was the goal. 214

The task assigned to the 4th Division, in the attack of October 4th, was to hold the northern edge of the Bois de Brieulles, to push through the Bois de Fays and the Bois de Malaumont into the Bois de Forêt, then to turn east and sweep the Bois de Forêt to the Meuse River. The open ground east of the Bois de Fays and the Bois de Malaumont was to be avoided because of the perfect observation and annihilating
fire that could be brought to bear on it by the Germans. The 214
success of this movement would have compelled the evacua-

tion of Brieulles by the Germans without the firing of a shot.
It was an ambitious plan but not beyond accomplishment.
The 33rd Division, on the right, held the line of the Meuse
and the northern edge of the Bois de la Côte Lémont as far as
215 Hill 280, so as to protect the right flank of the 4th Division.

Under the divisional plan the infantry attack was launched
without artillery preparation but preceded by an intense
barrage. A machine gun barrage was also laid on the south-
ern edge of the Bois de Fays by the 10th Machine Gun Batta-
lion and Companies "A" and "D" of the 12th Machine Gun
Battalion. The heavy French artillery placed destructive
fire upon the Bois de Fays and upon all known enemy artillery
216 positions within the Division sector. At 5:25 on the morn-
ing of October 4th the 58th Infantry and Companies "B" and
"C" of the 12th Machine Gun Battalion left their holes along
the northwestern edge of the Bois de Brieulles and proceeded
across the open ground in the direction of the southern edge
of the Bois de Fays. The advance was hidden from enemy
observation by a heavy morning mist. A smoke barrage had
been ordered and planned but the officers of the Gas Company
failed to reach the ground with their personnel and equip-
217 ment. The men advanced behind the artillery barrage, but
as the mist lifted and their lines were disclosed, they were
met with a storm of projectiles. Artillery from east of the
Meuse swept the whole front, deluging it with gas, shrapnel,
and high explosive. Machine guns poured a stream of lead
into their lines from every conceivable direction. There was
fire from the Meuse on the right, from the Bois de Fays in front
218 and from the Bois des Ogons on the left. The fields became
wreathed with smoke and gas. The men had to fight in their
masks. It was then realized that the struggle was to be bitter
and costly, that the Germans had taken full advantage of the
interlude between the first and second phases to reorganize
their troops and select the best positions. It was not on the
front of the 4th Division alone that this happened. The Ger-

mans were resisting along the entire Army front with a tenacity that was not altogether expected.

Despite the desperate character of the opposition, the 58th Infantry advanced into the Bois de Fays and gained a foothold on the southern edge. All the concentrated artillery fire of the Germans, from 77's up to the massive 210's, could not dislodge them. The fire came from so many different directions that the infantry thought their own artillery was firing short. Unfortunately, the 80th Division on the left was being
held at the Bois des Ogons which the Germans, with a true
appreciation of its tactical value, were stoutly defending. 219
Still our men went forward. The Bois de Fays was penetrated
and captured, the Bois de Malaumont passed and the troops
succeeded in crossing the Cunel–Brieulles road. The Bois de
Peut de Faux, to the east, was entered and the Bois de Forêt
was approached. The line, about 10:30 A. M., held along the 221
Cunel–Brieulles road. It was severe fighting every inch of 220
the way. The Germans had cleverly concealed their machine gun nests and poured a destructive fire into the advancing Americans. All that the men could do, amid the shells that were continually bursting overhead and among them, and the mixed gases that made the woods reek with their deadly fumes, was to seek out the machine gunners and maneuver around their positions until, surrounded, they were forced to surrender. The work called for accuracy of rifle fire but above all for fearlessness and endurance. But disappointment awaited the gallant infantry. After reaching the Cunel–Brieulles road the left flank of the 4th Division was exposed for an alarming distance. The 80th Division had been unable to get through the Bois des Ogons. The troops of the 58th were being decimated by fire from the direction of Brieulles on the right, from Cunel on the left and from the Bois de Peut de Faux in front. There was only one course open and that was to withdraw. With bitterness in their hearts the men retired, in the late afternoon of October 4th, from their hard won position to

the Fond de Ville aux Bois, half a kilometer in rear, where a line was established around three sides of the Bois de Fays
218 —a curious salient. About 5:30 P. M., a determined attack was made against the troops in the Bois de Fays by about 400 German infantry. The attack was launched from Brieulles, up the Ruisseau de Wassieu and the Ruisseau de Ponthieu and was preceded by a barrage that rolled through the Bois de Fays from east to west. The Germans were permitted to approach to within 300 yards; then such a withering fire from machine guns, rifles and trench mortars was poured upon them
222 that they broke and fled to the rear.

During tne night of October 3rd-4th the 59th Infantry had been relieved, at the northern edge of the Bois de Brieulles, by the 1st and 3rd Battalions of the 47th Infantry, reinforced by the 37 mm. guns and Stokes mortars of the regiment, two companies of the 11th Machine Gun Battalion and two guns
216 from the 16th Field Artillery. This force was to hold the woods, repel counter-attacks and neutralize the fire from Brieulles and a strong enemy position about 1½ kilometers southwest of Brieulles. During the day it harrassed the enemy by fire and by strong patrols that attacked his machine gun nests. The Germans left Brieulles itself unoccupied during the daytime but at night filled it with machine guns brought from east of the river. No American patrol could approach it without encountering heavy machine gun fire. Late in the evening of October 4th a patrol line was established from the east-
223 ernmost point of the Bois de Fays to Bonne Fontaine Farm.

The 59th Infantry, on being relieved from the front line, had been formed as reserve for the 8th Brigade in the south-
218 western portion of the Bois de Brieulles. During the afternoon one battalion was sent into the Bois de Fays to assist the 58th Infantry; the other two remained south of Bois de Fays, near the western edge of the Bois de Brieulles.

On the night of October 4th-5th the line of the 4th Division extended from Hill 280 along the northern edge of the Bois de Brieulles, eastern edge of the Bois de Fays, Fond de Ville aux Bois to the western boundary of the Division, thence south along that boundary to where it joined the right of the 80th Division. 224

The 80th Division, on the left, had, throughout the day encountered violent resistance from the Bois de Ogons. This Division, like the 4th had also suffered heavily from German artillery fire from east of the Meuse. Three gallant and determined attacks against the southern edge of the woods were repulsed with heavy loss to the attacking units. Their line that night extended from Hill 274, one kilometer south of the Bois des Ogons, northeast along the ridge to the 4th Division boundary. 225

The German communiqué of October 6th forms an illuminating comment on the tenacity of the men of the 4th Division. It read: "Army Group of von Gallwitz. Between the Argonne and the Meuse the Americans continued their powerful attacks. E. of Exermont they succeeded in pushing forward to the top of the wooded heights about a kilometer N. of this place, but gained nothing further by their renewed attacks in the afternoon. On both sides of the road from Charpentry to Romagne their attacks were once more completely shattered against the line of Westphalian and Alsatian-Lorrainian regiments. *Further east the enemy penetrated into the du Fays wood, but was everywhere else repulsed.*" 226

From now on, until the 80th Division could move forward and permit the resumption of the attack, the 8th Brigade was forced to consolidate its lines and defend the position it had gained.

Though the established lines remained stationary the troops were by no means inactive. On October 5th three attacks were repulsed, one made by more than 300 German infantry. 227
Each left machine guns and prisoners in the hands of the Americans. The artillery fire from east of the Meuse was incessant and deadly. Company "A," 10th Machine Gun

Battalion, lost all its officers on this day, when Captain Henry B. Keep and his four lieutenants were killed by this fire. Patrolling was constant. The area between the Fond de Ville aux Bois and the northern edge of the Bois de Peut de Faux became No Man's Land, a region infested with aggressive patrols from both combatants.

The infantry and machine gun units in the Bois de Fays were under a terrific strain. A scattered line of precarious holes, dug amid the seared and shattered trees of the wood, afforded the only possible shelter. It rained nearly every day and the dampness and cold, the mud and the darkness, made life almost unbearable. But they had to protect a salient open to attack from three sides and to suffer a continuous fire from every form of weapon that could throw a projectile. No fires could be built. Hot food could only be brought up from the Bois de Brieulles at night. Gas reeked through the woods. Aeroplanes bombed them at frequent intervals. The stench of dead bodies became unendurable, and as many as could be found in the thick underbrush were carried to the southern edge of the woods and later buried. Seldom has there been such a combination of horrors and seldom have the supreme qualities of endurance of mind and body and quiet heroism been in greater demand.

The Germans patrolled the front line and occasioned much trouble but it was their infiltration tactics that caused the most anxiety among the men of the Ivy Division. At night small parties of Germans would attempt to steal into the woods and fire into the backs of the front line infantry, in an endeavor to create consternation. These tactics demanded strict vigilance, particularly from the units guarding the flanks. On one occasion the Germans did succeed in penetrating into the lines by these methods but they were soon ejected. Then the hostile artillery, strong in the knowledge of all the springs and wells, the roads and trails through the woods, shelled them with a deadly accuracy that made it a journey of terror to get water or to deliver a message. The minds of the men were dulled and they became almost mechanical in their movements after a few days of this punishment. German attacks, which were frequent, were welcomed as a relief. The men wanted to go forward. They were tired of the Bois de Fays.

For sheer endurance there is little to equal, nothing to surpass the call on the Ivy Division during those days. That the positions in the Bois de Fays and the Bois de Brieulles were maintained, that fresh preparations were made for a new attack, is the finest testimony to the wonderful spirit of every branch in the Division.

Nor were the artillerymen without their sufferings. Gas shelling was so frequent that firing in gas masks became the order of the day. In the Bois de Septsarges, where some of the artillery was in position, shells fell continually and the losses in animals were so great as to threaten seriously the mobility of the units. In Nantillois dead animals lay everywhere. One four-line team, loaded with bread and hard bread, stood in the street. The driver lay dead in front of the mules and one of the mules was dead. The other three stood quietly in their harness as if nothing had happened. Fifteen water carts lay scattered about the town, their horses dead in harness.

During the night of October 5th-6th the battalion of the 318th Infantry (80th Division), which had been serving with the 8th Brigade, was relieved and returned to its own division. Its men had rendered valiant service and had acquitted themselves most creditably. 228

The 80th Division, after repeated efforts, had, after dark, managed to secure a foothold in the Bois des Ogons. On the evening of October 6th it had reached the northern edge. 229

October 6th, 7th and 8th saw no marked change in the situation. Under orders from the Corps commander the front lines organized their positions in anticipation of possible at-
tacks by the Germans. In the back area a second line of 230
defense, through the Bois de Septsarges, was prepared by the

View from Cuisy, looking north, showing long line of 4th Division supply trucks on the Cuisy–Gercourt road, approaching the 4th Division ration dump at the right of the picture. Field hospital, in tents, at left. The town, dumps, and hospital were daily shelled by the Germans. Taken Oct. 6, 1918

View of Septsarges from the southeast. The headquarters of the 7th and 8th Brigades were located here. The road crossing the sky-line in the centre of the picture is the Septsarges–Nantillois road

engineers. On October 6th the 3rd Battalion and the regimental machine gun company, 132nd Infantry (33rd Division), were attached to the 4th Division and that night joined the
231 59th Infantry, which was now in the Bois de Fays. The 58th Infantry and one battalion of the 59th were relieved from the Bois de Fays and withdrawn to the vicinity of Cuisy to
232 rest.

General Pershing now decided to make another big push. The attack of the First Army on October 9th was one in which the V Corps played the principal part. It was to capture the heights west of Romagne and those near Cunel. This included the penetration of the Kriemhilde Stellung south of the Romagne–Cunel road. The I Corps, on its left, and the III Corps, on its right, were to protect the flanks of the V Corps.
233 The attack of the V Corps was to begin at 8:30 A. M.

In the III Corps the 4th and 80th Divisions prepared barrage tables and elaborated complete plans for taking up the forward movement, in close cöoperation with each other, when
234 the line of the V Corps should arrive abreast of the III.

In the 4th Division, the 7th Brigade, reinforced by one battalion of the 59th Infantry and one battalion and the machine gun company of the 132nd Infantry (33rd Division), was designated as the unit to make the attack passing through the front line held on the Fond de Ville aux Bois. The remaining units of the 8th Brigade were withdrawn from the Bois de Fays to rest. The 47th Infantry was directed to hold the northern edge of the Bois de Brieulles, Hill 280 and the northern edge of the Bois de la Côte Lémont to the river, the divisional sector having been widened for this operation to include everything between the western boundary and the Meuse
235 River.

The 39th Infantry was designated as the assaulting unit. Its battalions were arranged in three lines, the 3rd Battalion in first line, the 2nd Battalion in support and the 1st in reserve. Each battalion had a machine gun company attached to it. General Poore directed the other attached units of his brigade to remain in the Bois de Fays. No definite hour could be fixed, in advance, for beginning the forward movement; that would depend on the progression of the V Corps. 236

At 10:30 A. M. the 3rd Battalion, 39th Infantry, passed through the lines of the battalion of the 132nd Infantry and settled down in fox-holes, south of the Fond de Ville aux Bois, to await the signal to advance. The march up to the 237
front line was a march of horror. Several paths led through the Bois de Fays and all were strewn on both sides with dead bodies. In one spot lay more than sixty men and in another over two hundred. Gas, artillery shells and ma- 238
chine gun bullets rained on them. In order not to crowd the Bois de Fays, which was a veritable death-trap, with too many troops the other two battalions of the 39th were held south of the woods. 239

It must be remembered that the front line of the 80th Division was, at this time, near the *northern* edge of the Bois des Ogons which is west of the *southern* edge of the Bois de Fays. In consequence, the 80th Division, moving behind its barrage at the prescribed rate, would not reach the front line of the 4th Division until an hour and forty minutes after it had started. It was not advisable for 240
the 4th Division to start ahead of the 80th; the barrage of the latter might easily fall on the flanks of the 4th Division and cause not only losses but dismay. Nothing is more trying to the infantry soldier than for him to know or feel that he is being fired on by his own artillery.

Hour after hour passed. Hour after hour the enemy shelling continued. The Division Headquarters attempted to obtain from the Headquarters of the 80th Division a definite time at which the barrage of the latter would begin. This the 80th could not furnish as its advance was contingent upon that of the 3rd Division, on its left. At last, at 3:00 P. M., came the coded message by radio "H hour is 3:30

241 P. M.; 15 minutes' preparation." The information was
promptly transmitted to the organizations. The leading
battalion of the 39th Infantry was to advance at 5:10
242 o'clock.

In the meantime this battalion, on the Fond de Ville
aux Bois, was undergoing a demoralizing experience. The
Germans knew that the 4th Division held the Fond de Ville
aux Bois; also that the 80th had started to advance under
a barrage. They doubtless believed that the 4th Division
would advance soon after the 80th had started. At 4:40 a tre-
mendous concentration of German artillery fire was placed
on the northern half of the Bois de Fays. Most of this
was gas. At the same time an exceedingly heavy machine
gun fire was opened from the Bois de Peut de Faux and the
Cunel–Brieulles road, to the east, the latter enfilading the
243 Fond de Ville aux Bois. Much of the artillery fire came
from east of the Meuse and, coming from their rear, led the
men to raise the cry that their own artillery was firing into
them. This was not true but it had its effect in diminished
morale. Gas masks had to be worn. With difficulty their
officers started to collect the now scattered men, to reform them
for the coming attack. A considerable number were gotten
together and at "H" hour, when darkness was approaching,
they advanced. In the failing light and amid the thick under-
brush little could be seen through the goggles of the masks.
The shell-fire seemed to increase with every passing minute.
The roar of heavy shells could be heard coming from the rear.
Soon the darkness in the thick woods became almost impene-
trable. The men in the front lines fell. The men behind
stumbled over their dead or wounded comrades as they ad-
vanced against their invisible enemy. They had nothing to
guide them but the sound of their officers' voices. They
could not see. They fell over the barbed wire which the
Germans had wound low through the underbrush and between
the trees. The advance stopped. The word was passed

along the line to withdraw and the men filtered back by ones and twos to the Fond de Ville aux Bois from which they had started. The attack had failed. 244

It had been a hopeless task. An attack in gas masks in open country and in daylight is by no means a perfectly simple matter. An attack in gas masks in the dusk of the evening, through a heavily wooded forest filled with barbed wire and tangled underbrush, promises little hope of success.

The attack was repeated at seven o'clock on the next morning. The 2nd and 1st Battalions of the 39th had been moved into the Bois de Fays after midnight. The 2nd Battalion was now placed in front line with the 1st Battalion in support. The 3rd Battalion, to which were added two companies of the 132nd Infantry, was placed in reserve. 245

Advancing behind a barrage, the 2nd Battalion crossed the Bois de Malaumont in spite of a repetition of the massed artillery, machine gun, and trench mortar resistance encountered on the previous day. In crossing the Bois de Peut de Faux it was subjected to direct rifle fire from the front and heavy machine gun fire from both flanks. Much of the latter appeared to come from the small group of trenches between the town of Cunel and the Bois de Forêt. The line fell back to the Bois de Malaumont and was reorganized. A barrage was called for and at nine o'clock the battalion advanced again. Once more it pushed up to the northern edge of the Bois de Peut de Faux and once more it was driven back, this time, however, only south of the Cunel–Brieulles road. Here the troops were again re-formed. The losses had been heavy; the lines were getting thin. One company was commanded by a corporal, every officer and sergeant having been killed or wounded.

As the regiment was now without majors, the Brigade Commander designated Major Floyd R. Waltz, 11th Machine Gun Battalion, for duty with the 39th; he was placed in 246
command of the leading battalion about noon.

Another barrage was now called for by the 2nd Battalion and behind it a third advance was begun about 2:30 P. M. Despite determined opposition the Bois de Peut de Faux was crossed. In the clearing between this woods and the Bois de Forêt the troops encountered such a destructive machine gun fire from the southern edge of the Bois de Forêt that the line fell back to the northern edge of the Bois
247 de Peut de Faux. Here it held. At 5:30 P. M. strong reconnoitering patrols were pushed out toward the Bois de Forêt. But the fire was too heavy. The patrols had to return. A platoon of 30 men cleared out the small trench system at the southwest corner of the Bois de Forêt and were later compelled to seek shelter in it against German
248 fire from Cunel and the Bois de Forêt.

The men dug in for the night. They were thankful to have left the Bois de Fays behind. They looked forward to the morrow with every confidence. The 80th Division, on the left, had reached a line running from a short distance south of the town of Cunel and the Cunel–Brieulles road to the
249 western edge of the Bois de Malaumont.

Plans of the First Army, for October 11th, contemplated
250 a general attack along the entire front. During the afternoon of October 10th many towns to the north of the American forces were reported to be in flames. Aviators also reported convoys moving to the north, indicating that the enemy
251 was making strenuous efforts to evacuate his materiel.

Early on the morning of October 11th, Colonel Parsons and his entire regimental staff were gassed. Lieut. Col. Troy H. Middleton, 47th Infantry, was now sent to the Bois de Fays to take command of the 39th Infantry and all troops in the
252 Bois de Fays and the woods north of it.

At 7:28 A. M., following closely behind a well-placed barrage, the attack was once more resumed. The disposition of the troops in the attacking line was as follows: 2nd Battalion, 39th Infantry, assaulting line; 1st Battalion in support, echeloned on the right rear of the 2nd Battalion to cover its right flank from attack from the eastern half of the Bois de Forêt; the 2nd Battalion of the 47th Infantry which, on October 10th, had been moved into the Bois de Fays, and the 3rd Battalion, 39th, now greatly diminished in strength, and two companies 132nd Infantry were in reserve in the northern edge of the Bois de Malaumont. 253

Under heavy machine gun and minenwerfer fire (there was little gas this morning) the troops crossed the open ground north of the Bois de Peut de Faux and entered the Bois de Forêt. The line of the narrow-gauge railroad in the northern part of the Bois de Forêt was reached about 8:15 A. M. Approximately two hundred machine guns and two hundred prisoners were captured during this advance. The Germans, from excellent firing positions and with excellent artillery support, fought coolly and resourcefully but all their experience and cleverness did not avail them against the savage onslaught of the Americans. Under the protection of the barrage the attack was pushed to the northern edge of the Bois de Forêt. About one o'clock the 1st Battalion of the 39th was withdrawn from the right flank to reinforce the rapidly thinning lines of the 2nd Battalion and the 2nd Battalion of the 47th Infantry was moved from its position in the Bois de Malaumont to the position formerly held by the 1st Battalion of the 39th. About 4 o'clock the 2nd Battalion 47th Infantry moved eastward, through the Bois de Forêt, about two hundred yards, preparatory to a farther advance under cover of a prearranged artillery barrage which was to roll east. The barrage came down as planned but, owing to the density of the underbrush, was permitted to pass eastward without being followed by the men —they believing that it was hostile shelling. Nothing further of importance occurred on the afternoon of the 11th with the exception of a German attack from the Bois de la Pultiére which was repulsed by the accurate fire of the infantry and machine gunners stationed near the western edge of the Bois

de Forêt. When darkness fell, on the evening of the 11th, the front line extended from the small trench system at the western extremity of the Bois de Forêt northeast along the western edge of the woods to a point 200 meters north of the Fontaines de Bouillon, then south to the southern edge of the Bois de Forêt and around the eastern extremity of the Bois de Peut de Faux. Patrols were sent out about 8 o'clock to Hill 299. The patrols stayed there until 4:30 on the following
253 morning.

On the morning of October 12th Company "L," 132nd Infantry (33rd Division), was sent into the front line to reinforce the left flank of the 2nd Battalion, 39th Infantry. A patrol of fifteen men from this company crawled across the open ground north of the Bois de Forêt to a small clump of trees (la Mi Noel) about 200 yards south of the top of Hill 299 and remained there in observation during the entire
254 day. About 11:00 A. M. another German attack was made from the direction of the Bois de la Pultiére, which was repulsed with ease. At about 12:30 P. M. a heavy and sustained artillery barrage was dropped on the front line then held by the 2nd and 1st Battalions of the 39th and one company of the 132nd Regiment, which caused harrowing losses. More than two hundred men were killed or wounded by this
255 artillery fire.

According to the unanimous testimony of the officers and men in the line that suffered, this artillery fire came from their left rear. There is, unfortunately, no other explanation than that the shells were fired by the artillery of another American division. Although a request was made that this firing be thoroughly investigated no report was received by the 4th Division concerning its origin. That it did not come from any unit of the 4th Division was established by suspending the firing of all the Divisional guns for a definite period, during which the shells continued to fall on the front line. The barrage lasted nearly two hours. It ceased about 2:30 P. M. Its temporary effect upon the morale of the men in the woods can be imagined. These men had been seventeen days under fire of high explosive and gas shells, of machine guns and snipers. For four days they had wrestled for this position. They had had nothing but cold food to eat and it had rained all the time. They had not had any sleep except what they could snatch during momentary lulls in the fighting when they lay in shell holes or against trees. They were physically exhausted. Their nerves were keyed up to the breaking point. To be shelled by American artillery was almost too much. Under command of their officers they fell back to the southwest and northeast road running through the middle of the Bois de Forêt until the artillery fire ceased and then, to their eternal glory be it said, returned and took up their old hard-won position on the edge of the woods. That, too, was the spirit of the 4th Division.

On October 12th it had been planned that, in the afternoon, the support battalion (2nd Battalion, 47th Infantry and one company 132nd Infantry) should, under protection of a barrage, clean up the rest of the Bois de Forêt—that is,
the eastern half. Unfortunately a misunderstanding arose 256
and, although the barrage was fired, the infantry never went forward. No other attack was attempted on October 12th by the troops, who had consolidated their position, because orders had been issued that the 5th Division would relieve the 4th that night. This order was later countermanded by the
III Corps. The relief did not take place although the of- 257
ficers of one battalion, 61st Infantry, made a preliminary reconnaissance on October 11th and two companies of the 61st Infantry arrived in the Bois de Peut de Faux on the morning of October 12th, and participated in some of the fighting. The troops of the 4th Division, in the Bois de Forêt, therefore, remained in line under fire. Outposts were shoved out to its most northern edge, east of Hill 299 and patrols to the ridge north of that. This was the "farthest north"

A battery of German 77 mm. guns captured by the 4th Division on Sept. 26, 1918, named Battery "Q," and used against the Germans

View in the Bois de Brieulles. This indicates better than words the nature of the woods through which our soldiers forced their way

View from a point 150 meters east of the old German power-house on the Ruisseau de Ponthieu, showing the stream and power-house in the foreground. In the background the Bois de Fays is on the left, the Fond de Ville aux Bois in the centre, and the Bois de Malaumont on the right. Some of the bitterest fighting in the World War took place in this area. Taken in March, 1919

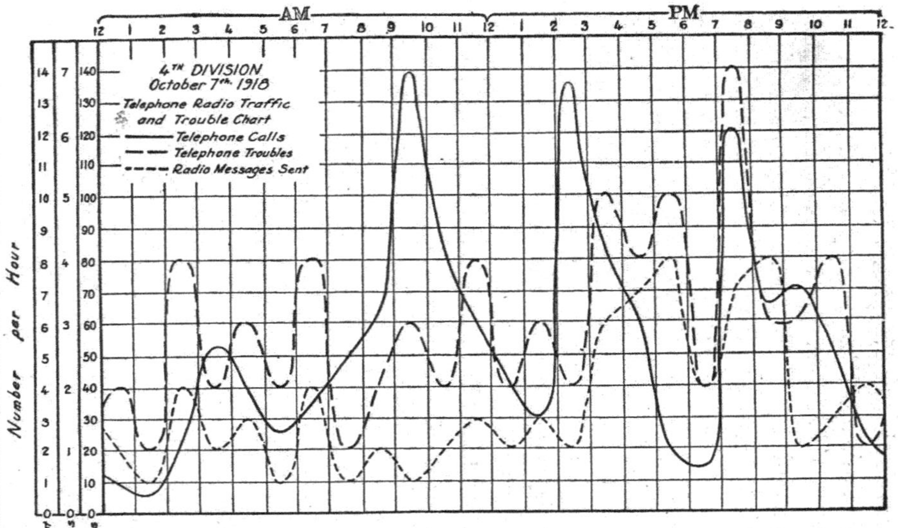

A graphic illustration of the telephone and radio traffic and of the telephone troubles handled by the signal troops of the 4th Division during a typical day in action. The peak loads in the telephone calls curve (solid line) show the heaviest telephone traffic from 8-9 A. M., 2-3 P. M., and 7-8 P. M.

reached by the 4th Division. About nine o'clock that night reports came in from the outposts that the Germans were attempting to infiltrate the position, and the patrols were withdrawn to the northern edge of the Bois de Forêt. A one-star rocket was fired over No Man's Land and by its light several crouching figures were seen approaching. As all the patrols had reported back at the outpost a heavy machine gun
fire was opened and maintained for about 25 minutes. Later 258
a second rocket showed nothing. The patrols were again sent out to the north. One patrol was also sent toward the Bois de la Pultiére. This patrol remained out till dawn on the 13th. During the night some shells fell in the area but did little damage. Casualties were mostly inflicted by machine gun fire which came from two directions, east and northwest. About dawn of the 13th the 11th Machine Gun Battalion repulsed an attack from the east through the Bois de Forêt by its own machine gun fire reinforced by the fire of
captured machine guns. 259

Between midnight and 8:00 A. M. on October 13th, the units of the 4th Division and the attached elements of the 132nd and 61st Infantry Regiments were relieved in the Bois de Forêt
by the 4th Infantry Regiment, 3rd Division. The 4th In- 260
fantry did not take over the advance line held by the 7th Brigade but occupied one that was shorter and less exposed—from the western corner of the Bois de Forêt along the northwestern edge for one kilometer, then along the unimproved road southeast through the forest to the northeastern corner of the Bois de Peut de Faux. The Commanding General, 5th Brigade, 3rd Division, stated that this shortened line was
made necessary by the depleted condition of his brigade. 261

When the units of the 7th Brigade in the Bois de Forêt were ordered to be relieved, the boundaries of the divisional sector were changed. The front now extended along the Meuse River. The western boundary ran from a point on the river one kilometer north of Brieulles, past the eastern cor-

ner of the Bois de Forêt to the northern corner of the Bois
de Fays, then south through the Bois de Fays to the western
edge of the Bois de Brieulles. The eastern boundary ran
almost due south from the northern point of the Bois de
262 la Côte Lémont. To the 47th Infantry was assigned the
duty of holding the front line. The 2nd Battalion of this regi-
ment and Company "M" were placed in the Bois de Fays to
hold its eastern edge. The remainder of the regiment con-
tinued to occupy the northern edges of the Bois de Brieulles
and the Bois de la Côte Lémont. The 39th Infantry was with-
drawn to the Bois de Septsarges. The battalion of the 59th
which had remained in the Bois de Fays was sent to join its
regiment south of Cuisy.

The 3rd Battalion, 132nd Infantry was relieved from duty
with the 4th Division on October 13th and directed to rejoin
263 the 33rd Division. To this battalion the 4th Division ex-
tends its sincere appreciation. Two of its companies partici-
pated in all the fighting in the Bois de Forêt; the other two and
the regimental machine gun company were in reserve in the
Bois de Fays. The Battalion Commander, Major John J.
Bullington, rendered most valuable aid to the Commanding
Officer, 39th Infantry, at a time when the latter was practically
without staff officers. The officers and enlisted men gallantly
fought side by side with their comrades of the 4th Division
and endeared themselves to all by their courage and confi-
dence. The battalion suffered heavily; 1 officer and 37
264 men were killed and 11 officers and 315 men were wounded.

Upon recommendation of the Division Commander, the
Commander-in-Chief, on October 14th, promoted Lieut. Col.
Middleton of the 47th Infantry to Colonel and Major Waltz
of the 11th Machine Gun Battalion to Lieut. Colonel in recog-
265 nition of their gallantry in action.

While the 7th Brigade had been fighting for the posses-
sion of the Bois de Forêt, the 47th Infantry (less the 2nd
Battalion) had been making adventurous incursions into the

German zone opposite the northern edge of the Bois de Brieulles. Patrols went out to the Tranchêe de Teton and even penetrated into the streets of Brieulles, crawling past the wrecked buildings in search of the enemy. Along the banks of the Meuse, Germans could be heard talking and, on one occasion, when a patrol was returning from Brieulles, it came across a German patrol making its way back to the town. On the morning of October 9th a 37 mm. gun or minenwerfer was seen mounted on the western end of the Tranchêe de Teton
but on the following morning it had disappeared. From that 266
time on, the trench was occupied, as an outpost position, by the 47th Infantry.

The new area, while by no means as active as the old, was far from being a quiet sector. The Bois de Brieulles and the Bois de Septsarges were constantly harrassed by enemy artillery. The 47th Infantry patrolled actively and held the front from October 13th to 19th. Then parts of the 6th Brigade, 3rd Division, arrived to relieve the regiment and during the night of October 18th-19th the front line was handed over to the incoming brigade, the relief being com-
pleted at 6:30 on the morning of the 19th. Meanwhile, on 267
October 14th, the 58th Infantry had been marched to the Bois de Beuge as reserve for the 5th Division. Here it remained
until October 18th when it marched back to Cuisy. 268

The 4th Field Artillery Brigade had acted during this time as part of a group composed of the artillery brigades of the 4th, 3rd, and 80th Divisions, and the 250th and 456th French Artillery regiments. Their mission was to support the attack of the 3rd and 5th Divisions. This whole group was commanded by General Babbitt, of the 4th Artillery Brigade. On October 16th the group was dissolved and the 4th Artillery
Brigade returned to support the 4th Division. 269

On the 19th all elements of the Division, with the exception of the 4th Artillery Brigade and the 4th Ammunition Train, turned their faces south and marched in the direction

269 of the Forêt de Hesse. It was not until the night of October 23rd-24th that the artillery was withdrawn for a rest of four days, after which it re-entered the action, to remain in it until the signing of the armistice.

On October 9th Lieut. Col. R. A. Wheeler, 4th Engineers, relieved Colonel A. A. Acher as Division Engineer, the latter having been transferred by G. H. Q. to similar duty with the 27th Division.

The 4th Division lost the valued leadership of General John L. Hines on October 10th. Once more it paid the inevitable price of having a great leader, for General Hines was selected by General Pershing to command the III Corps, General Robert L. Bullard having left to take charge of the Second Army. But the Division was happy, for General George H. Cameron, its first commander, returned to it from the V Corps.

CHAPTER XIII

WITH THE SECOND ARMY

THE 4th Division, less the Artillery Brigade and the Ammunition Train, was withdrawn from the battle line on October 19th. Twenty-four days previously its soldiers had 269
marched up to the front, in the closing days of September, as fine-looking a lot of men as any that had ever been seen in the World War. Their spirit was reflected in their bearing. They had taken part in two major operations and had acquitted themselves with credit. They were proud of themselves and of their division. Their uniforms were clean; their arms and accouterments were perfect; their faces glowed with the joy of life and the expectancy of battle. They were the pick of America's young manhood and they looked it.

Their appearance was quite different as they marched south, 24 days later, leaving behind them the scenes of horror and the stifling gases of the Bois de Forêt, the Bois de Fays and the Bois de Brieulles. During that period they had gone through unbelievable hardships, had suffered indescribable exposure and been subjected to a nervous strain beyond anything that men, unaccustomed to war, can imagine. They had seen their comrades wounded or stricken down in the hour of triumph. They had passed through the valley of the shadow of death. A keen observer could read in their eyes the memories of horrors that time could soften but only death destroy. Some of them talked and even laughed on their way. Most of them marched in silence, each absorbed in his own thoughts. Men who had never expected to come out alive smiled half cynically. Places where they had fought in their first advance were passed; how different they seemed now. Why here were men and wagons moving with absolute unconcern where, only a few days before, to stand upright meant death. To some it seemed like a tragedy, to others a tremendous joke.

As they marched south over the hills on their way to the Forêt de Hesse, these men were not the spick and span soldiers who had marched buoyantly to battle. The slime of the wood, the thick underbrush, the sear of the shrapnel had done their work. Their uniforms were torn and often stained with blood. Their cheeks were lean, their faces grim and a little haggard. They were tired and worn out but they marched well. They were shaved, their shoes were rubbed, their clothing cleaned of mud as well as could be. Their march discipline was perfect. And when, on the way, one regiment passed its old Division Commander, General John L. Hines, he was so impressed with the bearing of the men that he wrote General Cameron a letter in which he expressed his satisfaction and stated further that they looked more like soldiers
270 going into action than like soldiers coming out.

As kilometer after kilometer passed, as the packs grew heavier and heavier and feet responded less and less willingly to mental urgings, they began to ask themselves when they would reach a scene where the devastating hand of war had not spread. The monotonous sequence of upturned countryside and ruined houses told them that they had still far to go and they began to wonder, as soldiers wonder, if they had actually fought all this way. Why, to be sure, there was the old front line. "My God," someone exclaimed, "We certainly traveled."

Then the consciousness of success and the pride of achievement glowed in every man's mind and soul. They had fought and suffered, but it had not been in vain. They had driven the Germans back and made a great record for their Division. Something like natural cheerfulness returned to the men. Steps grew a little less weary. Sufferings lost something of their sharpness. They had fought in America's greatest battle, and had acquitted themselves worthily.

The men of the Ivy Division had, in truth, fought with splendid courage and devotion to duty. They had been steadfast and loyal. Their casualties had been heavy: 45 officers and 1120 men had been killed or had died of wounds, while 199 officers and 6048 men had been wounded, gassed, or were missing, and 2 officers and 45 men had died from other causes, making a total loss in strength of 7459 officers
and men. This was a heavy price to pay, but it was in- 271
evitable under the circumstances. A more than ordinarily difficult task had been accomplished. The Division had penetrated to a depth of 13 kilometers, over hills and through
woods, capturing 2731 prisoners, and taking 57 pieces of ar- 281
tillery, 4 minenwerfer, 228 machine guns, 2 tanks, and a vast
quantity of ammunition of all types. Above all it had re- 272
mained in action for twenty-four days, the longest consecu-
tive period any one division was in action. The determina- 273
tion of the enemy to resist the advance of the Division is well illustrated by the fact that he employed all of parts of nine different divisions against the Ivy men, one of them being the
famous 28th Prussian Division. 274

On October 14th the First Army, in conjunction with the Fourth French Army, made another great attack in which the 4th Division did not actively participate, merely holding
the front assigned to it. This joint attack enabled the 275
Fourth French Army to make definite gains. It reached the vicinity of Grandpré, where it established liaison with our 78th Division. The First Army met with desperate resistance. There was bloody fighting. The 82nd Division reached St. Juvin. The 32nd Division won Romagne and Hill 257, breaking through the enemy's main position on a front of one and a half miles. On October 21st the 5th Division, after four attacks, took the Bois des Rappes and the 78th

Looking northwest from the northeast corner of the Bois de Malaumont, showing the Cunel–Brieulles road and the Fond de Cunel in the centre of the picture. North of the road is the Bois de Peut de Faux

Narrow-gauge railroad track in the northwestern portion of the Bois de Forêt. Taken from a point 350 meters east of the western edge of the woods. The fighting in this immediate area, on October 11th and 12th, was of a most sanguinary character. 200 German machine guns were captured. Taken in March, 1919

276 carried Grandpré. On October 22nd the 3rd Division
277 cleaned out the eastern end of the Bois de Forêt. The
American front now extended from Grandpré to north of
Brieulles. The remainder of the month was spent by General
Pershing in preparation for another big drive to be launched
278 on November 1st. On October 30th the 89th Division made
a deep salient in the enemy lines by taking the Bois de Banthe-
ville and the 5th Division another by the capture of Aincre-
279 ville.

Preparations having been completed, General Pershing
launched his big attack on November 1st. The artillery had
been massed in great strength and in favorable positions, re-
serves were close to the front line and lines of communica-
tion well established. The First Army smashed its way
through the last hostile position, the Freya Stellung, taking
nearly 3000 prisoners. The Germans reeled before the blow
and began a retreat along the whole line but General Pershing
pursued them with unrelenting vigor. From November 1st to
November 4th the victorious American troops advanced nine-
teen kilometers, taking over 4000 prisoners and more than a
280 hundred guns. Then pivoting on the 5th Division, General
Pershing forced the Meuse on a wide front and, on November
9th the line extended from Bazeilles just south of Sedan to
Beaumont southeast of the Bois des Caures. Roughly
speaking, the American line ran from Verdun to Sedan. Not
a single German remained on the left bank of the Meuse. The
great Metz-Lille railroad, essential to German retirement, was
under fire. The enemy was in full retreat. The Battle of the
Meuse-Argonne, in which General Pershing had engaged
631,000 American troops and suffered 119,000 casualties,
was over. It had lasted 47 days, penetrated the enemy
lines to a depth of 55 kilometers, liberated 150 towns and
villages for France and freed 1650 square kilometers of
281 territory. Victory as complete as any in history had crowned
the American arms.

In the course of the Meuse-Argonne battle the German line, which on September 26th, the day of the initial attack, was held by only 6 divisions, was defended by 47 divisions, 15 of which were used twice or three times. Not only did 282
the Germans throw in their last reserves—literally their very last division—to stem the American tide but they depleted other parts of the front, thus enabling the British and the French to advance more rapidly.

After the battle, the Commander-in-Chief issued the following General Order: 283

> The enemy has capitulated. It is fitting that I address myself in thanks directly to the officers and soldiers of the American Expeditionary Forces, who by their heroic efforts have made possible this glorious result. Our armies, hurriedly raised and hastily trained, met a veteran enemy and by courage, discipline, and skill always defeated him. Without complaint you have endured incessant toil, privation and danger. You have seen many of your comrades make the supreme sacrifice that freedom might live. I thank you for the patience and courage with which you have endured. I congratulate you upon the splendid fruits of victory which your heroism and the blood of our gallant dead are now presenting to our nation. Your deeds will live forever on the most glorious pages of America's history.

During the entire time that the 4th Division was in action in the Meuse-Argonne battle, the weather was as bad as can possibly be imagined. It rained almost continually and even in the densest woods the soil was changed to a thick, sticky mud. Every shell hole was filled with water. Mist and fog made visibility very poor except for a few hours in the middle of the day and made liaison extremely difficult. The shattered state of the country and the absence of roads made it almost impossible to bring up supplies and evacuate the wounded. But the spirit of the Division inspired all its officers and men and enabled them to overcome difficulties, the true magnitude of which was not then realized.

To the success of the Division every branch of the service contributed its full part. The building of the Esnes-

Malancourt road by the 4th Engineers was perhaps the most important front-line engineering achievement of the American Army in France; it made possible the victory of the Meuse-Argonne. The infantry surpassed itself in valor, and proved not only its indomitable fighting spirit in attacking but equally splendid qualities of endurance in holding the woods under continuous and harrowing artillery fire. The artillery at last received a great opportunity to demonstrate its power and convinced all of its devotion to duty and accuracy in firing. The Signal Battalion, the Ammunition Train, the Medical Services, the Sanitary Train, the Supply Train whose trucks, under fire, moved the guns of the artillery, the runners and the staffs all worked with an energy and good will and suffered losses unflinchingly in a manner that reflected the greatest credit on themselves, and showed the pride felt by each man in the success of the Division.

That spirit may perhaps be shown by the story of a sergeant and 29 men. They were discharged from hospital in the early part of October and ordered to report to their regimental headquarters at Cuisy. They had all been wounded on the Vesle and had spent many unhappy days in hospital looking forward to the time when they could get back to their regiment, the 4th Engineers. With full packs and equipment they hiked on from the railhead, passing along the jammed roads, under intermittent shelling, until they reached Cuisy during one of those many periods when the Germans, with typical thoroughness, were searching with shrapnel and high explosives for Division Headquarters. An engineer officer standing on the side of a hill looking down into Cuisy noticed these men slowly making their way up toward him. They were weak from long confinement in hospital. During the minute intervals between shell bursts they would rush from one ditch or shell hole to another and drop for protection. It quickly became clear that they had a very definite mission in mind and, regardless of shell fire, were going to complete it.

When within saluting distance, the sergeant stepped up, saluted and said, "Sir, Sergeant Ducrot and 29 men report for duty." And then after a moment's pause, "Gee, Colonel, but it's fine to be back home again. That hospital was hell."

The battle served again to demonstrate the wealth of American ingenuity. The artillery, anxious to protect its horses from shell fire, particularly as the casualties were exceedingly heavy among animals, built corrals and palisades of saplings and filled up the interstices with clay. Behind this protection the horses, when unharnessed, were placed and although, of course, it was not sufficiently strong to withstand a direct hit, it would stop splinters that might otherwise maim. Then serviceable wheels were taken off the guns hit by enemy fire and fitted, so far as it was possible to fit them, to the guns whose wheels had been shattered by shell fire. Some of the guns of the Artillery Brigade have as many as six and seven wound stripes.

Of all the severe tasks, however, that fell to the Division, that of hauling food and ammunition was certainly one of the hardest. After September 27th the Ammunition and the Supply Trains had to go back a long distance to get their loads. Caught in the midst of traffic congestion and under fire for much of their journeys they were called upon to endure days and nights without sleep. They had to deliver their cargoes come what may. The infantry and the artillery must be munitioned and fed. That they did so under the very adverse conditions is the best testimony to their splendid services. To the M. P.'s also, who labored ceaselessly day and night to disentangle and direct the traffic on the Esnes–Malancourt road, all credit is due.

Above all is a special tribute due to the stretcher-bearers of the Division. The infantry spoke of these brave and valiant men with an admiration and sincere regard that had something of grief in it. To carry a man on a litter, under terrific shell fire, when the bearers cannot drop their load and "duck" is as severe a test of nerve as any, but it was done by the stretcher-bearers—15 to 20 men from each company. Naturally, their casualties were severe.

The 8th Brigade marched by day on the 19th to the Forêt
284 de Hesse, just north of Jouy-en-Argonne. The 7th Brigade
bivouacked around Cuisy on the evening of the same day and
285 arrived at the Forêt de Hesse on the 20th. Here the Division
Commander inspected all units. Headquarters was established
at Jouy-en-Argonne where, on October 22nd, General Cameron
left the Division to return to America, there to lend his
experience in the training of other divisions destined for
286 France. The command of the Division was taken over by
General Poore. While in the Forêt de Hesse the men tried the
delousing plant, but it was inadequate. There were no bar-
racks as the forest was just back of the old front line. The
water here was insufficient and of poor quality. After re-
peated protests by the Division Commander, reinforced by
those of the Corps Commander, against the retention of the
Division in the forest under these conditions, the Division
287 was assigned, on October 20th, to the Second Army. Orders
were received from the Second Army to move on October 21st, by marching, to the Toul area and take station in the vicinity of Vignot, just north of Commercy. The march occupied four days and was made in two columns. The 8th Brigade moved via Souilly, Issoncourt, Pierrefitte, Gimécourt and Lerouville. The 7th Brigade marched up the valley of the Aire River and then followed the route of the 8th Brigade. The resiliency of youth was shown by the manner in which the command had recovered from the strain of battle. Army inspectors closely observed the troops during the march and, upon its conclusion, letters were received by the Division Commander highly commending the march discipline. As soon as the Division arrived in the Vignot area the men were thoroughly deloused, training was resumed, and replacements, clothing, and animals received. Division Headquarters was

established at Vignot and the Division designated as Army
Reserve, Second Army. 269

On October 28th the 8th Brigade was moved about fif-
teen miles farther east with its headquarters at Avrain-
ville. Division Headquarters was transferred to Lucey and 269
training again resumed. It was here that, on October 31st,
General Mark L. Hersey joined the Division. General Her- 269
sey before being assigned to the 4th Division had successfully
commanded the 155th Brigade, 78th Division, which he had
organized, brought overseas and fought in the St. Mihiel
sector and during the bitter struggle of the Meuse-Argonne
campaign. As a military commander, his tireless energy and
knowledge of details was given to the Division at a time
when such knowledge was all-important. He was especially
insistent upon due attention to animals and transportation.
Alert, active, keen-eyed, he made his presence felt in every
unit. His observance of the "square deal," his ability and
his unselfish loyalty to the Division won for him the same
respect and devotion that had been given to his predecessors.

While at Lucey orders were received on November 3rd, re-
assigning the Division to the First Army, as part of the III
Corps, preparatory to its re-entry into the Meuse-Argonne
fight. These orders directed the Division to move by bus 289
back to Blercourt. The 8th Brigade, the 39th Infantry and
the 11th Machine Gun Battalion, the Signal Battalion and
part of the Sanitary Train moved on November 6th, in accord-
ance with these orders, there being only enough trucks pro-
vided to move half the Division. Before the trucks could re- 289
turn to move the 7th Brigade and the remaining auxiliary
troops, the order assigning the 4th Division to the III Corps
was revoked and the Division was returned to the control of the
Second Army to participate in what was to be the final blow,
on the Lorraine front. So the troops at Blercourt returned, 290
by bus, on November 9th. The 8th Brigade took station 269
in the vicinity of Void. During the absence of the 8th Brigade

in the First Army area, the 7th Brigade had been marched to
the Bois de la Belle Ozière, on November 10th, as reserve for
291 the IV Corps. The 39th Infantry, the 11th Machine Gun
Battalion and the 10th Machine Gun Battalion were directed
292 to join the 7th Brigade.

On November 11th the 8th Brigade was moved to the vicinity of Boucq and Division Headquarters established at that point. The Second Army was busy preparing to take part in what Marshal Foch had planned as the final great drive of the war. The blow was to be struck by a combined Franco-American army under the command of General de Castlenau, the French Army of twenty divisions being commanded by General Fayolle, and the American Army of six divisions, but nearly equal in number to the twenty French divisions, by
293 Lieut. General Bullard. The plan was to smash through the depleted German lines on the Lorraine front and, crossing the Moselle River, strike almost due east, passing to the south of Metz. The attack would have been supported by another
294 blow on the British front in Flanders. This drive, planned for November 14th, would undoubtedly have shattered the remnants of the German Army and exploited to the full the success then won by the First American Army in the Meuse-Argonne but the Germans, fearful of what was coming and conscious of their inability to face such a drive preferred to accept the terms of the armistice laid down on behalf of the Allies by Marshal Foch.

Meanwhile the 4th Artillery Brigade was still fighting on the Meuse-Argonne front. Withdrawn on the night of October 23rd-24th to the Forêt de Hesse for a five days' rest, it had suddenly been returned to the firing line in support of the III Corps. The 16th Field Artillery moved up to positions north of Romagne, the 77th between Cunel and the Farm de la Madeleine, the 13th to the vicinity of Cunel. The brigade participated in the great attack of November 1st and during the next ten days advanced some 30 kilometers with the infantry. The

16th Artillery was attached to the 90th Division and on November 2nd moved to position around Andevanne, where it supported that Division until November 7th, when the regiment's active operations terminated. The 13th and 77th Regiments served with the 3rd Artillery Brigade. On the nights of October 5th-6th and 6th-7th the 13th Artillery moved to the vicinity of Doulcon and the 77th to the east of Dun-sur-Meuse. On November 9th one battalion of the 77th took position northeast of Murvaux. On November 8th one battalion of the 13th Regiment took position about 1 kilometer west of Murvaux, and on the night of November 9th-10th the two remaining battalions moved to positions some 3 kilometers east of Murvaux directly behind la Sentinelle, well known in 5th Division history. On November 10th Headquarters of the 4th Artillery Brigade was established at Dun-sur-Meuse. The Mézières–Sedan–Montmédy railroad was under allied fire.

The last casualties in the Division, as the result of an enemy act, were suffered by the 1st Battalion, 13th Field Artillery, about 2:00 A. M., November 11th, when a heavy German shell wounded 4 men, one of them mortally. This regiment also, about this time, fired the last shot discharged by a 4th Division unit. The artillery units of the 4th Division rendered valiant service to the 5th Division in covering its crossing of the Meuse and later its advance in pursuit east of the Meuse.

On the eleventh hour of the eleventh day of the eleventh month peace settled over war-scarred Europe. The smoke from the cannon ceased, the machine guns ate up no more tape. Only the inevitable transport moved. It was the Sabbath of Mars. That night, all along the front, a weird display of Véry lights, rockets and burning powder turned darkness into day and gave the impression of some gigantic 4th of July celebration. In the woods the infantry lit fires and rejoiced that peace had returned to earth.

The losses of the artillery were 3 officers and 51 men killed,

American cemetery between the northwestern edge of the Bois de Forêt and Hill 299. The graves are those of members of the 39th Infantry. Taken March 25, 1919

A few of the German guns taken by the 4th Division on Sept. 26, 1918. These guns were assembled at the 4th Mobile Ordnance Repair Shop, to be remodeled and used against the Germans. Sivry-la-Perche, Oct. 11, 1918

2 officers and 9 men died from other causes, while 30 officers
271 and 274 men were wounded or gassed.

When the artillery units left the Argonne they had a 125-kilometer march in front of them to reach the 4th Divi-
295 sion Headquarters. They followed the same route taken by the foot troops. When they finally arrived at Pont-sur-Meuse the Brigade was re-equipped and prepared for the march into Germany.

During this period the infantry had been sending men to recreation points in central and southern France to recuperate from the strain of war. Here they lived in hotels like other guests, listened to entertainments provided exclusively for their ears but were not called on to pay any bills. It was all very strange after the turmoil of war but very acceptable. One sergeant who went to Nice got into bed on the day of his arrival at the hotel and stayed there during the whole week of his leave. He even ate his meals in bed.

CHAPTER XIV

THE MARCH INTO GERMANY

HOSTILITIES had ceased. Germany, in accepting the terms of the Armistice, had virtually surrendered unconditionally to the victorious Allied and American Forces. The terms of the Armistice were drastic in the extreme and not only broke Germany's military power but placed the Allied and American armies in a position from which they could, with ease, penetrate into Germany and if necessary occupy the entire country.

Briefly the terms of the Armistice provided for: 296

Immediate cessation of operations on land and sea.

Immediate evacuation of invaded countries and withdrawal of the German Army to the right bank of the Rhine.

Repatriation of all inhabitants of occupied countries and of prisoners of war, without reciprocity. No destruction of property or taking of prisoners upon evacuation.

Surrender of large quantities of equipment and war material of all kinds.

Maintenance personnel of all public utilities to remain at work.

Germany to reveal location of all mines, fuse traps, etc.

Right of requisition reserved to Allied troops. Requisitions by Germany to cease at once.

Evacuation of Russian territory and abandonment of treaties of Bucharest and Brest-Litovsk. Allies to have free access to evacuated territories by way of Dantzig and the Vistula.

Restoration of property; money indemnity to be paid by Germany.

Surrender of certain German ships of war. Allies to sweep up mine fields and have free access to the Baltic.

Blockade against Germany to remain in effect.

All German naval aircraft to be immobilized at once.

Captured and interned merchant vessels to be returned by Germany without destruction of material and without reciprocity.

No German shipping to be transferred to neutrals.

> Armistice to be in effect 30 days with option of extension. Termination by notification within 48 hours by either side permissible; in event of failure of the other to comply with the above terms, hostilities to commence 72 hours thereafter.

Under the terms of President Wilson's final reply to the German demand for a suspension of hostilities, all details were left to Marshal Foch who continued to be in supreme command of the military forces of the Allied and Associated Govern-
297 ments. Marshal Foch allotted the Aix-la-Chapelle sector to the Belgians, the Cologne sector to the British, the Coblenz sector to the Americans and the Mainz sector to the French. The assignments were logical. The Aix-la-Chapelle sector was in immediate proximity to Belgium and the Cologne sector admirably suited the British as it was the nearest to their base of supply. Two considerations prompted Marshal Foch to place the French in the southern sector. It was in prolongation of Alsace-Lorraine, from the whole of which, under the terms of the Armistice, the Germans had withdrawn and which was automatically taken over by the French and at once treated as part of France. The French General Headquarters had been moved to Metz. Besides these practical reasons the French were anxious to occupy a part of Germany which from 1792 to 1815 had belonged to France and where French sympathies were believed still to exist. This left the important centre section to the American Army. Facing Coblenz is the famous old fortress of Ehrenbreitstein, once considered impregnable but which to-day could be destroyed easily by heavy artillery.

To occupy this sector General Pershing created the Third American Army, composed of the III Corps under General Hines containing the 1st, 2nd, and 32nd Divisions and the IV Corps under General Charles H. Muir containing the 3rd, 4th, and 42nd Divisions. To this force was later added the VII Corps under General William G. Haan containing the 5th, 89th, and
298 the 90th Divisions. This Army was placed under the com-

mand of General Joseph T. Dickman, who as commander of the Third Division had won great distinction in staying the German drive toward Paris in July, 1918, and who, in the battle of the Meuse-Argonne, had been given the command of the I Corps of the First Army, when General Liggett was placed in command of the latter.

As soon as it was known that the American Army was going to occupy the Coblenz Bridgehead on the Rhine, there was great speculation in the various divisions as to which would be selected for the honor of representing American ideals in Germany. The men of the 4th Division who had fought so well, waited anxiously yet confidently, and when the good news came that the 4th was one of the six divisions selected to form the Third American Army the men cheered heartily. At once the Division set about getting new equipment, new uniforms and marching shoes, to make itself in every way representative of a conquering army.

The march to Germany was begun on November 20th. As 299
no general orders had been at that time issued, either by Marshal Foch or General Pershing, as to the conduct of the occupying troops toward the inhabitants, General Hersey, on November 19th, caused the following General Orders to be
read to all units before they started on the march: 300

> In the great offensive movements which have ended the present conflict, the men of this division have taken a conspicuous part. Your valor has been demonstrated. You have helped to change the map of Europe. You have made history.
>
> The American Army to-day is confronted with a greater task than that of defeating a martial foe. We are now to occupy enemy territory; we are to help build a new Government to take the place of the one we have destroyed; we must feed those whom we have overcome; and we must do all this with infinite tact and patience, and a keen appreciation of the smart that still lies in the open wound of their pride. If we, while animated by just pride of achievement, eliminate from our demeanor the overbearing of the conqueror; if we make them see that our perceptions include not only an appreciation of our sufferings but of theirs as well; if we convince them that our sole aim in entering the war was to make

the world a better place to live in, we shall succeed in implanting in each German breast a deep and abiding respect for America and American institutions.

The war is over. No useful purpose can be served by continuing, after war has ceased, the hatred that war engenders. As we have helped to destroy the old Germany, let us help now to build up a new Germany.

The work on which you are about to enter is an exceedingly difficult one. The humiliation and sting of defeat caused a proud people to feel a natural bitterness toward its conquerors. The feeling is intensified when the conquerors enter its territory. Women, children, old men, will show their resentment toward the troops of occupation. You must meet all manifestations of this bitterness in a dignified and soldierly manner. Show by discipline, self-control, generosity and helpfulness that the American soldier is as chivalrous as he is invincible.

The people of the United States are watching you to see how you will measure up to the requirements of this task. Meet the test as you have met all other tests.

The next day, November 20th, the great march began. The
III and IV Corps advanced abreast of each other, the latter
on the right. In the IV Corps the 3rd and 42nd Divisions led
the advance followed at a distance of two marches by the 4th
Division. The objective was the line Remich-Schengen, on
301 the Moselle River, along the eastern boundary of Luxemburg.

From the plain in the vicinity of Boucq, almost in the
shadow of Mont Sec, captured during the St. Mihiel drive,
the 7th and 8th Infantry Brigades, each with certain auxiliary
troops, moved in approximately parallel columns, with the
artillery following on both roads at about the intervals of a
day's march. The route of the right column, commanded by
General Booth, lay through Beney (a little west of Thiaucourt),
Chambley, Mars-la-Tour, Jarny, Moinville, Richemont,
Thionville, Garsch, Fixem, Schengen. The left column, under
General Poore, followed the route Apremont, Vigneulles, Jon-
ville, Labeuville, Conflans, Briey, Avril, Hayange, Oeutrange,
Rodemack, Remich. The artillery, under General Babbitt,
301 was one day's march in rear.

It was a cold morning well suited for marching and a wave

of old time gaiety and enthusiasm ran through the ranks as the order to move out was given. The Division presented a particularly fine appearance. The transportation was in the pink of condition. Ranks depleted in the Argonne fighting had been filled with replacements from the States. All evidences of fatigue had faded away. A wonderful transformation had indeed been effected since the day the men left the Bois de Forêt and the Bois de Brieulles. They were full of life and energy and stepped out in the pride of their young manhood, their eyes sparkling in anticipation of strange scenes, their lips responsive to quips and jokes, to mirth and sarcasm. The Ivy Division looked just what it was, a splendid body of American soldiers, smart and neat, steadfast and loyal, proud of itself and of its leaders.

Beyond ruined Thiaucourt the Division passed into the devastation of what had been No Man's Land after the wiping out of the St. Mihiel salient. Crossing the Paris-Metz road the troops entered Conflans, the great railroad centre in the midst of the Briey iron fields, where the Allied bombing aeroplanes had done extensive damage. In Conflans the people turned out en masse to welcome the Americans and to the great profusion of French tricolors, which had been preserved in nervous secrecy during the German occupation, were added many Stars and Stripes. These flags, made by the hands of patriotic Frenchwomen, had anywhere from ten to twenty stripes and from four to forty stars but the soldiers, hiding their smiles, showed their appreciation of the compliment to the United States and of the gratitude of the people by making great friends with the inhabitants. Then came the crossing into that part of Lorraine torn from France by the iron hand of Bismarck in 1871 but which was already restored to the French. Great was the greeting of the Lorrainers for the men of the Ivy Division as they marched under decorated arches hastily erected across the streets. Bands played and girls danced, while old folks wept. In the area Hayange, Schrémange,

Thionville, Hettange Grande, the Division halted from Nov-
302 ember 23rd to December 1st and this gave the good people of Lorraine an excellent opportunity to show their feelings toward America. Thanksgiving day was celebrated here with fervor and, although turkeys were unobtainable, an elaborate "spread" was enjoyed. But work during this week was not neglected; training proceeded regularly every day. At night the men had real beds to sleep on and enjoyed many of the comforts, which during the fighting, had seemed like dreams of a former existence. They lived happily, making friends with the townspeople, playing with the children, rocking the babies on their knees to the delight of proud mothers. The 4th Division knew all the joys and privileges of a liberating army.

The time set for the Allies to enter Germany now arrived. Army orders were issued to the troops explaining that they would find themselves among an enemy people and that their conduct must not tarnish the luster of their arms so bravely and victoriously borne on the field of battle. There was to be no "fraternizing." The German, whether an indulgent and kindhearted man or a coquettish and receptive woman, was to be left severely alone. Germany was still an enemy and no confidence was placed in the Government or the people. Prudence demanded that discipline and vigilance should not be relaxed.

The men of the Ivy Division contemplated the march into Germany with mixed feelings. Many wondered how the German people would behave when their houses were taken for billets, and visualized the necessity of sleeping with a rifle or revolver at hand, in case murder were attempted by the civilian population. Others thought that they would be segregated entirely from the Germans. All agreed that it was a great adventure that might hold many risks and many surprises, but one which they would not have missed for anything in the world.

On December 2nd the Ivy Division resumed the move-

ment. It marched into Luxemburg the next morning, cutting 299
across the corner of the tiny Grand Duchy near Remich, and
then the supreme moment came. On the 3rd of December
the leading elements crossed the Moselle River at Remich and
at Schengen and placed their feet on German soil. Without 299
the slightest ostentation, without any demonstration, with seemingly no emotion beyond curiosity, the Ivy Division made its entry into Germany where, that night, the men found billets and supper, having breakfasted in Lorraine and had dinner in Luxemburg.

The German people in the village viewed the sturdy figures and the military bearing of the Americans with all the admiration of a militaristic people taught from the cradle to worship the soldier. But it was the excellent appearance of the transportation and of the artillery equipment and particularly the field kitchens with their abundant supplies of good food that attracted the most attention. As was but natural the German children, many of them wearing forage caps similar to those used in the German Army, pressed round the soldiers with unabashed interest and contemplated the field kitchens with wonder in their eyes and desire in their stomachs. The soldiers made great friends with the children. Neither orders, threats nor punishment can keep a soldier from playing with children and the parents soon discovered that the Americans, although grim in battle, were not lacking in kindness now that hostilities had ceased.

In Lorraine the French children had quickly acquired a taste for American chewing gum. Their requests for it were made with a jovial camaraderie that presupposed acquiescence. "Got any gum?" became the usual form of greeting by children to the soldiers. In Germany the children wanted chocolate, of which they had seen none for three years; but their approach was different. An officer or soldier walking along the street would suddenly find a little, soft hand pushed into his own and, looking down, would find a pair of pleading blue eyes,

Major General Mark L. Hersey
Commanding 4th Division

Head of 8th Brigade passing under arch of welcome at Thionville (Diedenhofen), Lorraine, Nov. 23, 1918

Marching into Germany. 4th Engineers halted for the night on the bank of the Moselle River, Remich, Dec. 3, 1918

set in a rather pale, anemic face, gazing into his. Then "*Haben Sie Chocolade?*" came in a piping voice. She got it.

As the men of the Division moved through Germany they heard none of the welcoming cheers that had greeted them in every town and hamlet in France and Lorraine. There the warmth of their reception had stirred the men's emotions; now all was silence. The returning German Army had been welcomed by its own people with tears and cheers, but heads were turned aside or faces set as the conquering Americans passed by.

The men viewed the unfamiliar scenes around them with interest. The country, mostly agricultural, seemed prosperous and in striking contrast to the shell-torn fields of France. The villages were clean and the men sturdy and well fed; only the women and children showed signs of weakness and under-nourishment.

On December 6th the Army of Occupation was advancing to the Rhine by stages so arranged as to leave an interval of a few days between the withdrawal of the German troops and the arrival of the Americans. This was the day on which the armies of von Marwitz were to cross the Rhine at Coblenz and Bingen, completing the evacuation of the left bank of the river. The Americans were not due to arrive at the Rhine till December 11th. The citizens of Coblenz were afraid of what the revolutionists might do between the date of departure of the German Army and the arrival of the American Army of Occupation. A Soldiers' and Workmans' Council had been formed but this, instead of allaying the uneasiness, rather increased it. So after consultation with the representative of the American Bridgehead Commission, who had been sent forward to make necessary arrangements for the arrival of the Third Army Headquarters in Coblenz, the Bürgermeister of the city, with the approval of the German military authorities, asked the American Commander to send a battalion to Coblenz immediately to maintain order in the city and perform police duty. General Dickman honored the 4th Division by ordering it to provide a battalion for this purpose and General Hersey directed Colonel Bolles of the 39th Infantry to send one battalion, complete in every respect, to Coblenz. 303

The 2nd Battalion, then at Waldrach, was selected and, under Lieut. Colonel Lockett, marched at once to Rüwer-Paulin where it was inspected by the Division Commander. At 4:15 on the morning of December 8th, the men marched to the old Roman city of Trèves (Trier) and were put on board a long German train which was standing on a siding. After some little delay the German train crew got up steam and by 9 o'clock the train left Trèves for the Rhine. The battalion was accompanied by all the war correspondents accredited to the American army, who had come to tell the story of the first American troops to reach the Rhine. The journey was unusually long and tedious but the beauties of the scenery along the Moselle River, the up-to-date organization of the railroad, the approaches to Coblenz which looked not unlike an American city of the Middle West, all combined to make the trip interesting. There was no excitement. At 2:30 in the afternoon the train swung across the great iron bridge over the Moselle and stopped on a freight siding at the station in the heart of the city. 304

A German officer resplendent in full uniform, wearing a long gray cloak and pickelhaube, received the officers of the battalion. They returned his stiff salute and the men were detrained. Within a few minutes they had formed up and were marched to barracks before the people of Coblenz even knew they had arrived. There was no ceremony, no bands, no manifestation of any kind, nothing to suggest the entry of a conquering army. At dusk sentinels, two by two, were placed to guard the bridges over the Rhine and patrols were sent out to police the city. The socialist guards who, dressed in civilian clothes with white brassards on their arms and rifles slung

Near St. Gengoulph

General Pershing and Major General George H. Cameron at Lizy-sur-Ourcq, July 11, 1918

Hautevesnes in ruins

over their shoulders, were maintaining a semblance of order, were relieved.

The status of these American troops was a peculiar one; they were in Coblenz at the invitation of the German authorities and, until December 11th, technically the guests of the German city. The behavior of the men was beyond all praise. On the first day or two the people of the city were rather inclined to presume upon the mild and courteous manner in which they performed their duties. Coblenz had always
305 been a garrison town, the headquarters of the VII German
306 Army Corps and at one time the home of Von Moltke. The
seat of a royal palace, its civic pride and military prestige made it far from amenable to occupation. The women of Coblenz—widows, wives, or daughters of German soldiers for the most part—were particularly open in their hostility. Even the children showed aversion to the Americans, and one was heard to exclaim to an American officer who, standing at the foot of the giant statue of Kaiser Wilhelm der Grosse, was looking at the confluence of the Rhine and the Moselle, "Look well at our German Rhine. It is not yours; it will never be yours." Then the Germans, convinced that they were not beaten in the field and had signed the armistice merely to satisfy the cravings of both sides for peace, began to sneer at the small forces of the 39th Infantry that patrolled the streets. This was soon stopped. The restaurants and cafés, however, the shops and hotels, scenting an opportunity of making the occupation a source of profit, quickly settled down to cater to the Americans.

Meanwhile, the remainder of the Division was moving up to its area of occupation. The march was a severe strain on the men. They had to climb steep hills with heavy packs in driving rain. The new shoes made feet sore and the constant marching made legs weary. But still they went forward in splendid style. Their transport was kept in the best of condition. Their march discipline evoked the enthusiastic comments of corps and army inspectors. No complaint was re-

ceived at Division Headquarters, at any time during the march, that a soldier had deported himself in a manner unbecoming an American gentleman. But in the towns and villages where they rested there developed a familiar, friendly relationship between soldiers and inhabitants. This, being distinctly undesirable, began to worry the Division Commander and led him on December 11th to assemble the officers of the Division and clarify their ideas as to the status of the American Army in Germany. He pointed out to them that, for genera- 307
tions, the German civilians had looked upon an officer as a superior being and that the casual and amiable camaraderie which was creeping into the relations of American officers and soldiers with the German people must inevitably result in a lack of prestige and earn the contempt that follows familiarity. He made clear that the armistice which had been signed was merely a suspension of hostilities; that the United States was still at war with Germany; that the American officers and soldiers were in Germany, not as the friends of the Germans, but as their military governors. And he closed by saying, "You must be courteous but not familiar. You must never give nor receive hospitality or gifts. Whatever personal supplies you get you must pay for. The exact courtesy becoming an officer and a gentleman you must at all times punctiliously observe. At the same time you must demand the courtesy, consideration, and deference due you as an officer of the military governorship. It is, therefore, made your responsibility and your duty to see that the attitude of friendly fraternizing which has been adopted by some of our officers and soldiers ceases at once. You can expect to secure results only by constant correction on the spot. You will explain to your men that the German people are still controlled by active propaganda; that they are urged not to antagonize the Americans; that agitators are eager to insinuate themselves among our troops for the purpose of causing disaffection in our ranks. You will have them un-

derstand that their relationship with these people is to be confined to what is absolutely essential in the discharge of their duty. This means that, if necessary, disciplinary measures will be taken to break up these undesirable familiar relationships. It means that if you find one of your men or one of the men of this Division associating with German people, walking with them on the streets, or talking with them outside of the line of duty, eating or drinking with them, you will apply the proper disciplinary measures to cause the practice to cease."

Both sides of the Moselle River were used by the Division for its march. The 7th Brigade, followed by the 77th Field Artillery, marched on the left bank and the 8th Brigade, followed by the 13th Field Artillery, on the right. The route of the 7th Brigade lay through the picturesque town of Saarburg on the Saar River to ancient Trier, filled with crumbling monuments of Roman occupation. Beyond here the route followed the pleasant valley of the Moselle River to Schweich where it left the river and led northeast, across country, through Hetzerath and Clausen, again to strike the Moselle at Berncastel. Here the brigade rested for a day, reveling in the beautiful river scenery and the quaint architecture and streets of the old town of Cues, and examined with interest the ruins of the medieval castle perched like an eagle on the cliff overhead. On December 9th the march was resumed, via
308 Kappel, on Castellaun which was reached on December 10th.

Meanwhile, the 8th Brigade was moving toward the Rhine
308 by a route which lay south of the Moselle. Its march led through the heart of the Hunsrück, a terrain that was exceedingly difficult. All streams here run either northwest into the Moselle or southeast into the Rhine. In consequence the brigade, which was moving from southwest to northeast, was forced to cross an apparently endless succession of alternate high ridges and deep valleys. The weather was uniformly bad. Rain and sleet made marching a burden and turned the dirt roads, which had already been badly cut up by

the retreat of the Germans, into bogs. The artillery was frequently forced to push its guns up hill on the shoulders of the men. But when the sun did shine the beauties of the landscape repaid them for their efforts. Standing on a high ridge they could see, through the pine-tops, the white froth of a mountain stream far below them, hurrying to join the Moselle, its opposite bank covered with a park-like pine forest. In the distance, on the hillside, was a village, its red-roofed houses huddled closely together under the protecting tower of a church. Farther down the valley, where it grew wider, could be seen villages with their factories and tall chimneys and, beyond, the broad silver ribbon of the Moselle. The route lay through Saarburg, Schöndorf, Thalfang, Morbach, and Kirchberg to Simmern, which was reached December 10th.

Up to December 11th the whole of the 4th Division had been headed for the Rhine. It was to have had its headquarters at Boppard and its sector would have included the city of Coblenz and the area upstream as far as the junction with the French. But a change was made by Marshal Foch who enlarged the area assigned to the French Army of Occupation by diminishing that of the American Army; and so the 4th Division, when within thirteen miles of the river, was suddenly ordered to turn at right angles to its line of march. On December 13th the Division turned its shoulder on the Rhine and marched northward into the back area of the American
Army of Occupation. The 7th Brigade, the 77th Field 309
Artillery, and the 58th Infantry crossed the Moselle at Treis on a pontoon bridge, after it had been repaired by the 4th Engineers. The remainder of the Division crossed at Alf. On December 17th the march was completed and the Division
rested in its new area. 310

Under the change that was made necessary by the diminution of the American area the III Corps was placed east of the Rhine within the thirty-kilometer bridgehead. The IV Corps was located west of the Rhine; the 3rd and 42nd Divisions on the Rhine River and the 4th Division just southwest of them. The area of the VII Corps extended from the western border of the 4th Division back to the boundary of Luxemburg.

On December 18th the 2nd Battalion, 58th Infantry, was sent to Coblenz to assist the 2nd Battalion, 39th Infantry, in guarding the city. Lieut. Col. Lockett, as Military Commander of
311 Coblenz, assumed command of both battalions.

The march of the Third Army was the longest ever made by American troops in Europe. Approximately 330 kilometers were covered by the 4th Division in fifteen marching days. On November 21st, the 4th Engineers and the Engineeer Train marched 44 kilometers, the 39th Infantry and the 2nd Battalion of the 59th Infantry 40, the 11th and 12th Machine Gun Battalion 38, the 1st Battalion of the 16th Field Artillery 41, and the 2nd Battalion of the same regiment 38. On December 6th the 2nd Battalion of the 47th Infantry made 45 kilometers, the 3rd Battalion and the 4th Engineers 40, the 39th Infantry 39, and the 11th Machine Gun Battalion 38. Both of these days were preceded and followed by long marches. These were splendid feats, made possible only by the superb physical condition which the men of these units, in common with all the
312 others, had attained.

The medical records of the march also attest this fine condition of body, mind, and spirit, which sustained during the march the reputation made on the battlefield. It was hard even for trained men, and many of the 2,300 replacements that had joined the infantry regiments between November 8th and 16th had all too little chance to acquire the art of marching. The new men did bring the mumps with them, however, and shared them generously, causing a loss of 134 men to the marching column. In all, 2,197 men were evacuated to the Division Field Hospitals at an average daily rate of 86, of whom 312 were for influenza, 347 for bronchitis, and 326 for foot troubles. Nearly half of the foot troubles occurred in the first five days of the march and more than one third of them were caused by the British shoes issued just before the march began. In addition to those evacuated, 71 cases were treated in quarters, but the majority of the men carried their foot troubles with them without a word of complaint. 312

After the fourth day of the march there was little opportunity to send the sick to army or corps hospitals and from that time until just before the end of the march they had to be cared for within the Division Field Hospitals, as these moved forward with the troops. Of the 2,197 patients admitted, 732 were returned to duty during the march and 1,097 were evacuated to hospitals in the rear, 222 being the average number in the Division Field Hospitals each noon. 312

The daily collection of the sick was made difficult by the wide dispersion of the troops, poor communication, bad roads, short days, and dark, rainy nights. Ambulances found or tried to find each regiment every night and others were stationed at important road crossings and at centres of towns along the route of each column and remained there until after the troops had passed. The fear of not finding sick men and of leaving them behind uncared for—always the Division Surgeon's worst anxiety on the march—led to the utmost precaution in covering the routes and villages used, and each day every road was patrolled and every town visited through which troops had passed or where they had stayed on the previous day. There were occasional failures of co-ordination in the evacuation service but only seven cases failed to be taken care of promptly by the primary machinery of evacuation and of these all but one were picked up on the same or the following day by the supplementary patrol. The one case left behind was that of a soldier who, after the departure of his company from the village where it had been billeted overnight, was taken by the Bürgermeister to his home and placed in his guest-chamber. The good wife took such excellent care of him that when the ambulance finally called to take him to the hospital he protested violently against his removal.

The bridge over the Moselle at Schengen

Burg Cochem and the town of Cochem on the Moselle

"Pup tents" in which the men of the 4th Division slept for two nights prior to the inspection by General Pershing. Büchel, Mar. 17, 1919

CHAPTER XV

WITH THE ARMY OF OCCUPATION

THE area assigned to the 4th Division was bounded on the south by the Moselle River. It embraced the Kreises of Cochem and Adenau, the latter lying northwest of the former and the two together forming an irregularly shaped stretch of territory some thirty-four miles long from north to south and about twenty miles wide at the ends. In the centre, where the two Kreises joined, it narrowed to a width of about six miles. It was a huge area, necessitating wide dispersion of the troops with all the consequent difficulties of supply and communication. The country was a succession of steep hills and deep valleys. Wherever one went one had to climb. With the exception of the main highways, which were kept in condition by the state, the roads were poor, the accommodation such as one would expect among a mountain people. There were but two large towns—one of which was absorbed by the IV Corps as Headquarters—and comparatively few villages. To billet the men was a task of great difficulty for there were few desirable billets. Here the Division, feeling as if it were lost from the outer world, settled down to the work of occupation.

The Headquarters of the IV Corps, to which the 4th Division belonged, was established at Cochem, on the Moselle, in an imposing castle, the Burg Cochem, where the Prince of Wales was entertained when he visited the American Army of Occupation. The Headquarters of the 4th Division was located in the southwestern part of Kreis Cochem, at Bad Bertrich, a summer Spa almost on the edge of the Division's sector, which during the war had been used by the Germans as

a large convalescent hospital. The 8th Brigade was located in Kreis Cochem and General Booth established his Headquarters at Ediger. The 58th Infantry was at Moselkern, the nearest town to Burg Eltz, a famous castle in Germany, one of the only two which were not destroyed during and after the Thirty Years' War. The 59th Infantry was at Eller and the 12th Machine Gun Battalion at Sehl, just outside of Cochem. In this Kreis also was located the greater part of the Artillery Brigade. General Babbitt's Headquarters was at Bad Bertrich. The 77th Artillery was at Kaisersesch, the 13th Artillery at Ulmen. The 4th Engineers were sent to Dungenheim, the 10th Machine Gun Battalion to Hontheim, and the Signal Battalion to Müllenbach. The 7th Brigade was located in the Kreis Adenau, and General Poore established his Headquarters in Adenau itself where also was stationed the 47th Infantry. The 39th was at Kempenich, the 11th Machine Gun Battalion at Antweiler, the 16th Field Artillery at Kelberg. The Ammunition Train was placed at Klotten, Kreis Cochem, and the Supply Train divided between the railheads Adenau and Carden, one of which was in each Kreis. Field hospitals were established in each Kreis. The headquarters of the Sanitary Train was placed at Lutzerath. The northwestern part of Kreis 313
Adenau is in what is known as the Eifel, a mountainous region, associated in German folk-lore with dwarfs and gnomes, that fabled race of diminutive beings, the guardians of mines, quarries, etc.

A few days elapsed before the men became thoroughly settled in the area. As soon as they had entered permanent billets and accustomed themselves to the new surroundings the Division began to make arrangements for field sports, athletic contests, and all kinds of entertainments for the men. The organization commanders were keen on athletics and took great interest in developing a spirit of friendly rivalry between the various units. It was realized that the work of the Peace Conference would probably be prolonged and that no changes would be made in the Army of Occupation until peace was finally signed; so the Division settled down to make the best of its stay in Germany. Here training could be indulged in far more systematically than in France where only brief periods, between offensives, had been obtainable. A well-planned course was inaugurated with the object of bringing the personnel and materiel of the Division and of each unit in it to the highest state of perfection. But if more hours than previously were devoted to military training, equal attention was paid to making the most of the recreation period and, under the dual guidance of welfare officers and entertainment officers who provided moving pictures, boxing matches, and minstrel shows, life began to assume a color and variety which few would have thought possible amid the inhospitable woods, forbidding heights, and bleak rolling ground of the area.

The weather was cold, the men lived almost entirely out of doors, and naturally sports of all kinds were very popular. The 4th Division had one of the best football teams in the American Expeditionary Forces. There was plenty of material available for the athletic officers who were assigned to organize sports in the Division.

A number of schools were opened where men who did not possess a good working knowledge of the English language were given elementary instruction. This course was compulsory, but a number of others were established at which attendance was optional and the more studious-minded among the men derived lasting benefit from them. The large attendance at all the divisional schools was an earnest that the spirit of the Division would be maintained in peace as it had been in war.

But, of course, military training took first place in the activities of the Division. The hills and woods did not lend themselves to maneuvers and it was difficult to find suitable ground for mass movements of troops. The terrain, however, had certain advantages. The men soon became accustomed to climbing the hills, and the hard work and good food, the keen winter air and outdoor life, kept the Division at a high state of physical perfection and enhanced its morale. 314

During this time the 13th Field Artillery was motorized. The training of drivers and mechanics for the new service was completed in a very short time and soon the sight of the snorting tractors drawing their loads over the hills was one that astonished and impressed the Germans who looked on.

In the middle of January the Division suffered the loss of Brigadier General E. E. Booth, who was detached from the 8th Brigade and made G-1 of the Services of Supply and later its Chief of Staff. His tactical ability and resistless energy had been the animating force behind the 8th Brigade during the action in the Vesle River, the St. Mihiel Drive, and the Battle of the Meuse-Argonne. He was succeeded by Brigadier General Frank E. Bamford, who had been a regimental and brigade commander in the 1st Division. General Bamford was relieved on May 29th to rejoin the 1st Division and was succeeded by Brigadier General Francis C. Marshall.

On February 4th, the Division received notice from the Army that Colonel H. R. Richmond of the training and instruction branch of G. H. Q. would visit each division of the Army to inspect its animal-drawn transport. As his time was limited, only one regiment of the Division was to be inspected but the Division would be judged by the condition of that regiment. The 47th Infantry was selected as the unit for inspection. All the animal-drawn transport of the regiment was paraded on the drill ground at Adenau on February 7th. The wagons, animals, harness, and equipment were in splendid condition and elicited Colonel Richmond's highest praise. His report, made upon his return to G. H. Q., was to the effect that the 4th Division transport was the best in the Third Army. 315

The relation between the men of the 4th Division and the

German people among whom they now lived was characterized by courtesy and, when needed, by a firmness that proved to be the happiest of all methods for dealing with the population of the occupied area. When the Division first arrived in Germany the impression created upon the men by the attractions of a country untouched by the hand of war, by the well-kept towns and villages, the cleanliness and comforts they enjoyed in billets, was unquestionably very great. The contrast with the devastated and shell-torn areas of France, in which they had fought and lived, was impressive. Many things they saw gained their admiration; others their amused contempt. The manure piles under the bedroom windows afforded endless comment. Of course the replacements who had joined the Division since its withdrawal from the Meuse-Argonne, and who had not seen fighting, were more impressed by the attentions they received from the inhabitants than were those who had so recently seen death and suffering meted out to them by the Germans, and who could not forget. It is believed that the German Army in its retreat had left instructions with the various Landrats that efforts should be made to conciliate the Americans, as it was hoped that their friendship might be cultivated. So when the Americans occupied their country in force, the Germans, partly from fear, partly from policy, and partly from kindness, performed many small friendly acts which the soldiers were not slow to appreciate. There was little overcharging. This was in refreshing contrast to what had been experienced in France, where the merchants, generally, had been unable to withstand the temptation of doing what was characterized as "making the Americans pay for the war."

But with the passage of time and more intimate acquaintance with the informal, happy-go-lucky character of the American soldier came the inevitable friction. This was generally created by the younger German men for one of two causes—girls or politics. Clashes arose, not infrequently, because some local beau had found himself ousted from the affections of a "fräulein" by the superior physical attractions or longer purse of an American; then the disgruntled one would gather his cohorts to "do up" the soldier. These encounters were usually unfortunate for the Germans, who had not learned the scientific use of their fists. More serious, however, was the gradual infiltration of German ex-soldiers, their minds inflamed by radical ideas, who sought to create trouble and who had to be watched constantly. Then, too, the respect, amounting almost to awe, with which the Americans were at first treated had begun to wear away. So, as time went on, the really friendly feeling entertained for the German people by the soldiers during their earlier stay became only lukewarm. Certain it is that, aside from their desire to go home, many organization commanders received the order to return to America with a sigh of relief.

Officers in charge of civil affairs were appointed to deal
with the local authorities and attend to all questions of civil
administration which came within the province of the Army
of Occupation. The American policy was to interfere as 316
little as possible with the internal life of the German population. The administration of justice was allowed to continue as before and the Germans were left in unreserved control of public education and all such matter. But the Civil Affairs Officers saw to the enforcement of the regulations made by the Supreme Command for the ruling of the Germans. They issued passes for all movements of civilians from place to place. They settled all questions of billets and of requisitions. Then courts were established for the trial of Germans accused of violations of the rules and regulations promulgated by the military authority. Investigations had to be made and proceedings instituted on information lodged against alleged offenders; sanitary regulations had to be enforced and strict compliance with the "non-fraternization" order insisted upon. All this and many other things were placed in the hands of the

Civil Affairs Officers, the Assistant Provost Marshals, and the Military Police.

The splendid record of the Division, gained during the period of occupation,was largely due to its commander. General Hersey's long service as a civil administrator had given him experience that was invaluable in dealing with the situation. His treatment of the inhabitants, while firm and just, was tempered with kindness.

The two battalions of the 4th Division which furnished the garrison for the city of Coblenz, the 2nd Battalion of the 39th, and the 2nd Battalion of the 58th, had a much more difficult task than the rest of the Division. They had to secure the peace and good order of the city, furnish guards for the bridges and the public buildings in which were situated offices of the Third Army Headquarters, and provide details for all kinds of things. Their military band and guard mount always attracted a crowd of Germans, among whom it was possible to recognize former German officers. Those who came to see, with the secret hope that they could ridicule, went away impressed and surprised that a democracy in arms, which had in no sense been militaristic before the war, could present a guard mount that equalled the pre-war parades of the old German Army.

The Military Commander of Coblenz was Lieut. Colonel Lockett of the 4th Division who was later, on May 4th, suc-
317 ceeded by Colonel Fred R. Brown, 58th Infantry. The special pride that each took in the garrison was reflected in the amount of drill and other military duty which fell to the lot of the two battalions. But if forming the garrison of Coblenz meant plenty of hard work, there were compensations galore. The city was the amusement and leave centre for the entire Army of Occupation. In pre-war days it had been especially developed as a tourist resort and every possible convenience of life existed there. To the men in the small towns and little villages of the outlying areas life in Coblenz

seemed the next best thing to being ordered home. Trips to Coblenz were arranged by the Division for the men in the main area. At the Festhalle and other similar places excellent entertainments, both moving pictures and variety shows, were staged by the Y. M. C. A. for their benefit. The bouts between the various weight champions of the divisions in the Army of Occupation presented a wonderful opportunity for letting the world know that soldiers had lungs, and for practising much good-natured "kidding." The cheers from the Festhalle on boxing nights could be heard for half a mile. A large dance hall was built and a "flying squadron" of American girls who danced well brought to Coblenz to convince the soldier that the "made in America" trademark was the best and most desirable after all.

It was the success of the football team, however, that
claimed chief interest in the Division. Successively the
42nd Division and the IV Corps were defeated, the scores
being 7-0 and 3-0. Then came the turn of the 2nd Division, 317
the champion team of the III Corps. Proud of its achieve-
ments on the battlefield the 2nd Division was no less proud
of its football team. But on the banks of the Rhine by a
score of 10 to 7 the 2nd Division was defeated before a tre-
mendous gathering. There were hilarious scenes in Coblenz 317
that night. It was a famous victory. Then came the meeting
with the 89th Division, the champion team of the VII Corps,
which occupied a sector adjacent to that of the 4th Division.
This, too, was fought out at Coblenz on the same ground where
the 2nd had been defeated. The match was attended by Brit-
ish generals from Cologne and French generals from Mainz
who had come to see the final game of the football champion-
ship of the Army of Occupation. High were the hopes of the
Ivy men, but their ranks were depleted by injuries sustained
by some of their best players in previous games, and the
89th by a 14-0 score, won a splendid victory. It was a 317
severe disappointment, but the fact that the 89th after-

The Division in line

Passing in review

General Pershing addressing the Division at Büchel on March 18, 1919

Fourth of July on the Rhine, 1919

ward won the championship of the A. E. F. brought consolation.

But sports were neither the only nor the most important thing. The whole Division had been put through a thorough and systematic course of training. The results of this training were to be viewed by the Commander-in-Chief himself, on
317 March 18th. He had already visited the other divisions in the III and IV Corps and the turn of the Ivy Division had now come.

The terrain was so broken and wooded that the first and main difficulty was to find a suitable reviewing ground. At last a site near Büchel was selected, just northwest of where the road from Ulmen to Cochem crosses the road from Bad Bertrich to Kaisersesch, and orders were issued to the various regiments to prepare for the march. The distances were so great that some units of the 7th Brigade had to start on their journey two days before the review took place. Most of the troops were moved the day before the review. Through a blizzard the men swung over hills and through woods taking their kitchens with them. It was bitterly cold and the snow lay thick on the ground. The night before the review the men slept in their shelter tents in near-by woods. The field kitchens fed hot meals, and big fires were kept burning all night. The morning of March 18th dawned on a mist of whirling snowflakes but by the time of the review the snow had ceased. A white mantle, however, covered the plateau on which the reviewing ground stood and, when the infantry, artillery, and engineers with the other services were drawn up for inspection, the scene was as picturesque as it was impressive.

Owing to the nature of the ground the troops were drawn up on two sides of a square, the artillery on one side and the remainder of the Division on the other. The dismounted troops were arranged by brigades in line of regiments, each regiment being in column of companies. When faced to the flank for the march in review, each regiment was thus arranged in line of companies, each company in column of squads. The artillery regiments were drawn up in column of battalions. Each machine gun battalion was on the left of its brigade. On the third side of the square and so disposed as not to interfere with the march in review, was the horse-drawn transport of the Division, by regiments or separate organizations. The motor transport was drawn up on three roads which led to the Büchel crossing, as the reviewing ground was too soft to permit its evolution on the field.

General Pershing, with his staff, arrived, accompanied by the Corps and Army Commanders, Mr. Frank Hitchcock, former Postmaster General, and all the war correspondents accredited to the American Army. He was received by General Hersey and the Divisional Staff. Mounting his horse, General Pershing at once proceeded to inspect the transport. The wagons were in the best possible condition and the animals showed the care that had been lavished upon them. The drivers were all well groomed and the general appearance of neatness and efficiency met with the complete approval of the General. Still mounted, he next inspected the artillery, and here again the splendid condition of the guns and men won the highest praise. After completing his inspection of the artillery the General rode across the square, dismounted, and began his inspection of the troops, commencing with the 7th Brigade on the right.

In each company the front rank was faced about and the Commander-in-Chief passed between the front and rear ranks. Through line after line of soldiers standing at attention, General Pershing passed accompanied by Mr. Hitchcock, whose derby hat and civilian clothes brought many a smile to the eyes of the men. Not an individual was missed, so thorough was the inspection. As the Commander-in-Chief walked briskly between the men, they looked on him with unfeigned interest. It was the first time that the great majority of them had had a close view of him and his soldierly bearing, fine physique, and clear-cut features made a great impression. That the soldiers made an equal impression on him was eloquently expressed by his smiles and gestures. Whenever his eye caught a wound stripe he would stop and ask the man where he had been wounded and what he was doing at the time he was hit. He often inquired of the company commanders if all their men were wearing the wound stripes and service chevrons to which they were entitled. In one company the General's questions brought out vividly the truly national character of the Ivy Division. Three men standing next to each other all wore wound stripes and the General asked them in quick succession the names of their "home" States. Back came the replies "Vermont," "Kansas," "Texas." All the companies of a regiment having been inspected General Pershing passed to the next regiment. His inspection of the machine gun battalions was particularly searching, but everywhere the Commander-in-Chief found nothing to criticise and much to praise.

The inspection completed, the officers and men to be decorated were lined up in front of the centre of the dismounted troops facing the General. A large number of decorations were awarded, beginning with General Poore, who received both the D. S. M. and the D. S. C. The Commander-in-Chief addressed a few words of congratulation and shook hands with each man. When this brief ceremony was concluded General Pershing proceeded to the reviewing stand, accompanied by his staff, and all of the officers and men who had been decorated.

Alongside of the reviewing stand the Division band of 180 pieces was drawn up. The Division passed in review to the strains of "No Te Vayas de Zamboanga," an old Spanish march which was General Pershing's favorite when he was in command of all the troops in the Island of Mindanao during the Philippine campaign.

A more impressive scene could hardly be imagined. The pine-clad, snow-capped hills in the distance furnished a splendid background. Across the wind-swept field, on which patches of snow still drifted here and there, the long line of soldiers advanced, keeping perfect time with the band, their steel helmets surging forward like a disciplined wave. The sturdy figures of the men, their smart marching stride, the wave after wave of olive-drab that rolled by, gave an impression of tremendous power that filled all with pride who beheld it. At the head of each regiment were the Stars and Stripes and the regimental colors, their folds sweeping back as the bearers marched forward. As they approached the reviewing stand, the bugles sounded; officers hands went up to the salute; the regimental colors dipped; all eyes were turned to the General, and the moving mass passed on. Line followed line; the machine gun companies followed their regiments; the machine gun battalions followed their brigades; engineers, signal battalion, sanitary train. The condition of the ground did not permit the artillery to march past.

The thoughts of many of those who viewed this impressive scene turned to the Marne and the Ourcq, the Vesle and St. Mihiel, and to the scenes of the great struggle in the Meuse-Argonne. The pride of achievement was written on the faces of officers and men alike. If many were missing, if many had paid the last full price of devotion to their country and their country's cause, their sacrifice had not been in vain. Here the 4th Division stood on German soil, a part of the victorious American Army, its task well done, its purpose accomplished.

After the march past, the men were told to break ranks and group themselves as closely as possible about the reviewing stand. Speaking from the top of the stand, General Pershing addressed the officers and men of the Division. In clear, ringing tones he told the story of the entry of America into the war, at a time when the Allies needed help more than they had ever needed it. He told of the glorious part the American Army had played in winning the war and of the services rendered by the 4th Division. He reminded the men that when they returned home they would be looked upon as leaders and consulted by their fellow-citizens on many matters of national importance, that the prestige they, as veterans, would enjoy in their respective localities would give them additional privileges and responsibilities and urged them to prepare themselves to play as worthy a part in peace as they had played in war. Then he told them what a fine thing it was that amid all the temptations offered to them overseas they could go back to their mothers and sisters, their wives and sweethearts, happy in the knowledge that they were clean. He paid a special tribute to the women of America, so many of whom had worked for the army, and he spoke of their great influence with the men.

Then General Pershing did a thing which showed how much he was pleased with what he had seen. In all the addresses he had made to the troops after reviewing the other divisions the Commander-in-Chief had only made a passing reference to the appearance of the men. But to the 4th Division General Pershing said that he could not tell them how deep an impression their fine appearance had made upon him, and turning to General Hersey he congratulated him on the splendid body of men he had under his command.

Loud bursts of applause welled up from the tightly packed thousands of soldiers, but General Hersey stepped forward, and raising his hand for silence, said, "Men of the 4th Division, under the providence of an Almighty God there was given to the United States in her hour of need the Commander-in-Chief who has just addressed us. Take off your war helmets and with thanksgiving in your hearts give three cheers for General Pershing, our Great Commander."

Every helmet came off when the word rang out. There was a pause amid the bare-headed throng, but when the last words had left their leader's lips, the men of the 4th Division cheered General Pershing as he had perhaps never been cheered before. And when the men had ceased, the echoes and re-echoes came back from over the German hills. No man could have stood there unmoved by so heartfelt a demonstration of admiration. No man could hear those roaring cheers without feeling his blood run faster. If the men of the Ivy Division had won General Pershing's admiration on the field of battle, the opportunity had come at last for him to receive their tribute. And they gave it to the full.

CHAPTER XVI

HOMEWARD BOUND

EARLY in April the Division received orders to move up to the Rhine and take over the area formerly occupied by the 42nd Division which was returning to the
318 United States. This area followed the limits of the Kreis of Ahrweiler and was considered one of the most desirable in the whole Army of Occupation, as it contained many fine towns and health resorts which in peace times catered to numbers of tourists during the summer months. The billets were far superior to those in the Cochem and Adenau Kreises and every conceivable comfort was available. This area was within easy reach of Coblenz and immediately adjoined the sector occupied by the British. It was situated directly on one of the most picturesque and historic reaches of the Rhine.

With genuine regret the men of the 8th Brigade said goodbye to the Moselle River. The valley of the Moselle, with its miles upon miles of vineyards, its peaceful, slow-flowing stream, its pretty villages clinging to the banks beneath the majestic hills that rise, at times, sheer from the water's edge, is one of the most beautiful in Europe. Perched securely on the hilly crags are ruined castles once the home of robber barons and now the theme of inspiring legends. There is a quiet, old-world peacefulness about it which has an attraction all its own.

During the first half of April the Division moved, by marching and truck. The various units marched down the enchanting valley of the Ahr River. The road, which follows the windings of the stream, passes between high, verdant hills, thickly wooded, past ancient castles and splendid modern estates. Charming villages nestle amid the pine trees on the hills or lie in the narrow valley below. When the Division reached the Rhine, its Headquarters was established at Nieder-Breisig, in the summer mansion of a wealthy German cloth manufacturer. The Headquarters of the 7th Brigade was a little farther down stream, beyond the town of Remagen, in a hunting estate where the Kaiser had been a frequent visitor. Some of the trophies of the chase on the walls bear the signatures of the German Emperor and the Crown Prince. The 8th Brigade was located at Kripp and the 4th Artillery Brigade at Ahrweiler, the seat of the local government. The 39th Infantry had very desirable billets at Rolandseck at the extreme northern edge of the area and in immediate proximity to the British. Rolandseck is just opposite the ruins of the castle of Drachenfels, placed on the summit of one of the Seven Hills, famed in German song and story as the Siebengebirge. The 47th Infantry occupied the town of Remagen. The 10th and 11th Machine Gun Battalions were located at Brohl, between Nieder-Breisig and Andernach, at which latter place was situated the headquarters of the 3rd Division. Envied by all other organizations the 58th Infantry was in Coblenz. The 59th had its billets in Sinzig while the 12th Machine Gun Battalion was placed in Kripp. The artillery and the engineers were all in the valley of the Ahr. The 77th Artillery was at Neuenahr, one of the most luxurious "baths" in Germany. This city is a famous Spa and possesses fine medicinal springs and a number of large up-to-date hotels provided with every comfort. The streets are wide and well paved, lined with beautiful buildings and attractive shops. In addition to the 77th Artillery the Sanitary and the Supply Trains were also located there. The 13th Field Artillery occupied Ahrweiler, a very ancient fortified town still partly surrounded by an old Roman wall. The 16th was in two small towns, Heppingen and Heimersheim, just east of Neuenahr. The engineers were at Mays-

Field Day at Neuenahr, June 14, 1919

The Commander-in-Chief, with the Organization Commanders and Division Staff, at Niederbreisig.

1. General Pershing
2. Brig. Gen. Augustine McIntyre, Comndg. 4th F. A. Brig.
3. Col. Frank C. Bolles, Comndg. 39th Inf.
4. Col. Raymond A. Wheeler, Comndg. 4th Engrs.
5. Col. William B. Cowin, Comndg. Hq. Trains & M. P.
6. Maj. Alexander Whitworth, Comndg. 8th F. S. Bn.
7. Col. Christian A. Bach, Chief of Staff.
8. Col. Fred R. Brown, Comndg. 58th Inf.
9. Col. Troy H. Middleton, Comndg. 47th Inf.
10. Lt. Col. Donald D. Hay, Div. Inspector.
11. Col. Archibald F. Commiskey, Comndg. 77th F. A.
12. Lt. Col. Francis K. Newcomer, 4th Engrs.
13. Lt. Col. Charles I. Faddis, Comndg. 4th Ammn. Train.
14. Lt. Col. Thomas L. Sherburne, Div. Signal Officer.
15. Lt. Col. Wm. F. Robinson, Jr., A. C. of S. G-1.
16. Lt. Col. Sylvester Bonnaffon 3rd, Div. Welfare Off.
17. Col. Max B. Garber, Comndg. 59th Inf.
18. Maj. Henry W. Rogers, Adjt. 7th Brig.
19. Lt. Col. James A. Gillespie, 13th F. A.
20. Col. Paul Waterman, Division Surgeon.
21. Maj. Robert McAuslin, Div. Veterinarian.
22. Lt. Col. Wm. E. Wilmerding, Comndg. 4th San. Train
23. Capt. Dudley R. Tierney, Div. Chaplain.
24. Capt. Charles de Marenches, Personal Staff of C-in-C.
25. Maj. James H. Houston, Div. Q. M.
26. Maj. Charles H. Karlstad, Comndg. 11th M. G. Bn.
27. Maj. Edward M. Almond, Comndg. 12th M. G. Bn.
28. Col. John G. Quekemeyer, ADC to C-in-C.
29. Lt. Col. George W. Easterday, 16th F. A.
30. Lt. Col. Clarence Lininger, A. C. of S. G-3.
31. Maj. Frank A. Jones, Comndg. 10th M. G. Bn.
33. Lt. Col. Wm. H. Clendenin, A.C. of S. G-2.
34. Maj. John C. Hughes, ADC to C-in-C.
35. Capt. Gordon H. Michler, Div. Athletic Officer.
36. Capt. Wallace E. Cox, ADC to Div. Comndr.
37. Lt. Col. Henry W. Stiness, Div. J. A.
38. Col. Henry Beeuwkes, M.C., I.G.D.
39. 1st Lt. Edward G. Krieger, Asst. to G-2.
40. 1st Lt. Wells Hutchins, Asst. to G-3.
41. Maj. John T. McEwen, Adjt. 8th Brig.
42. Lt. Col. James M. Lockett, 39th Inf.
43. Mr. Henry Noble Hall, War Correspondent.
44. Maj. Edward S. Greenbaum, Asst. Div. J. A.
45. Lt. Col. Edward Bowditch, Jr., ADC to C-in-C.
46. Capt. Charles L. Bolté, Div. Adjutant.
47. Maj. Francis R. Fuller, Div. M.G. Officer.

choss and the Ammunition Train at Altenahr, both well back in the Ahr Valley. The Signal Battalion was at Bacheim. Thus the entire Division was grouped in a much smaller space
than before. Although this had obvious advantages for ad- 313
ministrative purposes, the space for maneuvers and parade grounds was much restricted.

As soon as the Division had become settled in its new area the athletic officers located baseball fields and training grounds for sports. The preparations for the Army of Occupation championship were in full swing but the move of the Division greatly interfered with the training of the men. Much time and energy had been expended, in the former area, on laying out athletic and baseball fields and getting the athletes of the Division together for careful training. All this work was lost. Men who had been selected for training had been returned to their units for the march to the Rhine, and all the work had to be done over again. As a result of this setback the Division failed to make the showing at the Army Field Day which, with the splendid material at its command, would otherwise have been certain.

It was in Germany that the Young Men's Christian Association, the Knights of Columbus, the Salvation Army, the Jewish Welfare Board, and other kindred organizations rendered the greatest service to the men of the Army. Prior to the time of the Armistice the benefits that accrued to the Division from the presence of these well-meaning war workers had been negligible. This was due to no lack of good intentions on their part but to the conditions under which they were operating. In the case of the Young Men's Christian Association, for instance, such motor transportation as they had brought with them had been commandeered, during the period of active operations, by the Army for bringing up absolutely necessary supplies. So, while they were willing to serve, they had nothing with which to work. It was not until after hostilities had ceased that the welfare associations

really came into their own. The largest of them all, the Young Men's Christian Association, suffered from the hasty expansion of its organization made at a time when there was an exceptional demand for able-bodied and efficient men for more essential war purposes. The nucleus of the organization and the great body of the added personnel were highly satisfactory, but the haste with which the new workers were chosen, the limitation of the field of choice, and the failure to probe the character of the applicants led, inevitably, to the inclusion in the Y. M. C. A. ranks of some individuals who brought grave discredit upon the organization. These are the cases that are remembered when the overwhelming mass of good work done is apt to be lost sight of. But the fact remains that the Y. M. C. A. was not popular with the men. Apart from actual cases of dishonesty, much discontent was caused by the fact that the Y. M. C. A. charged more than the army commissary for the same goods. On the other hand, virtually all the athletic supplies, moving pictures, and musical entertainments provided for the men were furnished by the Y. M. C. A. and furnished free of charge. The Knights of Columbus, a much smaller organization, and the Salvation Army, also did very good work and did not suffer
819 from the same criticism as did the Y. M. C. A.

The Red Cross had worked faithfully and unostentatiously throughout the whole period of active campaign as a part of the Medical Department, giving invaluable help at many times. During the period of occupation there was little need for its special services and the opportunities increased for the other welfare organizations to contribute to the well-being and pleasure of the troops.

On April 11th, General Babbitt was taken away from the
817 Division and ordered to America for important duty there. He had joined the 4th Artillery Brigade at Camp Greene and he commanded it, with distinction, through all the engagements in which it participated—the Vesle, the St. Mihiel

Offensive, and the Battle of the Meuse-Argonne. His departure left a distinct gap; his cheery personality and expert technical knowledge were greatly missed.

Fortunately the Division retained one of its fighting brigadiers. General Poore was permitted to remain with the 7th Brigade, much to its delight. He had brought it to Europe and had fought it in all the great American offensives of the World War. He took it back home. General Poore was recognized as a skillful tactician and had demonstrated his personal valor on the battlefield. Twice, for short periods, he ably commanded the Division.

On April 28th the two Infantry Brigades were paraded on the drill field at Remagen, where the Division Commander decorated the colors of the various regiments and the Signal and Machine Gun Battalions. The ribbons thus 317
tied to each of the regimental colors took the place of the silver bands which every American regiment is entitled to wear upon the staff of its colors, one for each battle in which it has fought. The bands themselves were to be attached on the return of the regiments to the United States. As the representative of the Commander-in-Chief General Hersey decorated a number of officers and soldiers with the Distinguished Service Cross and Croix de Guerre. Although the ceremony lacked something of the impressiveness of the review at Büchel, it had the glamor lent by the wonderful setting amidst which it was staged. On the edge of the field flowed the swift, still current of the Rhine. Securely embattled on the high crests of its eastern bank the ruins of ancient castles frowned down upon the presumptuous Americans who had dared to invade their domain. In the distance, on a high crag beyond Nieder-Breisig, loomed the castle of Rhineck, used as a post for the military police. This fastness forms part of the Von Bethmann-Hollweg estates, and in the grounds is the private burial place of the elder branch of the former Chancellor's family. The woods surrounding the

castle hide the grave of many a Jewish merchant or money-lender who, in the Middle Ages, had failed to satisfy the greed of the robber barons.

On this occasion Lieutenant Etienne Escudier, of the French Army, received the Distinguished Service Cross. Captain Jean L. Meurisse, of the French Army, was later similarly decorated. These officers had served since June and July, 1918, as liaison officers, with the 4th Division. By their splendid gallantry on the battlefield they had earned this honor. Hardly less notable were the services of Captains Bidaud, Carbonnel, Hommet, Bucher, and Rénard (who was killed Sept. 26, 1918) and Lieutenants Dael, Diconne, LeBrun, Limousin, and Thoumyre, all of whom had accompanied elements of the Division under fire, giving battalion commanders the benefit of their experience, especially during the early days. The Division acknowledges its debt to these gallant officers.

On May 10th General Hersey with a number of other division commanders left on a trip of inspection of the Services of Supply and Brigadier General Augustine MacIntyre who, on May 15, 1919, had taken command of the Artillery Brigade, acted as Division Commander. During this time General Pershing paid another visit to the Division on his way back from the British area of occupation. The Commander-in-Chief met all the commanders of units and the staff at the Division Headquarters in Nieder-Breisig, and was photographed with them.

On May 11th orders came, from the Third Army, to
320 prepare for return to the United States. The news was enthusiastically received. Proud as the men were of their Division, much as they enjoyed their experience of forming a part of the Army of Occupation in Germany, they were beginning to get homesick. They felt that their job was done; they had seen enough of Germany and longed to return to America. Everyone in the Division, officers and men alike,

talked of what they would do when they saw "God's country" again. The Division was placed under the control of the Services of Supply. The entrainment for St. Nazaire was scheduled for May 27th. The guns of the artillery, the machine guns and small arms of the infantry, the transport, were turned in and everything made ready to pull out of Germany and embark for home.

Suddenly, on May 20th, the Division was reassigned to
the Third Army and to the III Corps. All preparations 321
came to an abrupt end. Guns and transport were re-issued to the troops and once more the Division was placed on a war footing ready for any emergency. The Germans had declared their unwillingness to sign the peace terms drawn up in Paris. In order to compel their acceptance of the conditions fixed by the Allied and Associated Powers it might become necessary to move forward again. Marshal Foch had been instructed to advance on June 25th if the Germans did not announce their intention of signing the Treaty and he had ordered the American Army to hold itself in readiness to move, in conjunction with the French, British, and Belgian troops. The men were greatly disappointed at this
turn of events. 322

The crisis reached its height on June 17th. The old scenes of war were reënacted. The forward divisions moved up to the advanced bridgehead zones, artillery was brought up and placed in position, and a large amount of motor transportation concentrated east of the Rhine for the purpose of pushing the advance into Germany as fast as possible. The 59th Infantry was placed on the east bank of the river and 157 trucks of the 4th Division were temporarily assigned to the 1st Division. But on June 23rd the Germans announced their decision to sign the peace
and once more everybody's thoughts turned toward home. 322

The happy news of the signing of the peace at Versailles was received during the afternoon of June 28th. At last

the task for which the men had traveled so far and fought so well was accomplished. In the evening of the same day the Division was reassigned to the Services of Supply and ordered
323 to return home via Brest. Before leaving Germany, however, a strong feeling developed that some permanent association should be formed which would bind together, in after years, the officers and men who formed the 4th Division and who had felt the compelling influence of its undying spirit.

Division Headquarters was undecided whether or not an association formed along the lines of those that had been previously organized in other divisions would endure. To form an association which would not endure would have been a pure waste of effort. Other organizations had been formed, some for social, some for frankly political purposes. All of them had embraced to a greater or less degree a feature for the decoration of the graves of soldiers and the erection of monuments on the battlefield. Some contained a general statement showing that the object of the association was the betterment of American citizenship. None of these had elements which would indicate such a degree of permanence as was considered desirable.

In order to ascertain the opinions held by the enlisted men a circular was sent out which was read to each member of the
324 Division. After stating what might be done it invited the attention of the members of the Division to the possible objects. These were: the marking of the battlefields by suitable monuments; the marking of the graves of those men whose bodies were to be exhumed and taken back to the States; the formation of an endowment fund, the interest of which was to be used in creating scholarships for the children of such men as were financially unable to put their own children in school and college; the organization of a divisional paper or magazine to be issued monthly; the installation of a central office under the charge of a permanent secretary-treasurer;

and the activity of the Association in such questions of an entirely non-political nature as affected the military interests of the country at large. The Convention met at Neuenahr, on Sunday, June 29, 1919. All field officers of the Division were directed to be present together with the first sergeant of each company and likewise two enlisted men from each company elected as delegates by their own organization. From the way the enlisted men took hold of the meeting it was evident that they intended to have a voice in the proceedings.

A nominating committee, consisting of two thirds enlisted men, appointed for the purpose of selecting permanent officers for the Association, reported a slate so fair that it was a real inspiration and a tribute to our form of government. The active officers of the Association were about equally divided between officers and enlisted men. General Poore was nominated for President, and General Hersey, General Cameron, General Hines, and General Babbitt for Honorary Presidents. General Booth was nominated as Honorary Vice-President. The slate was unanimously elected, and nineteen directors, including the active officers of the Association, were chosen as a governing body. The conven- 325
tion lasted from 9:30 A. M. until past 5 o'clock. All the provisions recommended in the original circular were adopted, including the plan to publish a divisional magazine, *Ivy Leaves*, which has since attained wide success.

It was also decided that the 4th Division would erect monuments to the officers and men who had fallen, one monument on the Vesle, one at St. Mihiel, and one in the Meuse-Argonne, commemorating thus each major engagement in which the Division as a whole had participated in action. The monuments have since been erected. The design is very simple and dignified: each monument is a pyramid-topped shaft standing on a triple base pedestal. The monument is sixteen feet high. The Division insignia

One of the three 4th Division Monuments erected in France. This one is located on the Rouen–Rheims highway north of the Vesle River and a mile east of Bazoches

The 4th Division companies of the Composite Regiment on the Champs Elysées, after passing under the Arc de Triomphe. Paris, July 14, 1919

The Price of Victory. Romagne-sous-Montfauçon

in bronze is sunk into the face of the shaft and a large bronze tablet on the front of the pedestal is inscribed:

4TH DIVISION
UNITED STATES ARMY
1918

A party from the 4th Engineers was designated, before the Division left Germany, to erect the three monuments. There was considerable difficulty in obtaining the necessary materials and only one, that erected to the men who fell in the Meuse-Argonne, is made of granite, the other two being of very fine, durable, hard stone. The Vesle monument stands in a commanding position on the Rouen-Rheims road, on the north bank of the Vesle River about a mile east of the town of Bazoches. The St. Mihiel monument is on the Verdun-Metz road at its junction with the road from Fresnes-en-Woëvre and just east of the town of Manheulles. The Meuse-Argonne monument is erected on the Nantillois-Brieulles highway, about three fourths of a mile southwest of where the first houses of Brieulles once stood. The Division owns the ground on which these monuments stand.

The 4th Division began to entrain at Neuenahr, Sinzig, and
317 Coblenz on July 9th. There was much weeping among the fräuleins who crowded the stations, but the excellent deportment of the 4th Division in Germany was best shown by the large number of old men and women who came to the station to say good-bye and wish a safe return to the men who had been billeted in their houses.

Moving an entire division by rail from the Rhine to the Atlantic seaboard at Brest entails a large amount of labor which was rendered none the easier by the fact that the rolling stock in Europe, after nearly five years of war, was in the worst possible condition. The Division was moved mostly in German box cars, surrendered under the terms of tbe

Armistice, but the 4th Engineers had a great deal of work to do placing the cars in good condition for transporting the men, building into certain cars the necessary facilities for the preparation of food, such as ranges, water containers, etc. The main route of travel led through Cologne, Aix-la-Chapelle, Namur, Mons, Arras, Amiens, Rouen, Alençon, Laval, Rennes, St. Brieuc, to Brest. The journey was intensely interesting. The destroyed area of Belgium and northern France brought back vividly to the minds of all the scenes of devastation in which they had lived and fought. The names of towns and cities were familiar and recalled the early days of the war.

On its arrival at Brest the Division was quartered at Camp Pontanezen. In 1918, shortly after it was turned over to the Americans, this camp had been the source of much well-founded complaint. The mud was so deep, due to excessive rains, that duckboards had to be built all over the camp and the "duckboard" itself was adopted as the camp insignia. But when the 4th Division arrived there everything was in excellent shape. Good roads had been built and the men were comfortably situated. While at Camp Pontanezen word came that decorations which had been awarded to officers and men of the Division by the French Government would be presented. The necessary arrangements were made and on July 18th, in the main hall of the Hotel Continental where the Division Commander had his Headquarters, the French Admiral commanding at Brest bestowed a number of crosses of the Legion of Honor, Médailles Militaire and Croix de Guerre.

The Division remained at Brest awaiting transports to take it to America. During this period everybody had a final delousing and a final medical inspection and the men were issued fresh underclothing and such new uniforms as were necessary. The last detachment sailed on July 31st. Only a few days more than a year had elapsed since the Division first went into action in the Second Battle of the Marne.

Before leaving Germany, five hundred men of the 4th Division had been selected, pursuant to instructions from G. H. Q., to form two companies of a "composite" regiment which was to represent the American Expeditionary Forces in the various ceremonies and parades held in allied countries,
326 to celebrate the Victory. This regiment was entirely made up of men from "regular" divisions, there being twelve companies, two each from the 1st, 2nd, 3rd, 4th, 5th, and 6th Divisions. The two companies formed by the 4th Division were composed of selected men from the four infantry regiments and the three machine gun battalions. Officers and men alike were picked on the basis of appearance, conduct, and record. Most of them wore wound stripes; all of them had served in at least one major engagement. These men were collected at Remagen, trained for a short time, and then sent on to Coblenz to report to the Regimental Commander.

Their first public appearance was in Paris on July 4th, where they were reviewed by M. Poincaré the President of the French Republic, M. Clemenceau, Marshal Foch, and General Pershing. The review was held in the Place de la Concorde and, by a delicate attention, the regiment was led by a detachment with the colors of every French division with which American troops had been brigaded before they entered the line as separate units. The appearance of the men was a revelation to the Parisians. It was universally admitted to be the finest body of soldiers in Europe.

Of course these men had the honor of representing America in the great parade in Paris on July 14th, the national fête day of France which, by unanimous consent, was selected as the most fitting date for the celebration of victory by the Allied and Associated nations. No city in the world lends itself as well as Paris to a grand military spectacle. No people equal the French in the glory and glamor with which they surround patriotic displays; and never in the history of mankind had an occasion arisen better calculated to stir the imagination or more worthy of solemn and joyous celebration than Victory, won after a desperate and at times despairing conflict, by the genius of great generals, the heroism of their troops, and the inevitable weight of America's inexhaustible resources. The French rose to the occasion and Paris presented a never-to-be-forgotten spectacle when Marshal Foch and Marshal Joffre led the victors up the Avenue de la Grande Armèe, through the Arc de Triomphe, down the Champs-Elysées, by the Place de la Concorde, to the Boulevards and the Place de la Rèpublique. More than two million persons saw the pageant from every possible vantage point of the most beautiful city of Europe. And this great victory march, in which figured all the most famous military chiefs, all the generals commanding divisions at the front, with the colors of every regiment in the French Army and their colonels and color-guards, the colors of all the British regiments in France, picked detachments of the Italian, Belgian, Greek, Rumanian, and other allied troops, was headed by the American Army. Immediately behind Marshals Foch and Joffre and their staffs rode General Pershing, followed by his staff and the American Division Commanders. Then came the Composite Regiment and the massed colors of American regiments which had won undying fame in France. It was a fitting climax to the greatest of wars.

CHAPTER XVII

REVIEW

THE promise of America was fulfilled. She had entered the war for no selfish purpose but in defense of her own liberty and of civilization. In order to decide whether peace and justice should reign in the affairs of men, whether Right or Might should rule the destinies of mankind, America had employed "force, force to the utmost, force without stint or limit, the righteous and triumphant force" of a nation proud of its purposes, conscious of its strength, and wholly devoted to the task to which it had set its hand.

America entered the war at its most critical stage. At the beginning of 1917, the Allies had reached the highest point of their military power, a point they were almost certain not to exceed and which could not be long maintained. France had 2,965,000 men at the front with 5,800 pieces of light artillery and 20,000,000 shells, 3,650 pieces of heavy artillery and 7,000,000 shells. England had nearly 2,000,000 men in France with 3,000 pieces of light artillery and 11,000,000 shells, 2,000 pieces of heavy artillery and 4,000,000 shells, nearly all the British artillery being of the most modern type. Italy had added 20 per cent. to her armies during the winter of 1916–1917 and had some 60 divisions or a total force of more than 1,000,000 men. Belgium had 6 divisions numbering 156,000 men. The Russians had promised the Allies at the great Council of War, held on November 15, 1916, that they would have some 200 divisions in line and an additional 2,000,000 men for replacements. Finally, it was hoped that the Rumanian army would prove a factor of value. The British and French armies together could throw a hundred divisions into a general attack. 327

Great as were the preparations of the Central Powers, the balance of strength, in the beginning of 1917, lay wholly with the Allies on every front. As Colonel Réquin points out in his admirable work on America's share in the Victory, the European armies in the field, in April, 1917, when America entered the war, were as follows: on the western front at least 178 Allied divisions against 150 German divisions (Ludendorff states 190 Allied against 154 German); on the Italian front, 329
62 Italian divisions against 32 Austrian divisions; on the Russo-Rumanian front 200 Russian divisions and the Rumanian army against 127 divisions of German, Austrian, Bulgarian, and Turkish troops; and on the Macedonian front 24 Allied divisions against 15 enemy divisions, the latter, it is true, somewhat heavier than those of the Allies. But 330
important as was the numerical superiority of the Allies, they lacked the one great essential to victory—unity of command. Then, even before the attack could begin, the Russian Treaty of Peace, signed at Brest-Litovsk in March, 1917, freed a considerable number of German divisions on the eastern front. The battle on the western front began on April 14, 1917, a week after America had declared war on Germany. The French offensive was halted, after having failed to attain its objectives. Before the end of the month something akin to demoralization had set in and although General Pétain, who had replaced General Nivelle, succeeded in restoring the morale of his army and maintaining its strength, the rest of the fighting in France gave no very happy results. During the remainder of the year the Allied forces were gradually weakened by numerous small engagements. In 1917 the French and British lost very nearly half a million men in killed, died of wounds, missing, or prisoners, the figures being 231,000 for the British and 243,000 for the French. 328
The possibility of final and complete victory over the Cen-

tral Powers now depended upon the size of the army America could raise and the speed with which it could be thrown into the struggle.

In less than twenty months America raised, organized, trained, and equipped an army of 3,600,000 men, of which 2,048,000 were brought across the Atlantic and safely landed in Europe before hostilities ceased, without seriously interfering with the sending of war supplies of all kinds which the
331 United States was furnishing to the Allies. This astounding achievement, which far surpassed the wildest dreams of optimism, was made possible only by the splendid way in which the American people as a whole threw themselves into the struggle. Not only did the Congress of the United States devote all the energies of the nation and all its wealth to the purpose of the war, but men, women, and children in every calling supported the Army and cheerfully sacrificed not only comforts but even necessaries of life that the men who were upholding the honor of America might lack nothing. These freely yielded sacrifices, and self-imposed restrictions made possible the achievements of the army. The statement of this fact in no way detracts from the credit due the British for the coöperation of their navy and mercantile marine, without whose assistance it would not have been possible to transport 2,000,000 troops across the Atlantic; and to the French who supplied these troops with a large portion
333 of their artillery and ammunition and aeroplanes. It is too often lost sight of that, when the Armistice was signed, the American Army was still using French shells and that the French had furnished it with 1,871 75 mm. guns, 762 guns of 155 mm. short, 244 guns of 155 mm. long with tractors, 240 tanks, and 2,676 fully-equipped aeroplanes, not to mention 136,881 horses and a mass of other war materiel and
332 supplies.

America's part in the war can after all be most appropriately stated in the words used by Marshal Foch on March 11, 1919, at a banquet tendered to the American Delegation to the Peace Conference. The Commander-in-Chief of all the Allied Armies said: 334

"One year ago, on March 11, 1918, the American Army in France numbered only 300,000 men, that is, six infantry divisions, in training. Thirty thousand men were arriving every month. On March 21st the German offensive was launched against the junction of the Allied armies in the neighborhood of St. Quentin. You know what its effects were. It soon reached the Scarpe, moved up the Somme, which it crossed, and came down the Oise. The situation was grave. In those critical days, on March 28th, General Pershing and General Bliss came and generously asked me to lead them to battle. Both of them told me: 'All our troops are at your disposal; where shall we go?'

"Shortly afterward, on April 25th, at Sarcus, we met the two Generals again. On May 2nd, at Abbéville, in agreement with the Allied Governments, we asked the American Government to land in France, per month, 120,000 infantry or machine gunners, together with the necessary auxiliary troops. As a matter of fact, America sent us 60,000 in March, 94,000 in April, 200,000 in May, 245,000 in June, 295,000 in July, 235,000 in August. The American Army in France rose from 300,000 on March 11th to 954,000 in July and 1,700,000 at the beginning of October.

"On June 2nd the Supreme War Council at Versailles asked President Wilson to maintain the sending of troops at the rate of 200,000 or 300,000 per month and to prepare, for the spring of 1919, 100 American divisions. The President agreed to do so and said that if more men were needed they would be sent.

"But during this time American troops were not inactive. Already in May, two American infantry divisions were engaged with the First French Army in the Montdidier region, three in the Vosges, where they relieved French troops, and two others were in training.

"In June two American divisions were on the Marne where they contributed largely to the resistance against the enemy.

"On July 18th five American divisions took part in the glorious counter-offensive of the Tenth and Sixth French Armies and contributed greatly to its success.

"On July 24th, the First American Army was created under the command of General Pershing. Its task was to reëstablish direct communication between Paris and Nancy by forcing the enemy back from St. Mihiel.

"On September 12th, fourteen American divisions, eight in front line and six in reserve, captured the salient of St. Mihiel, taking 200 cannon and 15,000 prisoners.

"A few days later, fourteen American divisions were engaged, on September 26th, between the Aisne and the Meuse, in the difficult region of the Argonne, in a great offensive. On the second day Montfauçon was passed, on October 14th, Grandpré was taken; on the 21st, Chatillon; on the 30th, Bantheville; on November 1st, Buzancy; and the 4th, Beaumont; and on the 9th the whole line of the Meuse from Mauzon to Bazeilles was in our power.

"During the same time two American divisions were coöperating with the Fifth French Army toward Romaines, two others were with the British armies in the region of St. Quentin, two others, again in conjunction with the Fourth French Army, carried the formidable positions of Orfeuil; then two American divisions took part in the Flanders offensive on the Lys and the Scheldt. Finally, six others were ready with the French Army to attack in Lorraine on November 14th, when on November 11th, the Armistice came and disarmed us.

"Thus the American Army, backed by a Government thoroughly determined to carry on the struggle to the very end, had returned the visit of Lafayette to the new-born America. Thus the American Army, powerfully aided in securing Victory by the Armistice, which was tantamount to a surrender, an unconditional surrender.

"With these thrilling recollections and thinking of those days of anxiety and of success, I raise my glass in honor of President Wilson, who so gallantly supported the war, and in honor of my American comrades in arms, generals and soldiers alike, all equally glorious, who made decisive the victory of liberty."

Of the 1,960,000 American soldiers in France when the Armistice was signed about one and a quarter million had
332 been actually engaged in battle; about 70,000 were held in reserve; about 30,000 were casuals or unidentified, and
335 about 610,000 were engaged in the Services of Supply. A book could be written about the wonderful organization which American initiative and efficiency had built up in France with a view to supplying all the needs of the Army of 3,000,000 men which it was ultimately planned to land in Europe. Six principal points of debarkation, St. Nazaire, Bordeaux, Nantes, La Pallice, le Havre, and Brest had been selected, but their combined facilities did not exceed the handling of more than 10,000 tons of freight a day, whereas the minimum

necessary for an army of even a million men was nearly 60,000 tons a day. The Services of Supply enlarged the ports, built wharves, docks, railroads, warehouses, and gathered together immense stocks of supplies, equipment, and war material of all kinds necessary for the army. It is almost impossible to exaggerate the magnitude of the work performed or its importance and when, after the Armistice, the various division commanders and chiefs of staff were taken over the Services of Supply and shown the arrangements made for supporting the combatant forces and supplying their needs, they were dumbfounded. They saw miles upon miles of warehouses, filled with all kinds of stores, hundreds of miles of railroads laid by the Americans with quantities of American rolling stock, and engineering work of such immense scope, as for instance the docks and wharves of St. Nazaire, that it was almost impossible to believe they could have been constructed in so short a time. 332

Speaking generally, the Services of Supply functioned admirably. At no time was the 4th Division without food and ammunition, except in the rare cases where the exigencies of war made it impossible to bring up the supplies to the troops in the front line.

The 4th Division has the distinction of having served in every one of the American Army Corps which saw active service and of having formed part of the three American Armies created. It was with the First Corps in the Vesle; with the Second Corps on the British front; with the Third Corps in the Meuse-Argonne, and during its second short period of service in the Army of Occupation; with the Fourth Corps in Germany and when as part of the Second Army it was about to participate in the attack planned for November 14th; and with the Fifth Corps in the St. Mihiel operation. The 4th Division formed part of the First American Army in the Vesle fighting, at St. Mihiel and in the Meuse-Argonne; it formed part of the Second American Army when the

Armistice was declared, and it formed part of the Third American Army while in Germany. In addition to this the 4th Division served with one British division, the 16th, and with three French divisions, the 164th, the 33rd, and the 4th. It served in two French Corps, the VIIth and the IInd, and in one French Army, the Sixth.

Less than six months elapsed between the arrival of the 4th Division in France and the signing of the Armistice, but in that time it fought in all the great American offensives in France. During this fighting and the subsequent occupation of enemy territory the Division lost a total of 499 officers and 13,150 men, as follows: 69 officers and 2,000 men killed in action; 33 officers and 506 men died of wounds; 11 officers and 202 men died from other causes; and 386 officers and 10,442
336 men wounded. The 4th Division met and defeated sixteen
German divisions, two in the initial attack of July 18th, one
in the Sergy fight, two on the Vesle, two at St. Mihiel and nine
337 in the Meuse-Argonne. The Battle of France from July 18th
to November 11th lasted 117 days. Of this time all or part
of the Division was in action 69 days; in the front line, in
sector, 6 days, and in immediate reserve 12 days. It advanced
35 kilometers in action, took 61 guns, two tanks, a large num-
ber of minenwerfer and more than three hundred machine
guns beside 2,756 prisoners, not including those taken with the
338 French in July.

Although the 4th Division had little opportunity to train in the United States on account of the weather conditions and was flung into the struggle with less training in France than any other American division had received, its work at the Second Battle of the Marne, under French command, revealed the dash and courage of the Ivy men. The fight at Sergy showed some of those splendid qualities of endurance which were later to be emphasized on the Vesle. The slow torture of the river valley, the struggle there and the crossing of the Vesle under merciless odds had much to do in bringing out the full character of the Division, and when, after the exploit at St. Mihiel, the Ivy Division was given, in the great Battle of the Meuse-Argonne, a sector that, it was known, would try its fortitude to the utmost, it fully realized the high expectations of its commanders and showed itself worthy of the confidence General Pershing reposed in it. The 4th Division, under constant shell and machine gun fire, stayed continuously in action in the Meuse-Argonne for twenty-four days, a record unsurpassed by any other division, French, British, or American, in the whole war; and it also went as far as any other American division in the initial attack of the Meuse-Argonne battle reaching the Corps Objective by noon on the first day.

With such a record it was but natural that the Roll of Honor should be a long one. Most of the 4th Division's dead lie in the great cemetery of the Meuse-Argonne at Romagne-sous-Montfauçon, where they rest among the twenty-six thousand heroes buried on the gentle slopes of a verdant hill. When approaching this great cemetery from the Cunel road it looks in the distance like a vast field of white daisies. On a tall flag-pole of Oregon fir floats the Stars and Stripes. Beneath its folds and under the loving care of American officers and men our dead sleep in long well ordered rows as befits soldiers, united now as they were in life, each grave marked with a plain white cross. There lie men from every State in the Union, men of all the countless races which go to make up the "unique breed" of America, men of every station and calling in life; millionaires and paupers, college graduates and illiterates; but Americans all who crossed the seas to fight in a cause which they did not pretend was peculiarly their own, but which they knew was the cause of humanity. They made the greatest of all gifts, the gift of life, that others might be free. There let them rest.

This story of the 4th Division ceases with the departure from Brest. The Division, however, continues as an integral unit of the American Army; but of the officers and men who fought in the Great War most have returned to civil life. The 4th Division Association acts as a binding force and keeps alive that splendid spirit which, born of the regular army, inspired their great deeds. This history will have failed to accomplish its purpose if it does not fill the reader with admiration and respect for the men who so splendidly upheld the honor of their country. They were steadfast and loyal, their bravery and endurance was equalled only by the willingness with which they submitted to all sorts of hardships in the cause for which they fought. They were not only magnificent soldiers, but both on the field of battle and in the Army of Occupation they proved themselves by their treatment of the enemy to be chivalrous American gentlemen. Their pride in the 4th Division was a just and proper pride, and the Division itself will always be proud of them. The American Army was truly a democracy in arms. The men who fought were American citizens clad and armed in a particular manner for the performance of a specific duty; that duty accomplished, they were returned to the pursuits of peace and have mingled once more in the daily lives of their fellow citizens. Thus they proved that America has not changed, either in its temper or its institutions, since George Washington, at the close of the Revolutionary War, in General Orders addressed to the first of all American Armies, said:

"The generous task for which we first flew to arms being accomplished; the liberties of our country being fully acknowledged and firmly secured, and the characters of those who have persevered through every extremity of hardship, suffering and danger, being immortalized by the illustrious appellation of the Patriot Army, nothing now remains but for the actors of this mighty scene to preserve a perfect, unvarying consistency of character, through the very last act to close the drama with applause, and to retire from the military theater with the same approbation of angels and men which has crowned all their former virtuous actions."

BIBLIOGRAPHY

(Throughout the reference list the sources of information are abbreviated as indicated in the following list.)

BELLOC: "The Elements of the Great War; The First Phase," by Hilaire Belloc; Hearst's International Library Co., New York, 1915.

BRITAIN'S WHITE PAPER: "Great Britain's Parliamentary White Paper: Correspondence Respecting the European Crisis."

CHEVRILLON: "Britain and the War," by André Chevrillon; Hodder & Stoughton, London, 1917.

DIPLOMATIC CORRESPONDENCE: "Diplomatic Correspondence Between the United States and Belligerent Governments Relating to Neutral Rights and Commerce," Special Supplement to the Journal of International Law.

DOCUMENTS AND STATEMENTS: "Documents and Statements Relating to Peace Proposals and War Aims;" George Allen & Unwin, Ltd., London, 1919.

ENCY. BRIT.: Encyclopedia Brittanica, 11th Edition.

HAIG'S DESPATCHES: "Sir Douglas Haig's Despatches;" J. M. Dent & Sons, Ltd., London, 1919.

HINDENBURG: "Out of My Life," by Marshal Von Hindenburg; Cassell & Co., London, 1920.

LONDON TIMES HISTORY: The London Times History of the War.

LUDENDORFF: "My War Memories, 1914-1918"; by General Ludendorff; Hutchinson & Co., London, 1919.

MAURICE: "The Last Four Months," by Maj. Gen. Sir F. Maurice; Cassell & Co., London, 1919.

NIETZSCHE: "*Jenseits von Gut und Böse,*" "*Zur Genealogie der Moral,*" Friedrich Wilhelm Nietzsche.

N. Y. TIMES HISTORY: The New York Times History of the War.

PERSHING: Final Report of General John J. Pershing, War Dept., 1920.

PHILLIPSON: "Termination of War and Treaties of Peace," by Coleman Phillipson; T. Fisher Unwin, Ltd., London, 1916.

REPORT SECRETARY WAR: Annual Report of the Secretary of War for 1918.

REPORT SURGEON GENERAL: Annual Report of the Surgeon General of the Army, 1918.

RÉQUIN: "America's Race to Victory," by Lt. Col. E. Réquin; Frederick A. Stokes & Co., New York, 1919.

SCHLACHTEN UND GEFECHTE: "*Die Schlachten und Gefechte des Grossen Krieges, 1914-1918*," prepared by the German Great General Staff; Hermann Sack, Berlin, 1919.

SIMONDS: "History of the World War," by F. H. Simonds; Doubleday Page & Co., New York, 1919.

TREITSCHKE: "Politics," by Hermann von Treitschke; Constable & Co., London, 1916.

VON BERNHARDI: "Germany and the Next War," by F. Von Bernhardi; Longmans, Green & Co., London and New York, 1919.

VON DER GOLTZ: "The Nation in Arms," Von der Goltz; Hugh Rees, Ltd. London, 1913.

WAR WITH GERMANY: "The War With Germany," War Dept., 1919.

REFERENCES

NOTE: Items marked thus * refer to data which are contained in the files of the 4th Division records in the Historical Branch, War Plans Division, General Staff, in Washington. Items marked thus ** refer to data which are in the Historical Branch but in files other than those of the 4th Division.

Number *Reference*

1. War with Germany, p. 123.
2. Report Secretary War, p. 11.
3. Ency. Brit. Vol. XXVI, p. 677.
4. Ency. Brit. Vol. IV, p. 420.
5. Ency. Brit. Vol. IV, p. 421.
6. Ency. Brit. Vol. XI, p. 861.
7. Ency. Brit. Vol. XI, p. 52.
8. Ency. Brit. Vol. XI, p. 864.
9. Ency. Brit. Vol. XI, p. 54.
10. Von der Goltz, Chapter V.
11. Phillipson, Appendix A.
12. Ency. Brit. Vol. XXVIII, p. 55.
13. Ency. Brit. Vol. III, p. 677.
14. Ency. Brit. Vol. IV, p. 6.
15. Ency. Brit. Vol. IV, p. 339.
16. Ency. Brit. Vol. XI, p. 873; Vol. XXIV, p. 714.
17. Ency. Brit. Vol. IV, p. 8.
18. Ency. Brit. Vol. XI, p. 874; Vol. XI, p. 11.
19. Von Bernhardi, Chapter V.
20. Nietzsche.
21. Treitschke.
22. Ency. Brit. Vol. XI, "Germany;" Belloc, Part I; Simonds, Vol. I, Chapter I.
23. Chevrillon, pp. vii-57.
24. Britain's White Paper.
25. Speech, German Chancellor to Reichstag, Aug. 4, 1914.
26. Ludendorff, Vol. II, Chapter I.

Number *Reference*

27. London Times History, Vol. I, p. 320
28. Diplomatic Correspondence, Vol. XI.
29. Diplomatic Correspondence, Vol. IX.
30. Diplomatic Correspondence, Vol. X.
31. N. Y. Times History, Vol. XII, p. 72
32. N. Y. Times History, Vol. X, p. 968.
33. Cablegram, Gen. Pershing to War Dept. July 6, 1917.**
34. Letter, Capt. R. T. Heard, 6th F. A., Feb. 12, 1919.**
35. Cablegram, Gen. Pershing to War Dept. Nov. 4, 1917.**; Pershing, p. 8, p. 23.
36. Letter, The A. G. O. to C. G. 4th Div.
37. Organizations of the Army.**
38. War Diary, G-3 4th Div.*
39. Report Surgeon General.
40. Simonds, Vol. II, Chapter VI.
41. Simonds, Vol. II, Chapter VIII.
42. Simonds, Vol. II, Chapter XIII.
43. London Times History, Vol. VIII, pp. 73, 75.
44. Simonds, Vol. III, Chapter X.
45. London Times History, Vol. IX, p. 88.
46. Ludendorff, Vol. II, Chapter, I.
47. Réquin, Chapter I.
48. Haig's Despatches, pp. 81-84.
49. Maurice, Chapters I-II.
50. Ludendorff, Vol. II, pp. 527-588.
51. Haig's Despatches, pp. 177-241.
52. Maurice, Chapter I.
53. Some Accomplishments of the S. O. S.**, p. 10.
54. Files of U. S. State Department.
55. Speech, Marshal Foch to Peace Delegation, Mar. 11, 1919.
56. Report Secretary War, p. 31.
57. War Diary, G-3 4th Div.*
58. History 4th Engineers.*
59. Cablegram, Gen. Pershing to War Dept. May 24, 1918.**
60. History 4th F. A. Brigade.*
61. Ludendorff, Vol. II, pp. 589-678; Hindenburg, pp. 356-369.
62. Réquin, Chapter VIII.

Number *Reference*

63. Summary of Information, G.H.Q.A.E.F., June 1918.**
64. Report Secretary War, p. 75.
65. F. O. 1, 4th Div. June 8, 1918.*
66. G. O. 29, 4th Div. June 13, 1918.*
67. F. O. 3, 4th Div. June 14, 1918.*
68. F. O. 4, 4th Div. June 16, 1918.*
69. F. O. 9, 4th Div. July 4, 1918.*
70. Hindenburg, Chapter XXI.
71. Schlachten und Gefechte, p. 376; Ludendorff, pp. 665-666.
72. Maurice, pp. 95-96.
73. Hindenburg, p. 378.
74. Pershing, p. 35.
75. Summary of Information, G.H.Q.A.E.F., July 18, 1918.**
76. Réquin, pp. 177-179.
77. History 7th Inf. Brig., p. 19,* History 39th Inf. p. 25.* F. O. 8, 7th Inf. Brig. July 16, 1918.*
78. Plan of Engagement, Hq. 164th Fr. Inf. Div. July 17, 1918.*
79. Report 1st Bn. 58th Inf.*
80. Report 2nd Bn. 58th Inf.*
81. Affidavit Capt. T. E. Roderick, 58th Inf.*
82. Statement Maj. L. Farrell, 59th Inf.*
83. Report Gen. Booth on Chézy Operations.*
84. G. O. of Operations, 164th Fr. Inf. Div. July 18, 1918.*
85. S. O. 3rd Bur. No. 604 / S. OP. 164th Fr. Inf. Div. July 20, 1918.*
86. History 59th Infantry.*
87. S. O. 3566, 6th Fr. Army, July 22, 1918.*
88. Order 2/PC 3rd Bur. II ACFr. July 22, 1918.*
89. F. O. 13, 4th Div. July 24, 1918.*
90. F. O. 13½, 4th Div. July 24, 1918.*
91. Order 3597, 6th Fr. Army, July 25, 1918.*
92. Order 3615, 6th Fr. Army, July 27, 1918.*
93. Letter, C.G., I A.C.U.S. to C.G. 4th Div. July 28, 1918.*
94. F. O. 28, I A.C.U.S., July 28, 1918.*
95. F. O. 17, 4th Div. July 29, 1918.*
96. Letter, C.G., I A.C.U.S. to C.G. 4th Div. July 30, 1918.*
97. F. O. 21, 4th Div. July 30, 1918.*

Number	*Reference*
98.	F. O. 22, 4th Div. July 31, 1918.*
99.	Summary of Intelligence 25, I A.C.U.S., July 29, 1918.*
100.	Report of Capt. G. J. Braun, 47th Inf.*
101.	History 42nd Division.**
102.	History 47th Infantry,* pp. 39-40.
103.	History 39th Infantry, p. 35.*
104.	F. O. 33, I A.C.U.S., Aug. 1, 1918.*
105.	Summary of Intelligence 30, I A.C.U.S., Aug. 3, 1918.*
106.	Cable Communiqué Aug. 3, 1918, G.H.Q.A.E.F. to War Dept. re 4th and 42nd Divs.** History 39th Inf. pp. 41, 43.* History 59th Inf.* Report Gen. Booth on Vesle Operations.* Statement Lt. Col. M. C. Mitchell, 39th Inf. Mar. 5, 1920.*
107.	Message C.O. 1st Bn 59th Inf. to C.G 8th Brig. 10:40 A. M. Aug. 3, 1918.* Report Gen. Booth on Vesle Operations.*
108.	Message Gen. Booth to Hq. 4th Div. Aug. 3, 1918.*
109.	Memo C.G. I A.C.U.S. to C.G. 4th Div. Aug. 3, 1918.*
110.	F. O. 24, 4th Div. Aug. 3, 1918.*
111.	History 39th Infantry, pp. 41-42.*
112.	Message C.O. 59th Inf. to C.G. 8th Brig. 11:10 P. M. and 11:40 P.M. Aug. 3, 1918.*
113.	F. O. 6, 8th Brig. Aug. 4, 1918.*
114.	Message C.O. 39th Inf. to C.G. 7th Brig. 8:35 A. M. Aug. 4, 1918.* Carte des Operations, 6th Fr. Army.* Map, Aisne-Marne Offensive, G-3 GHQ.** History 7th Brigade, p. 22.* History 39th Infantry, p. 44.*
115.	Report Gen. Booth on Vesle Operations.*
116.	Message Gen. Booth to C.O. 58th Inf. Aug. 4, 1918.* Message Gen. Booth to Hq. 4th Div. 11:20 A. M. Aug. 4, 1918.*
117.	History 2nd Bn 58th Inf.*
118.	Memo 226/G-3, I A.C.U.S., Aug. 4, 1918.* F. O. 26, 4th Div. Aug. 4, 1918.
119.	Message C.O. 39th Inf. to C.G. 7th Brig. 2:50 A. M. Aug. 5, 1918.*
120.	History 2nd Bn 58th Inf.*
121.	Message from C.O. 1st Bn 58th Inf.*
122.	Message Maj. Fuller to Hq. 4th Div. 2 P. M. Aug. 5, 1918.* Message Gen. Booth to Hq. 4th Div. 2:10 P. M. Aug. 5, 1918.*

Number *Reference*

123. History 39th Infantry, p. 45.*
124. Files of 4th Division records.*
125. Message Capt. Jones to C.O. 4th Engrs. 1:50 P. M. Aug. 5, 1918.*
126. History 4th F. A. Brigade.*
127. G. O. 350, 62nd Fr. Inf. Div. Aug. 6, 1918.*
128. F. O. 27, 4th Div. Aug. 6, 1918.*
129. F. O. 2, 58th Inf. Aug. 6, 1918.*
130. Message C.G. 8th Brig. to Hq. 4th Div. 9:35 P. M. Aug. 6, 1918.* Message C.O. 1st Bn. 58th Inf. to C.O. 58th Inf. 7:45 P. M. Aug. 6, 1918.*
131. Message 7th Brig. to Hq. 4th Div. Aug. 6, 1918.* Message 39th Inf. to 3rd Bn. 58th Inf. Aug. 6, 1918.*
132. History 39th Infantry, p. 47.*
133. Letter C.O. 10th M. G. Bn. to C. of S. Aug. 6, 1918.*
134. Messages Maj. Fletcher to Hq. 4th Div. 7:40 P. M. and 7:55 P. M. Aug. 6, 1918.* Message C.G. 8th. Brig. to Hq. 4th Div. 8 P. M. Aug. 6, 1918.* Message Art. L. O. at Hq. 8th Brig. to Hq. 4th Div. 8:15 P. M. Aug. 6, 1918.*
135. Messages C.O. 58th Inf. to C.G. 8th Brig. 8:45 P. M. and 10:03 P. M. Aug. 6, 1918.*
136. Message C.G. 8th Brig. to C.O. 58th Inf. 10:35 P. M. Aug. 6, 1918.* Report Gen. Booth on Vesle Operations.*
137. Operations Report 39th Inf. Aug. 7, 1918. F. O. 19, 7th Brig. Aug. 6, 1918.*
138. History 47th Infantry, pp. 47-48.*
139. Message 7th Brig. to Hq. 4th Div. 6 P. M. Aug. 7, 1918.* Message L. O. 7th Brig. to Hq. 4th Div. 4:15 P. M. Aug. 7, 1918.* History 47th Infantry, p. 48.*
140. Operations Report 4th Div. Aug. 8, 1918.* Message C.G. 8th Brig. to Adjt. 8th Brig. 7:40 A. M. Aug. 7, 1918.*
141. Messages C.O. 58th Inf. to 8th Brig. 6:20 P. M. and 6:55 P. M. Aug. 7, 1918.* Message C.O. 1st Bn. 58th Inf. to C.O. 58th Inf. 6:30 P. M. Aug. 7, 1918.*
142. Message C. O. 58th Inf. to C. G. 8th Brig. Aug. 7, 1918.*
143. Message Adjt. 7th Brig. to Hq. 4th Div. 4:35 P. M. Aug. 8, 1918.*

Number *Reference*

144. Operations Report 59th Inf. Aug. 8, 1918.* Message 8th Brig. to Hq. 4th Div. 3:20 P. M. Aug. 8, 1918.*
145. Message C.G. 7th Brig. to Hq. 4th Div. 7:40 P. M. Aug. 9, 1918.* Messages C.O. 47th Inf. to Hq. 4th Div. 3:45 P. M. and 8 P. M. Aug. 9, 1918.
146. Message C.G. 8th Brig. to Hq. 4th Div. 10:50 P. M. Aug. 9, 1918.*
147. F. O. 30, 4th Div. Aug. 10, 1918.*
148. F. O. 31, 4th Div. Aug. 11, 1918.*
149. History 4th F. A. Brigade.*
150. Operations Report 4th Div. Aug. 12, 1918.*
151. F. O. 34, Aug. 14; F. O. 35, Aug. 15; F. O. 36, Aug. 16; 4th Div. 1918.*
152. F. O. 10, 4th F. A. Brig. Aug. 18, 1918.*
153. History V A.C.U.S.**
154. F. O. 38, 4th Div. Aug. 31, 1918.*
155. Pershing, pp. 37-40.
156. Pershing, pp. 41-42.
157. Pershing, p. 39.
158. F. O. 41, 4th Div. Sept. 5, 1918.*
159. F. O. 43, 4th Div. Sept. 5, 1918.*
160. F. O. 44, 4th Div. Sept. 29, 1918.*
161. F. O. 18, V A.C.U.S., Sept. 11, 1918.*
162. War with Germany, p. 110.
163. Operations Report 4th Div. Sept. 15, 1918.*
164. Message C.G. 8th Brig. to Hq. 4th Div. 10:30 P. M. Sept. 14, 1918.*
165. F. O. 47, 4th Div. Sept. 12, 1918.*
166. F. O. 48, 4th Div. Sept. 13, 1918.*
167. F. O. 49, 4th Div. Sept. 13, 1918.*
168. Pershing, p. 42.
169. F. O. 13, 8th Brig. Sept. 14, 1918.*
170. History 4th F. A. Brig.*
171. See appendix.
172. F. O. 50, 4th Div. Sept. 19, 1918.*
173. Pershing, p. 41
174. Pershing, Plate 3.

Number *Reference*

175. Hindenburg, pp. 426-427.
176. Hindenburg, p. 408.
177. Hindenburg, p. 417.
178. Pershing, p. 43.
179. Pershing, p. 46.
180. F. O. 18, III A.C.U.S., Sept. 21, 1918.*
181. Pershing, p. 44.
182. Graphic Summaries of Intelligence, 1st Army.** Enemy order of battle Sept. 25-26, 1918.**
183. F. O. 54, 4th Div. Sept. 24, 1918.*
184. Annex "D" F. O. 54, 4th Div., Sept. 24, 1918.*
185. Annex "B" F. O. 54, 4th Div., Sept. 24, 1918.*
186. History 39th Infantry, p. 58.*
187. Message C.G. 7th Brig. to Hq. 4th Div. 2:20 P. M. Sept. 26, 1918.*
188. Pershing, Plate 4.
189. Message C. of S. III A.C.U.S. to C.G. 4th Div. 4 P. M. Sept. 26, 1918.*
190. Message C.G. 4th Div. to C.G. 8th Brig. 8:45 A. M. Sept. 26, 1918.*
191. Message C.G 8th Brig. to C.G. 4th Div. 4:25 P. M. Sept. 26, 1918.*
192. Annex 5, F. O. 18, III A.C.U.S., Sept. 21, 1918.*
193. History 16th Field Artillery.*
194. History 77th Field Artillery.*
195. History 13th Field Artillery.*
196. Message C. of S. 4th Div. to 4th Div. 12:30 A. M. Sept. 27, 1918.*
197. Message C. of S. 4th Div. to Div. P. C., 10:35 A. M. Sept. 27, 1918.*
198. Message C. of S. 4th Div. to III A.C.U.S., 12:50 P. M. Sept. 27, 1918.*
199. Operations Report 4th Div. Sept. 28, 1918.* Message C.O. 47th Inf. to C. G. 7th Brig. Sept. 27, 1918.*
200. Message Adjt. 7th Brig. to C. G. 7th Brig. 3:05 P. M. Sept. 27, 1918.*
201. Report C. O. 13th F. A. to C. G. 4th F. A. Brigade.*

Number *Reference*

202. History 16th Field Artillery.*

203. History 13th Field Artillery.*

204. F. O. 19, III A.C.U.S., Sept. 27, 1918.*

205. F. O. 57, 4th Div. Sept. 28, 1918.*

206. Statement Col. J. K. Parsons.* Report Capt. Ralph Smith, 5:20 A. M. Sept. 28, 1918.* Message C. O. 59th Inf. to Hq. 4th Div. 2:05 P. M. Sept. 28, 1918.*

207. Message C. O. 39th Inf. to Hq. 4th Div. 4:10 P. M. Sept. 28, 1918.*

208. F. O. 56, 4th Div. Sept. 28, 1918.*

209. Report Lt. Marshall, 58th Inf. and all messages received and sent by him.*

210. Report Capt. P. E. Engel, 58th Inf.*

211. History 4th Sanitary Train.*

212. History 3rd Division, pp. 15, 67-68.**

213. G-3 Order No. 120, III A.C.U.S., Sept. 28, 1918.*

214. F. O. 22, III A.C.U.S., Oct. 3, 1918.*

215. Pershing, Plate 4.

216. F. O. 60, 4th Div. Oct. 3, 1918.*

217. Message C. G. 8th Brig. to C. G. 4th Div. 6:15 A. M. Oct. 4, 1918.*

218. Report Gen. Booth on Meuse-Argonne Operations.*

219. Messages L. O. with 80th Div. to Hq. 4th Div. 9:15 A. M. and 9.25 A. M. Oct. 4, 1918.*

220. Message L. O. with 8th Brig. to Hq. 4th Div. 10:50 A. M. Oct. 4, 1918.* Message 11th M. G. Bn. to Hq. 4th Div. 11:30 A. M. Oct. 4 1918.* Message C. G. 7th Brig. to Hq. 4th Div. 12:15 P. M. Oct. 4, 1918.* Message C. G. 8th Brig. to Hq. 4th Div. 4:30 P. M. Oct. 4, 1918.*

221. Message L. O. with front line to C. G. 8th Brig. 12:50 P. M. Oct. 4, 1918.*

222. History 2nd Bn. 58th Inf.* Message C. G. 8th Brig. to C. G. 4th Div., 7:20 P. M. Oct. 4, 1918.*

223. Message Maj. Smith to Hq. 4th Div. 5:55 P. M. Oct. 4, 1918.*

224. Message C. G. 8th Brig. to Hq. 4th Div. 4:30 P. M. Oct. 4, 1918.* History 47th Infantry, p. 65.* Report Gen. Booth on Meuse-Argonne Operations.*

Number *Reference*

225. History 318th Infantry, pp. 67-68.** Messages L. O. with 80th Div. to Hq. 4th Div. 3:40, 3:50, 3:58, 4:10, 4:23, 4:37, and 4:50 P. M. Oct. 4, 1918.*
226. Summaries of Information G.H.Q.A.E.F., October, 1918.*
227. History 2nd Bn. 58th Infantry.*
228. G-3 Order No. 145, III A.C.U.S., Oct. 5, 1918.*
229. Message Inf. Brig. of 80th Div. to one of its regts. 3:30 P. M. Oct. 5, 1918.* Message L. O. with 80th Div. to Hq. 4th Div. 9:30 P. M. Oct. 5, 1918.* Summary of Operations, 1st Army U. S.**
230. F. O. 23, III A.C.U.S., Oct. 6, 1918.* F. O. 63, 4th Div. Oct. 6, 1918.*
231. G-3 Order No. 156, part I, III A.C.U.S., Oct. 6, 1918.* F. O. 62, 4th Div. Oct. 6, 1918.* F. O. 22, 8th Brig. Oct. 6, 1918.*
232. F. O. 22, 8th Brig. Oct. 6, 1918.*
233. F. O. 24, IIIA. C.U.S., Oct. 7, 1918.* Memo G-3/713, III A.C.U.S., Oct. 7, 1918.* Memo G-3 1st Army U. S., Oct. 8, 1918.*
234. F. O. 25, III A.C.U.S., Oct. 7, 1918.* F. O. 64, 4th Div. Oct. 8, 1918.* F. O. 18, 4th F. A. Brig. Oct. 8, 1918.*
235. F. O. 64, 4th Div. Oct. 8, 1918.*
236. Message G-3 4th Div. to C. G. 8th Brig. Oct. 9, 1918.*
237. Message C. G. 7th Brig. to Hq. 4th Div. 10:50 and 11:55 A. M. Oct. 9, 1918.* Message C. G. 8th Brig. to Hq. 4th Div. 10:45 A. M. and 12:30 P. M. Oct. 9, 1918.* Message C. O. 59th Inf. to Hq. 4th Div. thru C. G. 8th Brig. 1 P. M. Oct. 9, 1918.*
238. History 2nd Bn 58th Inf.* History 39th Infantry, p. 76.*
239. History 39th Infantry, p. 76.*
240. Message L. O. with 80th Div. to 7th Brig. 5:45 P. M. Oct. 9, 1918.*
241. Message 80th Div. to Hq. 4th Div. 3:02 P. M. Oct. 9, 1918.*
242. History 39th Infantry, p. 75.*
243. Message C. O. 11th M. G. Bn. to Hq. 4th Div. 5 P. M. Oct. 9, 1918.*
244. Message C. G. 7th Brig. to Hq. 4th Div. 6:30 P. M. Oct. 9,

Number *Reference*

1918.* Message C. O. 59th Inf. to Hq. 4th Div. 6:45 P. M. Oct. 9, 1918.* History 39th Infantry, p. 76.* Message, C. G. 8th Brig. to Hq. 4th Div. 6:35 P. M. Oct. 9, 1918.* Message Capt. Wood to C. O. 39th Inf. 8:15 P. M. Oct. 9, 1918.*

245. Message C. G. III A.C.U.S. to C. G. 4th Div. 9.45 P.M., Oct. 9, 1918.* F. O. 31, 7th Brig. Oct. 9, 1918.* History 39th Infantry, p. 76.*

246. Message C. O. 39th Inf. to Hq. 4th Div. 2:05 P. M. Oct. 10, 1918.* History 39th Infantry, p. 76.* Message C. G. 7th Brig. to Hq. 4th Div. 4:25 P. M. Oct. 10, 1918.*

247. History 39th Infantry, p. 78.*

248. Statements Maj. F. R. Waltz and Capt. R. W. Norton.*

249. Message C. G. 8th Brig. to Hq. 4th Div. 6:40 P. M. Oct. 10, 1918.* Message III A.C.U.S. to 4th Div. 6:05 P. M. Oct. 10, 1918.* Message C. G. 159th Inf. Brig. to Hq. 4th Div. 9:15 P. M. Oct. 10, 1918.*

250. F. O. 56, 1st Army U. S., Oct. 10, 1918.* F. O. 27, III, A.C.U.S., Oct. 10, 1918.*

251. Report aviator to 4th Div. 9:30 A. M. Oct. 10, 1918.* Message 88th Aero Squadron to 4th Div. 5:15 P. M. Oct. 10, 1918.*

252. Par. 2, S. O. 24, 7th Brig., Oct. 11, 1918.* Message Maj. Sibert to Hq. 4th Div. 10:03 A. M. Oct. 11, 1918.*

253. History 39th Infantry, pp. 78-79.* History 7th Brigade, p. 47.* Statement Capt. R. W. Norton, 39th Inf.*

254. Statements Maj. F. R. Waltz and Capt. R. W. Norton.*

255. Messages Maj. Waltz to C. O. 39th Inf. 12:05, 2:55 and 4:30 P. M. Oct. 11, 1918.* Message C. G. 7th Brig. to Hq. 4th Div. 3:50 P. M. Oct. 11, 1918.* Statements Maj. Waltz and Capt. Norton.* History 39th Infantry, p. 81.*

256. History 7th Brigade, p. 47.* Statmeent Capt. R. W. Norton.*

257. G-3 Order No. 174, III A.C.U.S., Oct. 11, 1918.* Orders No. 31, III A.C.U.S., Oct. 11, 1918.* F. O. 66, 4th Div. Oct. 11, 1918.* Message C. G. III A.C.U.S. to C. G. 4th Div. 2:35 A. M. Oct. 12, 1918.* History 5th Division, p. 133.**

Number *Reference*

258. History 39th Infantry, p. 82.* Statement Capt. R. W. Norton.*
259. Statement Capt. R. W. Norton.*
260. History 39th Infantry, p. 82.* Statements Maj. Waltz and Capt. Norton.* History 3rd Division, pp. 70-71.** War Diary G-3 4th Division.*
261. Statement Gen. B. A. Poore, June 26, 1919.*
262. F. O. 66, 4th Div. Oct. 11, 1918.* F. O. 68, 4th Div. Oct. 13, 1918.*
263. Par. 5, F. O. 66, 4th Div. Oct. 11, 1918.*
264. Report Maj. John J. Bullington, 132nd Inf. Nov. 14, 1918.**
265. Telegram C. G. III A.C.U.S. to A.G.G.H.Q.A.E.F. Oct. 13, 1918.*
266. Intelligence Summary 47th Inf. Oct. 9, 1918.* Summary of Intelligence 4th Div. No. 31, Oct. 9, 1918.* History 47th Infantry, p. 66.*
267. F. O. 70, 4th Div. Oct. 18, 1918.* Message C. G. 7th Brig. to Hq. 4th Div. 6:45 A. M. Oct. 19, 1918.*
268. F. O. 68, 4th Div. Oct. 13, 1918.* G-3 Order No. 182, III A.C.U.S., Oct. 13, 1918.*
269. War Diary G-3 4th Division.*
270. Letter C.G. III A.C.U.S. to C. G. 4th Div. Oct. 20, 1918.*
271. Casualty Tables 4th Division.*
272. Files of 4th Division records.*
273. Graphic Chart of Service in Meuse-Argonne, 1st Army U. S.**
274. Combined Daily Order of Battle, G-2 A-1 G.H.Q.A.E.F.**
275. Pershing, p. 49. Summary of Operations 1st Army U. S.**
276. Réquin, p. 189.
277. History 3rd Division, p. 35.**
278. Pershing, pp. 50-51.
279. Pershing, Plate 4.
280. Pershing, p. 51 and Plate 4. Réquin, p. 190.
281. Meuse-Argonne Battle, G-1 A. F. in F. p. 6.**
282. Pershing, Plate 7.
283. G. O. 203, G.H.Q.A.E.F., Nov. 12, 1918.*

Number *Reference*

284. F. O. 24, 8th Brig. Oct. 18, 1918.*
285. History 7th Brigade, p. 48.*
286. S. O. 293, G.H.Q.A.E.F., 1918. War Diary G-3 4th Division.*
287. F. O. 79 and S. O. 460, 1st Army U. S., Oct. 20, 1918.*
288. F. O. 71, Oct. 20; F. O. 72, Oct. 21; F. O. 73, Oct. 22; 4th Div. 1918.*
289. S. O. 535, 1st Army U. S., Nov. 3, 1918.* F. O. 76, 4th Div. Nov. 8 1918.*
290. F. O. 79, 4th Div. Nov. 8, 1918.* F. O. 109, 1st Army U. S., Nov. 9, 1918.*
291. F. O. 81, 4th Div. Nov. 10, 1918.*
292. F. O. 80, 4th Div. Nov. 10, 1918.*
293. Pershing, p. 53.
294. Maurice, pp. 234-235.
295. History 4th F. A. Brigade.*
296. London Times History, Vol. XX, pp. 450-453.
297. Documents and Statements, pp. 258-259.
298. Pershing, p. 56.
299. War Diary G-3 4th Division.*
300. G. O. 90, 4th Div. Nov. 19, 1918.*
301. F. O. 85, 4th Div. Nov. 17, 1918.*
302. F. O. 90, Nov. 23; F. O. 93, Dec. 1; 4th Div. 1918.*
303. F. O. 99, 4th Div. Dec. 5, 1918.*
304. History 39th Infantry, pp. 121-122.*
305. Ency. Brit. Vol. XI, p. 825.
306. Ency. Brit. Vol. XVIII, p. 678.
307. Address C. G. 4th Div. to Field Officers, Dec. 11, 1918.*
308. F. O.s 95-106, 4th Div. 1918.*
309. F. O. 107, 4th Div. Dec. 12, 1918.*
310. F. O.s 108-111, 4th Div. 1918.*
311. F. O. 112, 4th Div. Dec. 17, 1918.*
312. Report Division Surgeon, 4th Division.*
313. Station Lists 4th Division.*
314. History 13th Field Artillery.*
315. Memo Gen. H. B. Fiske, G.H.Q.A.E.F., Feb. 17, 1919.*
316. Pershing, p. 87.

Number *Reference*

317. War Diary G-3 4th Division.* G-3 Memo No. 83, 4th Div. Mar. 19, 1919.*
318. F. O. 3, Mar. 29; F. O. 5, Apr. 9; 4th Div. 1919.*
319. Reports I. G. 4th Div. and I.G. G.H.Q. A.E.F., on investigation Y. M. C. A.**
320. Operations Order 158, 3rd Army U. S., May 11, 1919.*
321. Operations Order 205, 3rd Army U. S., May 20, 1919.*
322. History Third Army, p. 42.**
323. Operations Order 279, 3rd Army U. S., June 28, 1919.*
324. Memo 69, 4th Div. June 22, 1919.*
325. Memo 72, 4th Div. Jul. 1, 1919.*
326. Letter C. G. 3rd Army to C. G. 4th Div. Apr. 21, 1919.* G. O. 46, 3rd Army U. S., May 19, 1919.*
327. Réquin, Chapter I.
328. Réquin, p. 200 note 1.
329. Ludendorff, Vol. II, p. 405.
330. Réquin, p. 8.
331. Some Accomplishments of the S. O. S.**
332. Some Accomplishments of the S. O. S.**
333. Pershing, pp. 75-76.
334. Réquin, p. 191.
335. Pershing, p. 75.
336. Casualty Tables 4th Division.*
337. B. R. 6th Fr. Army, Jul. 18-22, 1918.* Summaries of Intelligence I A.C.U.S., Jul. 28-31, 1918.* Order of Battle G.H.Q.A.E.F.** Graphic Summaries of Intelligence, 1st Army U. S.**
338. Files of 4th Division records.*

APPENDIX A

DIVISION, BRIGADE, AND ORGANIZATION COMMANDERS

From the inception of the 4th Division to the termination of its overseas service on August 1, 1919

DIVISION COMMANDERS

Major General George H. Cameron . .	Dec. 3, 1917 to Aug. 16, 1918
*Brigadier General Benjamin A. Poore .	Aug. 16, 1918 to Aug. 27, 1918
Major General John L. Hines	Aug. 27, 1918 to Oct. 11, 1918
Major General George H. Cameron . .	Oct. 11, 1918 to Oct. 22, 1918
*Brigadier General Benjamin A. Poore .	Oct. 22, 1918 to Oct. 31, 1918
Major General Mark L. Hersey . . .	Oct. 31, 1918 to Aug. 1, 1919

*As senior brigade commander.

CHIEF OF STAFF

Colonel Christian A. Bach	Dec. 12, 1917 to Aug. 1, 1919

A. C. of S. G-1

Lieut. Col. Edward Canfield, Jr. . . .	May 17, 1918 to Apr. 16, 1919
Lieut. Col. Wm. F. Robinson, Jr. . . .	Apr. 16, 1919 to Aug. 1, 1919

A. C. of S. G-2

Captain Elbert E. Farman	Dec. 19, 1917 to Mar. 29, 1918
Captain Max S. Murray	Mar. 29, 1918 to Jun. 3, 1918
Lieut. Col. Emmertt W. Savage . . .	Jun. 3, 1918 to Jan. 24, 1919
Major T. J. Hayes	Jan. 24, 1919 to Mar. 17, 1919
Lieut. Col. Wm. H. Clendenin	Mar. 17, 1919 to Aug. 1, 1919

A. C. of S. G-3

Major Charles H. Rice	Jan. 26, 1918 to Jun. 9, 1918
Lieut. Col. Henry L. Cooper	Jun. 9, 1918 to Aug. 2, 1918
Lieut. Col. Hugh A. Parker	Aug. 14, 1918 to Nov. 4, 1918
Lieut. Col. Charles S. Caffery	Nov. 5, 1918 to Jan. 31, 1919
Lieut. Col. A. C. Tipton	Jan. 31, 1919 to Apr. 20, 1919
Lieut. Col. Clarence Lininger	May 3, 1919 to Aug. 1, 1919

7TH INFANTRY BRIGADE

*Colonel Leon S. Roudiez	Dec. 9, 1917 to Jan. 4, 1918
Brigadier General John S. Mallory . .	Jan. 5, 1918 to Apr. 2, 1918
Brigadier General B. A. Poore. . . .	Apr. 3, 1918 to Aug. 14, 1918
*Colonel Frank C. Bolles	Aug. 14, 1918 to Aug. 21, 1918
Brigadier General B. A. Poore. . . .	Aug. 21, 1918 to Oct. 21, 1918

7TH INFANTRY BRIGADE—*Continued*

*Colonel Frank C. Bolles	Oct. 21, 1918 to Oct. 31, 1918
Brigadier General B. A. Poore	Nov. 1, 1918 to May 12, 1919
*Colonel Frank C. Bolles	May 13, 1919 to May 17, 1919
Brigadier General B. A. Poore	May 18, 1919 to May 23, 1919
*Colonel Frank C. Bolles	May 24, 1919 to Jun. 14, 1919
Brigadier General B. A. Poore	Jun. 15, 1919 to Aug, 1, 1919

*As senior regimental commander.

39TH INFANTRY REGIMENT

Colonel William C. Bennett	Jun. 23, 1917 to Apr. 9, 1918
Colonel Frank C. Bolles	Apr. 9, 1918 to Sept. 28, 1918
Colonel James K. Parsons	Sept. 28, 1918 to Oct. 11, 1918
Lieut. Col. Troy H. Middleton	Oct. 11, 1918 to Oct. 14, 1918
Lieut. Col. O. P. M. Hazzard	Oct. 14, 1918 to Nov. 12, 1918
Colonel Frank C. Bolles	Nov. 12, 1918 to May 11, 1919
Lieut. Col. James M. Lockett	May 11, 1919 to May 20, 1919
Colonel Frank C. Bolles	May 20, 1919 to Aug. 1, 1919

47TH INFANTRY REGIMENT

Colonel Leon S. Roudiez	Sept. 30, 1917 to Jul. 29, 1918
Lieut. Col. Robert H. Peck	Jul. 30, 1918 to Oct. 28, 1918
Colonel Troy H. Middleton	Oct. 28, 1918 to Jul. 6, 1919
Lieut. Col. Manton C. Mitchell	Jul. 6, 1919 to Aug. 1, 1919

11TH MACHINE GUN BATTALION

Captain Harvey H. Fletcher	Dec. 8, 1917 to Mar. 25, 1918
Major William G. Murchison	Mar. 25, 1918 to Jul. 4, 1918
Major Floyd R. Waltz	Jul. 4, 1918 to Oct. 31, 1918
Major Manton S. Eddy	Oct. 31, 1918 to Mar. 5, 1919
Major Robert M. Littlejohn	Mar. 5, 1919 to May 2, 1919
Major Charles H. Karlstad	May 2, 1919 to Aug. 1, 1919

8TH INFANTRY BRIGADE

*Colonel Benjamin W. Atkinson	Nov. 23, 1917 to Jan. 13, 1918
Brigadier General Samuel W. Miller	Jan. 13, 1918 to Mar. 24, 1918
Brigadier General Frank D. Webster	Mar. 24, 1918 to Jul. 25, 1918
*Colonel Carroll F. Armistead	Jul. 25, 1918 to Jul. 27, 1918
Brigadier General Ewing E. Booth	Jul. 27, 1918 to Jan. 9, 1919
*Colonel Fred M. Wise	Jan. 9, 1919 to Feb. 7, 1919
Brigadier General Frank E. Bamford	Feb. 7, 1919 to May 29, 1919
Brigadier General Francis C. Marshall	May 29, 1919 to Aug. 1, 1919

*As senior regimental commander.

58TH INFANTRY REGIMENT

Colonel Lutz Wahl	Aug. 18, 1917 to Feb. 3, 1918
Captain George T. Everitt	Feb. 3, 1918 to Mar. 16, 1918
Colonel Carroll F. Armistead	Mar. 16, 1918 to Aug. 25, 1918
Colonel George H. McMaster	Aug. 25, 1918 to Oct. 2, 1918
Lieut. Col. Leonard T. Baker	Oct. 2, 1918 to Oct. 30, 1918
Major Gilbert R. Cook	Oct. 30, 1918 to Nov. 1, 1918
Lieut. Col. Max B. Garber	Nov. 1, 1918 to Jan. 3, 1919
Colonel Fred R. Brown	Jan. 3, 1919 to Aug. 1, 1919

59TH INFANTRY REGIMENT

Colonel Benjamin W. Atkinson	Jun. 5, 1917 to Jul. 25, 1918
Lieut. Col. Charles F. Andrews	Jul. 26, 1918 to Aug. 7, 1918
Lieut. Col. Max B. Garber	Aug. 8, 1918 to Aug. 21, 1918
Colonel George H. McMaster	Aug. 21, 1918 to Aug. 25, 1918
Lieut. Col. Max B. Garber	Aug. 25, 1918 to Sept. 3, 1918
Colonel Fred M. Wise	Sept. 4, 1918 to Jan. 30, 1919
Colonel James S. Parker	Jan. 31, 1919 to Apr. 23, 1919
Colonel Max B. Garber	Apr. 24, 1919 to Aug. 1, 1919

12TH MACHINE GUN BATTALION

1st Lieut. Edward M. Almond	Dec. 1, 1917 to Mar. 1, 1918
Captain Francis R. Fuller	Mar. 1, 1918 to Mar. 5, 1918
Captain H. J. M. Smith	Mar. 5, 1918 to Apr. 14, 1918
Major Max B. Garber	Apr. 14, 1918 to Aug. 11, 1918
Captain Benjamin J. Holt, Jr.	Aug. 11, 1918 to Aug. 15, 1918
Major John B. Duckstad	Aug. 15, 1918 to Sept. 5, 1918
Captain Vic K. Burriss	Sept. 6, 1918 to Sept. 25, 1918
Major Harvey H. Fletcher	Sept. 25, 1918 to Oct. 1, 1918
Major Edward M. Almond	Oct. 1, 1918 to Aug. 1, 1919

4TH FIELD ARTILLERY BRIGADE

Brigadier General Edwin B. Babbitt	Jan. 5, 1918 to Apr. 11, 1919
*Colonel Wright Smith	Apr. 11, 1919 to May 5, 1919
Brigadier General Augustine McIntyre	May 5, 1919 to Jun. 25, 1919
*Colonel Wright Smith	Jun. 25, 1919 to Aug. 1, 1919

*As senior regimental commander.

13TH FIELD ARTILLERY REGIMENT

Colonel Wright Smith	Jun. 1, 1917 to Aug. 1, 1919

16TH FIELD ARTILLERY REGIMENT

Colonel Henry L. Newbold	Dec. 1917 to May 8, 1918
Major Harold H. Bateman	May 8, 1918 to Jun. 7, 1918
Lieut. Col. Albert T. Bishop	Jun. 7, 1918 to Jun. 9, 1918
Colonel Henry W. Butner	Jun. 9, 1918 to Aug. 17, 1918
Major Harold H. Bateman	Aug. 17, 1918 to Aug. 27, 1918
Lieut. Col. Wm. H. Rucker	Aug. 27, 1918 to Oct. 18, 1918
Lieut. Col. Harold H. Bateman	Oct. 18, 1918 to Nov. 7, 1918
Colonel Henning F. Colley	Nov. 7, 1918 to Nov. 23, 1918
Lieut. Col. Wm. H. Rucker	Nov. 23, 1918 to Dec. 9, 1918
Colonel O. P. M. Hazzard	Dec. 9, 1918 to Jan. 1, 1916
Lieut. Col. Wm. H. Rucker	Jan. 1, 1919 to Feb. 22, 1919
Colonel O. P. M. Hazzard	Feb. 22, 1919 to Mar. 2, 1919
Lieut. Col. Wm. H. Rucker	Mar. 2, 1919 to Aug. 1, 1919

77TH FIELD ARTILLERY REGIMENT

Colonel Guy H. Preston	May 23, 1917 to May 1, 1918
Major Frederick E. Shnyder	May 1, 1918 to May 8, 1918
Lieut. Col. A. F. Commiskey	May 8, 1918 to Aug. 25, 1918
Colonel Creed F. Cox	Aug. 25, 1918 to Oct. 15, 1918
Colonel A. F. Commiskey	Oct. 15, 1918 to Jan. 6, 1919
Lieut. Col. Kenneth P. Lord	Jan. 6, 1919 to Mar. 16, 1919
Colonel A. F. Commiskey	Mar. 16, 1919 to Aug. 1, 1919

4TH TRENCH MORTAR BATTERY

Captain Robert H. Ennis	Dec. 1, 1917 to Oct. 15, 1918
Captain E. Clinton Le Gette	Oct. 15, 1918 to Aug. 1, 1919

4TH ENGINEER REGIMENT

Colonel Jay J. Morrow	Aug. 21, 1917 to Jun. 18, 1918
Colonel Albert H. Acher	Jun. 19, 1918 to Oct. 9, 1918
Colonel Raymond A. Wheeler	Oct. 10, 1918 to Jan. 14, 1919
Colonel Elliott J. Dent	Jan. 15, 1919 to Feb. 23, 1919
Colonel Raymond A. Wheeler	Feb. 24, 1919 to Aug. 1, 1919

8TH FIELD SIGNAL BATTALION

Major Girard L. McEntee	Aug. 24, 1917 to Feb. 18, 1918
Major Resolve P. Palmer	Feb. 18, 1918 to Aug. 8, 1918
Major John H. Stutesman	Aug. 8, 1918 to Aug. 24, 1918
Major Alexander E. Whitworth	Sept. 3, 1918 to Aug. 1, 1919

10TH MACHINE GUN BATTALION

Major Albert Hardman	Nov. 22, 1917 to Feb. 27, 1918
Major Franklin C. Sibert	Feb. 27, 1918 to Aug. 21, 1918
Major Francis R. Fuller	Aug. 21, 1918 to May 8, 1919
Major Frank A. Jones	May 8, 1919 to May 15, 1919
Major Francis R. Fuller	May 15, 1919t o Aug. 1, 1919

4TH TRAIN HEADQUARTERS

Colonel James J. Hornbrook	Dec. 27, 1917 to Jul. 20, 1918
Major James M. Lockett	Jul. 20, 1918 to Jul. 27, 1918
Major H. J. M. Smith	Jul. 28, 1918 to Sept. 1, 1918
Lieut. Col. Arthur Hixon	Sept. 1, 1918 to Nov. 1, 1918
Major H. J. M. Smith	Nov. 1, 1918 to Dec. 29, 1918
Colonel William B. Cowin	Dec. 29, 1918 to Jun. 12, 1919
Major H. J. M. Smith	Jun. 13, 1919 to Jun. 20, 1919
Captain Perley B. Hartwell	Jun. 20, 1919 to Jul. 12, 1919
Major H. J. M. Smith	Jul. 13, 1919 to Aug. 1, 1919

4TH MILITARY POLICE COMPANY

1st Lieut. Okla E. Smith	Dec. 10, 1918 to Jan. 1, 1919
1st Lieut. William F. O'Brien	Jan. 1, 1919 to Feb. 1, 1919
1st Lieut. Alvin R. Baker	Feb. 1, 1919 to Mar. 1, 1919
Captain Harrison S. Beecher	Mar. 1, 1919 to Aug. 1, 1919

4TH SANITARY TRAIN

Lieut. Col. Wm. E. Wilmerding	Mar. 5, 1918 to Aug. 1, 1919

4TH AMMUNITION TRAIN

Lieut. Col. Alden Trotter	Jan. 1, 1918 to Jun. 30, 1918
Major Ben. F. Ristine	Jul. 1, 1918 to Aug. 1, 1918
Captain Milton L. Knowlton	Aug. 2, 1918 to Aug. 22, 1918
Major Donald Markle	Aug. 23, 1918 to Oct. 11, 1918
Lieut. Col. Charles I. Faddis	Oct. 12, 1918 to Aug. 1, 1919

4TH SUPPLY TRAIN

Captain Jack L. Meyer	Dec. 10, 1917 to Aug. 23, 1918
Major Milton L. Knowlton	Aug. 23, 1918 to Dec. 3, 1918
Captain Donald Clement	Dec. 3, 1918 to Jan. 13, 1919
Major Carl H. Seale	Jan. 13, 1919 to Feb. 9, 1919
Captain Charles Montooth	Feb. 9, 1919 to Feb. 15, 1919
Captain Samuel McClellan	Feb. 15, 1919 to Aug. 1, 1919

4TH ENGINEER TRAIN

1st Lieut. Clifford E. Chase	Dec. 1, 1917 to Mar. 18, 1918
1st Lieut. Hardy J. Story	Mar. 18, 1918 to Apr. 19, 1919
1st Lieut. Harold J. Carniglia	Apr. 19, 1919 to Aug. 1, 1919

DIVISION HEADQUARTERS TROOP

Captain Paul R. Frank	Dec. 18, 1917 to Feb. 27, 1918
1st Lieut. A. B. C. Smith	Feb. 27, 1918 to Mar. 27, 1918
Captain H. J. M. Smith	Mar. 27, 1918 to Jul. 26, 1918
Captain Leo McCusker	Jul. 26, 1918 to Dec. 12, 1918
Captain Richard T. Schlossberg	Dec. 12, 1918 to May 12, 1919
1st Lieut. Thomas Mather	May 12, 1919 to Jun. 4, 1919
Captain W. F. Stimax	Jun. 4, 1919 to Aug. 1, 1919

DIVISION QUARTERMASTER

Lieut. Col. Edward Canfield, Jr.	Dec. 15 1917 to Apr. 18, 1918
Lieut. Col. Frederick Gilbreath	Jun. 23, 1918 to Oct. 29, 1918
Major William P. Murphy	Oct. 21, 1918 to Apr. 22, 1919
Major Paul W. Butin	May 26, 1919 to Jun. 27, 1919
Major James L. Houston	May 23, 1919 to Aug. 1, 1919

DIVISION SIGNAL OFFICER

Lieut. Col. George E. Kumpe	Dec. 17, 1917 to Mar. 3, 1918
Major Girard L. McEntee	Mar. 3, 1918 to Jun. 13, 1918
Lieut. Col. Thomas L. Sherburne	Jun. 13, 1918 to Aug. 1, 1919

DIVISION ADJUTANT

Major James G. McIlroy	Dec. 19, 1917 to Apr. 2, 1918
Major Jesse D. Elliott	Apr. 2, 1918 to Jun. 18, 1918
Major Howard J. Savage	Jun. 19, 1918 to Oct. 6, 1918
Major Lon S. Haymes	Oct. 6, 1918 to Jan. 20, 1919
Lieut. Col. Max B. Garber	Jan. 20, 1919 to Apr. 4, 1919
Captain Charles L. Bolte	Apr. 4, 1919 to Jun. 4, 1919
Major J. Shapter Caldwell	Jun. 4, 1919 to Aug. 1, 1919

DIVISION INSPECTOR

Lieut. Col. Marcellus G. Spinks	Dec. 23, 1917 to Mar. 23, 1918
Lieut. Col. James M. Petty	Mar. 24, 1918 to Jun. 18, 1918
Lieut. Col. Arthur G. Hixon	Sept. 12, 1918 to Oct. 31, 1918
Major Walter P. Tyler	Nov. 1, 1918 to Mar. 1, 1919
Lieut. Col. Vincent M. Elmore	Mar. 1, 1919 to May 12, 1919
Lieut. Col. Donald D. Hay	May 12, 1919 to Aug. 1, 1919

DIVISION JUDGE ADVOCATE

Lieut. Col. Charles C. Teare	Feb. 2, 1918 to Mar. 25, 1919
Lieut. Col. Henry W. Stiness	Mar. 21, 1919 to Aug. 1, 1919

DIVISION ORDNANCE OFFICER

Major W. P. Baker	Jan. 1, 1918 to Mar. 1, 1918
Major Lyman N. Hine	Mar. 1, 1918 to Nov. 16, 1918
Major Louis Waefelaer	Nov. 16, 1918 to Apr. 6, 1919
Major Edward G. Flaig	Apr. 15, 1919 to Aug. 1, 1919

DIVISION ENGINEER

Colonel Jay J. Morrow	Dec. 30, 1917 to Jun. 18, 1918
Colonel Albert H. Acher	Jun. 18, 1918 to Oct. 9, 1918
Colonel Raymond A. Wheeler	Oct. 9, 1918 to Jan. 14, 1919
Colonel Elliott J. Dent	Jan. 14, 1919 to Feb. 23, 1919
Colonel Raymond A. Wheeler	Feb. 23, 1919 to Aug. 1, 1919

DIVISION SURGEON

Colonel Robert H. Carswell	Dec. 20, 1917 to Oct. 10, 1918
Colonel Paul Waterman	Nov. 15, 1918 to Aug. 1, 1919

DIVISION DENTAL SURGEON

Major Harry E. Smalley	Jan. 1, 1918 to Aug. 1, 1919

DIVISION VETERINARIAN

Major Robert McAuslin	Mar. 16, 1918 to Aug. 1, 1919

APPENDIX B

AWARDS OF DECORATIONS

DISTINGUISHED SERVICE CROSS

August Aibner	Sergeant	Co. M. 58th Infantry.
Samuel P. Adkisson	1st Lieutenant	39th Infantry.
James P. Behan	Private	Btry. A, 13th F. A.
Harold V. Beal	Corporal	Btry. A, 13th F. A.
George N. Bringham	Corporal	Co. I, 47th Infantry.
Fred E. Billman	Priv., Med. Corps	47th Infantry.
Gustav J. Braun	Captain	47th Infantry.
Frank C. Bolles	Colonel	39th Infantry.
William B. Beach	Sergeant	Co. C., 4th Engineers.
Harold Bedolfe	Sergeant, 1st Class	Co. B., 4th Engineers.
Lawrence Boup	Private	Co. A,. 59th Infantry.
Raymond Buma	Corporal	M. G. Co., 39th Infantry.
Joseph Bassi	Private	Co. I., 59th Infantry.
George Brown	Private	Hq. Co., 59th Infantry.
Olaf Brekke	Private	Co. C., 58th Infantry.
Tom F. Barto	Corporal	Co. D., 4th Engineers.
Guy L. Boardman	Private	Co. A., 59th Infantry.
Elwyn L. Berwick	Corporal	Btry. C., 13th F. A.
Homer J. Bleau	Sergeant	Co. A., 59th Infantry.
Walter Currie	Private	Co. A., 59th Infantry.
Frank B. Cook	2nd. Lieutenant	4th Engineers.
Charles F. Carbaugh	Sergeant	Co. F., 47th Infantry.
Marion H. Cardwell	Captain	58th Infantry.
James B. Carpenter	Private	Co. H., 47th Infantry.
Claude F. Cherry	Sergeant	Co. B., 11th Mach. Gun Bn.
Willis M. Campbell	Sergeant	Co. B., 59th Infantry.
Earl W. Curtis	Private	Co. B., 59th Infantry.
Fred E. Cullen	Corporal	Co. D., 12th Mach. Gun Bn.
William J. Cahill	Private	Co. D., 59th Infantry.
Joseph H. Carvo	Private	Co. I., 47th Infantry.
James Conway	Private	Co. C., 58th Infantry.
Alberis Callewaert	Private	Hq. Co., 58th Infantry.
Arthur I. Clark	Sergeant	Co. C., 39th Infantry.
Charles T. Dunbar	Corporal	Co. F., 4th Engineers.
Leslie C. Dill	Private	Co. B., 47th Infantry.
Charles E. DeLew	Captain	4th Engineers.
Walter Detrow	Sergeant	Co. B., 47th Infantry.
Albert Dietz	Sergeant	Co. I., 59th Infantry.
Clinton Day	Private	Co. C., 58th Infantry.
James Dilworth	Private	Co. A., 59th Infantry.
Frank J. Downs	Sergeant	Co. B., 58th Infantry.
Charles B. Duncan	Captain	77th Field Artillery.
Harold W. Enright	Private	Co. I., 47th Infantry.

PETER W. EBBERT	1st Lieutenant	58th Infantry.
EMIL J. EKLUND	2nd Lieutenant	58th Infantry.
ETIENNE ESCUDIER	1st Lieut.	79th Infantry (French).
CHARLES H. EVANS	Private	Co. B., 39th Infantry.
CHARLES E. EPLER	Private	Co. A., 59th Infantry.
PIETRO FORMICA	Private	Co. A., 59th Infantry.
EARL R. FRETZ	Sergeant, 1st Class	Co. B., 4th Engineers.
ERNEST FOSNAS	Corporal	Co. A., 59th Infantry.
DAVIS V. FORD	Corporal	Co. C., 4th Engineers.
ISAAC GATAINO	Corporal	Co. I., 47th Infantry.
JAMES P. GROWDEN	Captain	4th Engineers.
ARTHUR J. GOETSCH	Sergeant	Co. D., 4th Engineers.
CHARLES J. GLEN	Private	Co. D., 4th Engineers.
GLEN M. GROVE	Sergeant	Co. D., 11th Mach. Gun. Bn.
HENRY J. GARST	Corporal	Co. H., 47th Infantry.
FRANK B. GRESHAM	Sergeant	Co. G., 39th Infantry.
LEONARD E. GUY	Sergeant	Co. C., 58th Infantry.
CORNELIUS T. GLYNN	Corporal	Co. K., 59th Infantry.
RUBIN L. GEORGE	Corporal	Co. A., 59th Infantry.
ARTHUR M. HAMILTON	Sergeant	Co. E., 58th Infantry.
ROY HARRIS	Private	Co. F. 4th Engineers.
HAROLD DELLA MONTE HALL	Private	Co. A., 59th Infantry.
MATHIAS W. HANEY	Captain	39th Infantry.
SAMUEL H. HANNA	Sergeant	Co. B., 12th Mach. Gun Bn.
EDWARD G. HASKEW	Sergeant	Ambulance Co., No. 33.
SAMUEL H. HOUSTON	Major	58th Infantry.
WILLIAM HERREN	1st Sergeant	Mach. Gun Co., 58th Inf.
WILLIAM H. HAMMOND	1st Lieutenant	39th Infantry.
WILLIAM B. HOOK	Sergeant	Co. B. 4th Engineers.
MORRISON HAYES	Corporal	Co. D., 12th Mach. Gun Bn.
HENRY HOWARD	Sergeant	Co. A., 39th Infantry.
WILLIAM HUNTER	Private	Co. D., 58th Infantry.
ALBERT L. A. IHRKE	Private	Co. B., 47th Infantry.
FRANK JAWORSKI	Corporal	Co. F., 4th Engineers.
HOMER J. JARVIS	1st Lieutenant	11th Mach. Gun Bn.
REUBEN L. JOHNSON	Private	Co. B., 47th Infantry.
MAX S. KOSS	Private	Co. K., 47th Infantry.
CHARLES KELLY	Sergeant	Co. C., 12th Mach. Gun Bn
ORVAL KLINE	2nd Lieutenant	11th Mach. Gun Bn.
JACOB KREIS	Private	Co. I., 47th Infantry.
THOMAS W. KEARNS	1st Lieutenant	47th Infantry.
PAUL VON KREBS	1st Sergeant	Co. M., 47th Infantry.
LATTHER E. LINDAHL	Sergeant	Co. I., 47th Infantry.
FRED A. LIEUALLEN	Captain, M. C.	47th Infantry.
JOHN LEGNOSKY	1st Sergeant	Co. L., 58th Infantry.
JOE LIMON	Private	Co. M., 47th Infantry.
EDWARD K. LAWLESS	Sergeant Major	39th Infantry.
CLYDE H. LINDSAY	Private	Co. A., 59th Infantry.
JOSEPH LINGAWSKI	Private	Co. L., 59th Infantry.
CHARLES J. LOVE	Private	Co. K., 59th Infantry.
EARL M. MCKINLEY	1st Lieutenant	11th Mach. Gun Bn.
JOHN J. MADORE	Private	Co. C., 47th Infantry.
MURRAY K. MACKALL	Captain	4th Engineers.
ARNO S. MCCLELLAN	1st Lieutenant	47th Infantry.
GEORGE C. MCCELVY	1st Lieutenant	47th Infantry.
RICHARD MARCELLA	Private	Mach. Gun Co., 47th Inf.
ROBERT H. MURDOCK	1st. Lieut. M. C.	47th Infantry.

Howard C. McCall	Captain	59th Infantry.
Hans E. Morgan	Private	Co. B., 47th Infantry.
Roy E. Mathews	Private	Co. E., 58th Infantry.
James Manning	Corporal	Co. C., 4th Engineers.
Forrest L. Martz	Private, 1st Class	Co. E., 58th Infantry.
Nelson Martin	Corporal	Co. H., 58th Infantry.
Manton C. Mitchell	Lieut. Colonel	39th Infantry.
Arnot L. McArthy	Private	Co. D., 59th Infantry.
William H. McGinnis	Corporal	Co. D., 12th Mach. Gun. Bn.
Arthur M. Miller	Private, 1st Class	Co. B., 47th Infantry.
Joseph McCollum	Wagoner	Co. B., 10th Mach. Gun Bn.
Robert G. Marshall	Captain	58th Infantry.
Jean L. Meurisse	Captain	27th Infantry (French)
Robert A. Madden	Private	Co. I., 47th Infantry.
Cecil N. Martin	Private	Co. I., 47th Infantry.
Edward McAndrews	Sergeant	Co. B., 12th Mach. Gun Bn.
Francis K. Newcomer	Lieut. Colonel	4th Engineers.
Robert W. Norton	Captain	39th Infantry.
John H. Norton	Captain	47th Infantry.
John W. Norton	Sergeant	Co. I., 39th Infantry.
Morton Osborn	Sergeant	Co. H., 47th Infantry.
Cornelius J. O'Brien	Corporal	Co. D., 4th Engineers.
Richard G. Plumley	Captain	39th Infantry.
James J. Pirtle	1st Lieutenant	59th Infantry.
Arthur Paulson	Sergeant	Co. A., 59th Infantry.
Benjamin A. Poore	Brig. General	7th Infantry Brigade.
John H. Pratt, Jr.	2nd Lieutenant	47th Infantry.
Paul J. Pappas	Private	Co. M., 39th Infantry.
Donald H. Pegg	Private	12th Mach. Gun Bn.
J. K. Parsons	Colonel	39th Infantry.
Ernest R. Potter	1st Sergeant	Co. D., 39th Infantry.
Oscar W. Peterson	Sergeant	Co. A., 59th Infantry.
Carl Rasmussen	Private	Co. B., 39th Infantry.
James V. Rice	Private, 1st Class	Co. C., 8th Field Sig. Bn.
Leo D. Roberts	Sergeant	Co. A., 11th Mach. Gun Bn.
Raymond D. Robertson	Sergeant	Co. F., 4th Engineers.
Lowell H. Riley	2nd Lieutenant	58th Infantry.
Edwin D. Ritchie	Private	Co. M., 47th Infantry.
Stephen Riggio	Private	Co. K., 39th Infantry.
Lee M. Ray	Sergeant Major	39th Infantry.
Charles C. Rismiller	Private	Med. Detach., 4th Engineers.
Fred N. Rapp	Corporal	Mach. Gun Co., 59th Inf.
James Roberts	Sergeant	Co. K., 39th Infantry.
Raymond R. Smith	Corporal	Co. C. 11th Mach. Gun Bn.
Rutherford H. Spessard	Major	58th Infantry
Otto A. Schwanke	Private, 1st Class	Co. B., 47th Infantry.
Albert B. Simpson	1st Lieutenant	11th Mach. Gun Bn.
Louis Scionti	Sergeant	Co. F., 47th Infantry.
William A. Shea	Sergeant	Mach. Gun Co., 39th Inf.
Ralph Slate	Captain	39th Infantry.
William Shemin	Sergeant	Co. G., 47th Infantry.
Anthony F. Shedlock	Sergeant	Co. H., 58th Infantry.
David Stearns	Corporal	Co. E., 4th Engineers.
Wallis H. Sturtevant	Corporal	Co. D., 12th Mach. Gun Bn.
Joe Smith	Private	Co. C., 39th Infantry
Ford D. Smith	Corporal	Co. D., 4th Engineers.
Henry G. Turdury	Private	Co. C., 12th Mach. Gun Bn.

Henry D. Turner	Sergeant	Co. B. 10th Mach. Gun Bn.
Clark O. Tayntor	1st Lieutenant	47th Infantry.
John C. Vann	2nd Lieutenant	47th Infantry.
Thomas Vanderveen	Private	Co. C., 11th Mach. Gun Bn.
Arthur H. Warfield	Sergeant	Co. B., 47th Infantry.
Stephen J. Weston	Sergeant	Co. I., 47th Infantry.
Emmet W. Waltman	Corporal	Co. F., 4th Engineers.
William J. Wood	Sergeant	Co. D., 4th Engineers.
Columbus Whipple	Private	Co. H., 47th Infantry.
Robert L. Worden	Wagoner	Ambulance Co. No. 21.
John S. Weimer	Private	Co. M., 47th Infantry.
Merle R. Windsor	Corporal	Co. D., 12th Mach. Gun Bn.
Harrison B. Webster	Major, Med. Corps	47th Infantry.
Carroll B. West	Sergeant	Co. B., 12th Mach. Gun Bn.
Gilbert W. Wilcox	Private	Hq. Detach., 4th Engrs.
Joseph Waskiewic	Private	Co. A., 11th Mach. Gun Bn.

DISTINGUISHED SERVICE MEDAL

Albert H. Acher	Colonel	4th Engineers.
Edwin B. Babbitt	Brig. General	4th Field Arty. Brig.
Christian A. Bach	Colonel	Chief of Staff.
Frank C. Bolles	Colonel	39th Infantry.
Ewing E. Booth	Brig. General	7th Infantry Brig.
Edward Canfield	Colonel	A. C. of S., G-1
Archibald A. Commisky	Colonel	77th Field Arty.
Max B. Garber	Colonel	59th Infantry.
Mark L. Hersey	Major General	U. S. A.
John L. Hines	Major General	U. S. A.
Troy H. Middleton	Colonel	47th Infantry.
Benjamin A. Poore	Brig. General	7th Infantry Brig.
William H. Rucker	Colonel	16th Field Arty.
Wright Smith	Colonel	13th Field Arty.
Fred M. Wise	Colonel, U. S. M. C.	59th Infantry.
Raymond A. Wheeler	Colonel	4th Engineers

FRENCH

COMMANDEUR DE LA LÉGION D'HONNEUR

George H. Cameron	Major General	U. S. A.
John L. Hines	Major General	U. S. A.

OFFICIER DE LA LÉGION D'HONNEUR

Christian A. Bach	Colonel	Chief of Staff.
Frank C. Bolles	Colonel	39th Infantry.
Ewing E. Booth	Brig. General	8th Infantry Brigade.
Mark L. Hersey	Major General	U. S. A.
Benjamin A. Poore	Brig. General	7th Infantry Brigade.

CHEVALIER DE LA LÉGION D'HONNEUR

Mathias W. Haney	Captain	39th Infantry.
Robert W. Norton	Captain	39th Infantry.
William Hammond	1st Lieutenant	39th Infantry.
Orval Kline	2nd Lieutenant	11th Mach. Gun Bn.

MÉDAILLE MILITAIRE

William Beach	Sergeant	4th Engineers
Walter H. Detrow	Private	47th Infantry.
Clyde H. Lindsay	Private	59th Infantry.

CROIX DE GUERRE

Christian A. Bach	Colonel	Chief of Staff
William Beach	Sergeant, 1st Class	4th Engineers.
William Beach	Sergeant, 1st Class	4th Engineers. (2nd Dec.)
John P. Bells	Captain	59th Infantry.
Arthur Bickstein	Bugler	39th Infantry.
Howard W. Bitner	Corporal	47th Infantry.
Leslie H. Blank	1st Lieutenant	59th Infantry.
Elwyn L. Berwick	Corporal	13th Field Arty.
Max F. Berngardt	Private	4th French Mor. Bty.
Claude Burgess	Private	16th Field Arty.
Frank C. Bolles	Colonel	39th Infantry.
Ewing E. Booth	Brig. General	8th Inf. Brigade.
Lawrence S. Boup	Private	59th Infantry.
William C. Boynton	Sergeant	16th Field Arty.
Henry Brewer	Sergeant	16th Field Arty.
Henry Brosemer	Color Sergeant	58th Infantry.
William J. Cahill	Private	59th Infantry.
Kenneth E. Caldwell	1st Lieutenant	59th Infantry.
Marion H. Cardwell	Captain	58th Infantry.
Charles C. Champowich	Sergeant	16th Field Arty.
Joseph Clement	Major	39th Infantry.
Morris N. Cohen	Private	4th Trench Mor. Bty.
Arthur Clark	Sergeant	39th Infantry.
Harry A. Conway	Private, 1st Class	58th Infantry.
Gilbert R. Cook	Lt. Colonel	58th Infantry.
Frank B. Cook	2nd Lieutenant	4th Engineers.
Earl W. Curtis	Private	59th Infantry.
Phillipp J. Davidson	1st Lieutenant	39th Infantry.
Claude Day	Sergeant	16th Field Arty.
John Dean	Sergeant	39th Infantry
Walter H. Detrow	Private	47th Infantry.
Vernon D. Elliott	Sergeant	16th Field Arty.
William B. Ely	Private, 1st Class	16th Field Arty.
Ben F. Ellis	Sergeant	4th Trench Mor. Bty.
Albert W. Emmens	2nd Lieutenant	39th Infantry.
Charles E. Enler	Private	59th Infantry.
Robert H. Ennis	Captain	4th Trench Mor. Bty.
Harold W. Enright	Private	47th Infantry.
Charles H. Evans	Sergeant	39th Infantry.
Thomas H. Evans	Sergeant	39th Infantry.
Harold E. Fife	1st Lieutenant	11th Mach. Gun Bn.
Simon Finn	Sergeant	16th Field Arty.
Hugh J. Fitzsimmons	Private	39th Infantry.
Isaac Gataino	Corporal	47th Infantry.
Robert George	Private	4th Trench Mor. Bty.
Cornelius Glynn	Corporal	59th Infantry.
Arthur J. Goetsch	Sergeant	4th Engineers.
Archibald R. Gordon	2nd Lieutenant	39th Infantry.
David S. Grant	2nd Lieutenant	39th Infantry.

Frank B. Gresham	Sergeant	39th Infantry.
Glenn M. Grove	Sergeant	11th Mach Gun. Bn.
James P. Growden	Captain.	4th Engineers.
Henry Hahn	Private	58th Infantry.
William Hammond.	1st Lieutenant.	39th Infantry.
Mathias W. Haney	Captain.	39th Infantry.
Roy Harris.	Private	4th Engineers.
Harry Hartman	Private	16th Field Arty.
Daniel R. Healy	Sergeant	39th Infantry.
William Herren	1st Sergeant	58th Infantry.
Mark L. Hersey	Major General.	U. S. A.
Richard M. Hicks.	Private	16th Field Arty.
Llewellyn J. Hartman	Sergeant	Co. A., 58th Infantry.
Samuel H. Houston	Major	58th Infantry.
William B. Hook	Sergeant	4th Engineers.
John L. Hines	Major General.	U. S. A.
Harry T. Hoffman	Sergeant	59th Infantry.
Henry Howard	Sergeant	39th Infantry.
Fred Hubbell	Private	58th Infantry.
Theodore Jacquet	Private	1st Bn. 58th Inf.
Harry C. Jones	Sergeant	59th Infantry.
Peter P. Jones.	Private, 1st Class	Med. Detach. 39th Inf.
Christian P. Jorgenson	Sergeant	16th Field Arty.
Orval Kline	2nd Lieutenant	11th Mach. Gun Bn.
Max S. Koss	Private	47th Infantry.
Robert Kostelak	Sergeant	39th Infantry.
Ralph E. Ladue	2nd Lieutenant	11th Mach. Gun Bn.
Earl S. Leaman	Sergeant	16th Field Arty.
Clyde H. Lindsay	Private	59th Infantry.
Clyde H. Lindsay	Private	59th Infantry (2nd Dec.).
Claudius L. Lloyd	Captain.	58th Infantry.
Siberius Lonnohan	Corporal	16th Field Arty.
Frank de Luca.	Corporal	39th Infantry.
Walter T. Mann	Corporal	39th Infantry.
Richard Marcella	Bugler	47th Infantry.
Joseph C. Martin	Bugler	58th Infantry.
Roy E. Mathews	Private	58th Infantry.
Arnot L. McArthey	Private	59th Infantry.
George C. McCelvy	1st Lieutenant.	47th Infantry.
Joseph McCuen	Private	59th Infantry.
James J. McGuinness.	Private	12th Mach. Gun Bn.
Robert F. Meyer	Sergeant	16th Field Arty.
Byron T. Miller	1st Lieutenant.	16th Field Artillery.
Geoffrey W. Miller	Private	39th Infantry.
Manton C. Mitchell.	Lieut. Colonel	39th Infantry.
Harry A. Musham.	Major	59th Infantry.
James Newberry	Sergeant	59th Infantry.
John W. Norton	Sergeant	39th Infantry.
Robert W. Norton	Captain	39th Infantry
George F. Pearsall	Private	58th Infantry.
Robert H. Peck	Colonel	39th Infantry.
Benjamin A. Poore	Brig. General	7th Inf. Brigade.
Clark A. Putnam	1st Lieutenant.	59th Infantry.
Paul J. Pappas	Private	39th Infantry.
Emmett M. Ratliff	Major	59th Infantry.
Mark Reid	Corporal	39th Infantry.
Bernard J. Remas.	Sergeant	4th Trench Mor. Bty.

Stephano Riggio	Private	39th Infantry.
Eugene S. Ripley	1st Lieutenant	16th Field Artillery.
Edward D. Ritchie	Private	47th Infantry.
William H. Rucker	Colonel	16th Field Arty.
Otto A. Schwanke	Private	47th Infantry.
Louis Scionti	Sergeant	47th Infantry.
Thomas L. Sherburne	Lieut. Colonel	Div. Sig. Officer.
Ralph Slate	Captain	39th Infantry.
Ralph Slate	Major	39th Infantry (2nd. Dec.)
Bert R. Smith	Sergeant	16th Field Arty.
James Smith	Sergeant	16th Field Arty.
Raymond R. Smith	Corporal	11th Mach. Gun Bn.
Eugene G. Snedaker	Captain	4th Engineers.
Rutherford H. Spessard	Major	58th Infantry.
Paul Strickland	2nd Lieutenant	39th Infantry.
Joseph Tamisier	Interpreter	16th Field Arty.
Henry Terrell, Jr.	Major	39th Infantry.
George Tippery	Private	16th Field Arty.
Horace G. Toole	Sergeant	16th Field Arty.
Henry G. Turdury	Private	12th Mach. Gun Bn.
John C. Vann	2nd Lieutenant	47th Infantry.
George Van Santwood	Sergeant	39th Infantry.
Sylvester W. Wadden	2nd Lieutenant	4th Trench Mor. Bty.
Emmett W. Waltman	Corporal	4th Engineers.
Arthur H. Warfiels	Sergeant	47th Infantry.
Stephen J. Weston	Sergeant	47th Infantry.
Merle R. Windsor	Corporal	12th Machine Gun Bn.
Robert F. Winters	Sergeant	39th Infantry.
William J. Wood	Sergeant	4th Engineers.
Everett D. Woods	1st Lieutenant	12th Mach. Gun Bn.
Joseph Wypychoski	Private	4th Trench Mor. Bty.
Nicholas Yourkewich	Private, 1st Class	12th Mach. Gun Bn.
James H. Woodson	Sergeant	16th Field Arty.

BELGIAN

CHEVALIER DE L'ORDRE LEOPOLD II

Raymond D. Robertson	Sergeant	Co. F., 4th Engineers.
Murray F. MacKall	Captain	4th Engineers.

CHEVALIER DE L'ORDRE DE LA COURONNE

Charles E. DeLew	1st Lieutenant	4th Engineers.

DÉCORATION MILITAIRE

Charles T. Dunbar	Corporal	Co. F., 4th Engineers.
Frank Jaworski	Corporal	Co. F., 4th Engineers.
Emmet W. Waltman	Corporal	Co. F., 4th Engineers.

CROIX DE GUERRE

Roy Harris	Private	Co. F., 4th Engineers.

ITALIAN

ORDER OF DEI SANTE MAURIXIO E LAZZARO (UFFICIALE)

Louis A. Falligant	Major	Hq. 4th Division.

CROCE DI GUERRA

Albert Dietz	Sergeant	Co. I., 59th Infantry.
Charles T. Dunbar	Corporal	Co. F., 4th Engineers.
Harold W. Enright	Private	Co. I., 47th Infantry.
Allen Fletcher	Major	59th Infantry.
Davis V. Ford	Private	Co. C., 4th Engineers.
Leonard E. Guy	Sergeant	Co. C., 58th Infantry.
Henry J. Garst	Corporal	Co. H., 47th Infantry.
Albert L. A. Ihrke	Private	Co. B., 47th Infantry.
Frank Jaworski	Corporal	Co. F., 4th Engineers.
Luther E. Lindahl	Sergeant	Co. I., 47th Infantry.
Hans E. Morgan	Private	Co. B., 47th Infantry.
Forrest L. Martz	Private	Co. C., 12th Mach. Gun Bn.
Benjamin A. Poore	Brig. General	7th Inf. Brigade.
Paul J. Pappas	Private	Co. M., 39th Infantry.
Edwin D. Ritchie	Private	Co. M., 47th Infantry.
William A. Shea	Sergeant	Mach. Gun Co., 39th Inf.
Joe Smith	Private	Co. C., 39th Infantry.
John S. Weimer	Private	Co. M., 47th Infantry.

APPENDIX C

CITATIONS AND COMMENDATIONS

7° Corps d'Armée
État-Major
3me Bureau

P. C., le 21 Juillet 1918
Heure de la signature: 21.15

ORDRE GÉNÉRAL No. 407

Le Général Commandant le C. A. transmet aux État-Majors, troupes et Services sous ses ordres les félicitations du Président du Conseil Ministre de la Guerre, du Général Commandant en Chef les Armées Alliées et du Général Commandant les Armées du Nord et du Nord-Est à l'occasion des succès remportes depuis le 18 Juillet.

Il y joint ses félicitations personnelles avec l'assurance que ces succès se poursuivront jusqu'à la déroute complète de l'adversaire.

Le Général MASSENET Commandant le 7° C. A.
Signé: MASSENET.

P. A.: le Chef d'État-Major
HOURRUAN
Destinataires:
4me Division U. S.

2° Corps d'Armée
État-Major
1er Bureau
No. 2972 C.

Au Q. G. le 23 Juillet 1918.

ORDRE No. 262

La 7° Brigade Américaine cesse de faire partie du 2° Corps d'Armée.

Le Général tient à lui exprimer ses remerciements pour l'aide précieuse qu'elle lui a apportée et à lui addresser au nom de tous ses meilleurs vœux pour la poursuite de sa glorieuse carrière.

Le Général et les troupes du 2° Corps n'oublieront pas le bel entrain et la bravoure de leurs camarades américains au cours de la Bataille; ils saluent les officiers et les soldats tombés au Buisson de Cresnes et à Noroy.

Les fatigues et les dangers courus en commun ont fait de nos alliées d'hier des compagnons d'armes et le souvenir de la Brigade Poore doit rester au 2° Corps.

Le Général Commandant le 2° Corps d'Armée.
PHILIPOT.

P.A.
Le Chef d'État-Major
Rousseau

164th Inf. Div.,
General Staff.
3rd Bureau
No. 7520-3. HEADQUARTERS, July 28, 1918.

OPERATIONS OF THE 164TH INF. DIV.

From the 17th to 27th of July, 1918.

Map 1-80000.

(EXTRACT)

JULY 18th

At 6:00 A. M. the Dussauge Group has attained its first objective. The American Battalion* attached to this Group in a splendid dash took the village of Chevillon.

GENERAL GAUCHER,
Commanding 164th Division.

HQ. July 30, 1918.

7th Army Corps
General Staff
1st Bureau
No. 11372/P

LETTER OF CONGRATULATION

General MASSENET, commanding the 7th Army Corps, sends his congratulations to the officers and men of the 8th Infantry Brigade, Engineers, and Signal Corps of the 4th American Infantry Division, who by their bravery and spirit of duty, contributed in a large measure to the success of the offensive operations during the period from the 18th to the 28th of July. These young troops have proved themselves equal to the best and such a début promises them new successes in the near future.

The Commanding General of the 7th Army Corps likewise congratulates the personnel of the Motor Supply Train of the 4th U. S. Infantry Division, who night and day, with untiring good will, rendered valuable assistance to the French troops during the battle.

GENERAL MASSENET.

G. A. R.
État-Major HQ. August 4th, 1918.
3rd Bureau
No. 4.190.

GENERAL ORDER

The second battle of the MARNE ends, like the first, in a victory. The CHATEAU-THIERRY "pocket" exists no more.

The VIth and Xth Armies, also the allied troops fighting at their side, have taken a glorious part in that battle.

Their swift and powerful entrance into the battle, on July 18, broke up entirely, as a first result, the offensive of the enemy, and compelled him to retreat across the MARNE.

Since that time, owing to our strong attacks, chased night and day without stop, he has been forced to fall back across the VESLE, leaving in our hands 25,000 prisoners, 600 guns, 4,000 machine guns, and 500 minenwerfer.

*(Note: The 2nd Battalion, 58th Infantry, is the American Battalion referred to in this order).

We owe these results to the energy and skill of the Chiefs, and to the extraordinary valor of the troops, who, for more than 15 days, had to march and fight without rest.

I am sending to the Commanders of the Xth and VIth Armies, Generals Mangin and DeGoutte, to the Commanders of the British and American units, and to all the troops, the token of my admiration for their knowledge, their courage, and their heroic tenacity.

They may all be proud of the work accomplished! It is great, because it has greatly contributed to secure the final victory for us, and to hasten the day of its accomplishment.

Signed: Fayolle.

Official:
The Chief of Staff:
Signed: Paquette.

VIth Army
État-Major
3rd Bureau
No. 2.564/3

P. C., August 8, 1918.

P. A.
The Chief of Staff:
G. Brion

Copies sent to:
1st Army Corps U. S.
3rd Army Corps U. S.
Artillery
Engineers
Aeronautics
Telegraphy
1st, 2nd, and 4th Bureaux, S. R.

Copy to:
4th Division

MEMORANDUM
HEADQUARTERS, FIRST ARMY CORPS
AMERICAN E. F.

August 9, 1918.

From: Chief of Staff, 1st Army Corps, Am. E. F.
To: Brig. General Ewing E. Booth, Commanding 8th Brigade,
(Thru Commanding General, 4th Division).

Subject: Commendation.

1. The Corps Commander directs me to inform you that the character of your work in pushing to the Vesle, crossing it, reaching your assigned objective and maintaining your position there, excites his personal admiration and he takes this method of informing you thereof as well as of his official commendation.

2. Prompt and certain carrying out of his orders by you and the officers and men of your Brigade will be made a matter of record by the Corps Commander in his official report of present operations.

By command of Major General Liggett:
Malin Craig,
Chief of Staff.

Sixth Army.

General Order P. C., August 9th, 1918.

Prior to the great offensive of July 18th, the American troops, forming part of the Sixth French Army, distinguished themselves in wresting from the enemy the Bois de la Brigade de Marine and the village of Vaux, in halting the offensive on the Marne and Fossoy.

Since then they have taken a most glorious part in the second battle of the Marne, vying in ardor and valor with the French troops. During twenty days of incessant combat, they liberated numerous French villages and achieved, across a most difficult terrain, an advance of 40 kilometers, which carried them to the Vesle.

Their glorious deeds are marked by the names which will illuminate, in the future, the military history of the United States:

Torcy, Belleau, Plateau d'Êtrèpilly, Épieds, Le Charmel, l'Ourcq, Seringes-et-Nesles, Sergy, La Vesle, et Fismes.

The young divisions, who were under fire for the first time, showed themselves worthy of the old warlike traditions of the regular army. They had the same ardent desire to fight the Boche, the same discipline that always executes the order of the commander no matter what difficulties are to be conquered or what sacrifices to be made.

The magnificent results attained are due to the energy and ability of the commanders and to the bravery of the soldiers.

I am proud to have commanded such troops.

The General commanding the Sixth Army:

DEGOUTTE.

HEADQUARTERS FOURTH DIVISION
AMERICAN E. F.

France, August 14, 1918.

General Orders:
No. 46.

OFFICERS AND MEN OF THE FOURTH DIVISION:

After twenty-seven days of marching and fighting, our Division has been withdrawn from the front for a hard-earned and well deserved rest, and for the first time during that period it is now possible suitably to record our achievements.

With our training period still unfinished, our infantry and machine guns were rushed into the line on the night of July 17-18 to take part in Marshal Foch's now famous drive from the MARNE. Under the able commanders of the 164th Division, 7th Army Corps, VI French Army, and side by side with our gallant allies, battalions of the 8th Brigade drove the enemy from HAUTEVESNES, ST. GENGOULPH, CHÉZY, CHEVILLON, PRIEZ and COURCHAMPS, with such pluck and vigor that over four hundred prisoners, eighteen guns and many mortars and machine guns fell into the hands of General Gaucher, who commended our troops for "splendid dash." At the end of two days' fighting, the 164th Division was the most advanced in the VI Army.

Further north, our 39th Infantry, under its own colonel, took over a sector at night and was later cited in orders of the 33rd French Division for "magnificent ardor" and for the capture of the woods of CRESNES, the village of NOROY, an enemy battery, and a great number of trench mortars and machine guns, as well as over one hundred prisoners, including two officers. On July 23rd our troops were withdrawn and concentrated to resume a status of training, but under a sudden change of orders, were immediately marched to join the 1st Corps U. S., and placed in second line behind the 42nd Div., U. S., on the

OURCQ. Here two battalions of the 47th Infantry suffered heavy losses when pushed up to reinforce an exhausted brigade of the front line. On August 3rd the Division passed through to the front and, operating for the first time under its own commanders, continued the drive as far as the VESLE River. Here the enemy had established himself in force and successfully resisted further Allied advance.

The IVY DIVISION, baptized in full battle, has been christened a fighting unit. It has been tried out and has stood the test. With no preliminary experience in front sector, it took its full share in the greatest attack that has yet been launched by the Allied Forces. No soldiers have ever been called upon to stand a more gruelling grind upon their fortitude, endurance, and morale, than has been imposed upon the men of this Division by the fighting in the valley of the VESLE.

The Division Commander desires to express, in equal measure, his appreciation of the splendid coöperation of all branches:

To the Engineers, who, under heavy fire and with heavy losses constructed bridges over the VESLE, and, under shelling maintained roads that made supply possible;

To the Signal Corps, who labored day and night to maintain our lines of communication;

To the Medical Service, whose units were pushed forward to the firing line, working without rest to alleviate the sufferings of the wounded; and

To our Chaplains with their faithful parties, who carefully buried our dead.

He desires to commend in no less degree the personnel of the Trains, who, from the opening of the campaign, have not failed on a single occasion to furnish the combat elements with food and ammunition, and the Military Police, who, by intelligent traffic regulation, made it possible to supply a division in a congested area over devious and difficult roads.

The Division Commander is justly proud to command officers and men who have measured up to the highest standards of Americanism.

We mourn our dead. For the living, there is the work of to-morrow.

By command of MAJOR GENERAL CAMERON:

C. A. BACH,
Lieut. Col., General Staff,
Chief of Staff.

Official:
HOWARD J. SAVAGE
Capt. A. G. D., N. A.,
Acting Adjutant.

HEADQUARTERS 7TH INFANTRY BRIGADE
AMERICAN EXPEDITIONARY FORCES

France, 14 August, 1918.

General Orders:
No. 12

1. The Commanding General takes pleasure in publishing to the Brigade the following General Order of the 33rd French Infantry Division:

July 20, 1918.

33rd DIVISION HQ.,
NO. 42/0 GEN. ORDER.

The General commanding the 33rd Infantry Division cites in the ORDER OF THE DIVISION:

"*The 39th regiment of Infantry, U. S.*

"Attached to the Division to hold the sector, was called to take part in the battle the day after its arrival. Under the command of Colonel Bolles gave proof in receiving its baptism of fire, of admirable bravery.

"Took the thicket of Cresnes and the village of Noroy; captured an enemy battery, a great number of minenwerfer and machine guns, and made more than a hundred prisoners.

"The General Tanant commanding the 33rd Division."

2. Companies "A" and "C," 11th Machine Gun Battalion were at the time covered by the order, attached to the 39th Infantry, and are entitled to share in the citation with that regiment.

By command of General Poore:
A. D. Falconer,
Major, Q. M. C.
Adjutant.

G. H. Q.
AMERICAN EXPEDITIONARY FORCES

France, August 28, 1918.

General Orders:
No. 143.

It fills me with pride to record in General Orders a tribute to the service and achievements of the First and Third Corps, comprising the 1st, 2nd, 3rd, 4th, 26th, 28th, 32nd, and 42nd Divisions of the American Expeditionary Forces.

You came to the battlefield at the crucial hour of the Allied cause. For almost four years the most formidable army the world has yet seen had pressed its invasion of France, and stood threatening its capital. At no time had that army been more powerful or menacing than when, on July 15th, it struck again to destroy in one great battle the brave men opposed to it and to enforce its brutal will upon the world and civilization. Three days later, in conjunction with our Allies, you counter-attacked. The Allied Armies gained a brilliant victory that marks the turning point of the war. You did more than give our brave Allies the support to which as a nation our faith was pledged. You proved that our altruism, our pacific spirit, our sense of justice, have not blunted our virility or our courage. You have shown that American initiative and energy are as fit for the test of war as for the pursuits of peace. You have justly won the unstinted praise of our Allies and the eternal gratitude of our countrymen.

We have paid for our success in the lives of many of our brave comrades. We shall cherish their memory always, and claim for our history and literature their bravery, achievement, and sacrifice.

This order will be read to all organizations at the first assembly formation after its receipt.

John J. Pershing,
General, Commander-in-Chief.

Official:
Robert C. Davis,
Adjutant General.

Washington, September 14.

To General John J. Pershing,
American Expeditionary Forces,
France.

Accept my warmest congratulations on the brilliant achievements of the army under your command. The boys have done what we expected of them and done it the way we most admire. We are deeply proud of them and of their chief. Please convey to all concerned my grateful and affectionate thanks.

Signed: WOODROW WILSON.

Waterfall, Sept. 14-15, 1918.

The Commander-in-Chief is pleased to transmit to the command the following telegram which he has just received: "My dear General, The First American Army under your command on this first day has won a magnificent victory by a maneuver as skillfully prepared as it was valiantly executed. I extend to you as well as to the officers and troops under your command my warmest compliments. Marshal Foch." The Army Commander directs that the foregoing telegram be distributed to the forces of your command.

DRUM.

HEADQUARTERS
FIFTH ARMY CORPS
AMERICAN E. F.

General Orders: France, Sept. 16, 1918.
No. 17.

1. The following telegram from the Commander-in-Chief, American E. F., is published to the command:

"Sept. 15th, 1918.

"MAJOR GENERAL GEORGE H. CAMERON,
"Commanding General, Fifth Corps.

"Please extend to the officers and men of the Fifth Corps my sincere congratulations for the part they have taken in the first battle of the American Army. Our successes have thrilled our countrymen and evoked the enthusiasm of our Allies. Will you convey to the command my cordial appreciation of their work. I am proud of the accomplishment.

PERSHING."

By command of MAJOR GENERAL CAMERON:
W. B. BURTT,
Brigadier General,
Chief of Staff.

Official:
HARRY C. KAEFRING,
Adjutant General.
Distribution:
4th Division.

HEADQUARTERS FIFTH ARMY CORPS
AMERICAN EXPEDITIONARY FORCES

France, September 19, 1918.

General Orders:
No. 20.

1. The following telegram from the Commanding General, First Army, is published to the command:

"Hq. First Army, Sept. 17, 1918.

COMMANDING GENERAL, Fifth Army Corps:

The Commander-in-Chief is pleased to transmit the following telegram which he has just received from General Pershing's Headquarters, American E. F.:

'All ranks of the British Army in France welcome with enduring admiration and pleasure the victory which has attended your personal command. I beg you to accept and convey to all ranks my best congratulations and those of all ranks of the British Army under my command.

HAIG.'

H. A. DRUM.

Chief of Staff."

By command of MAJOR GENERAL CAMERON:

W. B. BURTT,

Brigadier General,

Chief of Staff.

Official:

HARRY C. KAEFRING,

Adjutant General.

Distribution:

4th Division.

HEADQUARTERS THIRD ARMY CORPS

AMERICAN EXPEDITIONARY FORCES

France, October 5, 1918.

General Orders:

No. 29.

Extract

2. The Corps Commander, in General Orders, cites the gallant conduct of the 4th Division, especially the 7th and 8th Infantry Brigades, in the seizure against great difficulties, of the BOIS DE FAYS, and the holding of it against repeated and determined counterattacks, between September 26th and October 5th. You are there; stay there!

By command of MAJOR GENERAL BULLARD:

A. W. BJORNSTAD,

Brigadier General, G. S.,

Chief of Staff.

Official:

DAVID O'KEEFE,

Adjutant General.

HEADQUARTERS THIRD ARMY CORPS

AMERICAN EXPEDITIONARY FORCES

France, 20 Oct., 1918.

From: Commanding General, Third Corps.

To: Commanding General, 4th Division.

Subject: Appearance of men of 58th Infantry.

1. Yesterday, the 19th instant, I passed the 58th Infantry of your Division on the road between ESNES and MALANCOURT. The superficial inspection of this regiment made from an automobile in passing, disclosed the fact that these men, although they had been in a very active sector fighting for twenty-five days, presented a clean and orderly appearance. Most of the men were cleanly shaven. Their clothing was free from mud, and they presented altogether a very creditable appearance.

2. I wish to congratulate you upon having such an organization in your Division and I desire also that you congratulate the Colonel of this regiment. I very much desire that my remarks be made known to the individual members of this organization.

J. L. HINES,
Major General, U. S. A.

GRAND QUARTIER GÉNÉRAL
des Armées
Du Nord et du Nord-Est
Etat-Major
Bureau du Personnel
(Décorations)

Ordre No. 10.887 "D" (Extrait)

Aprés approbation du Général Commandant en Chef les Forces Expéditionnaires Américaines en France, le Général Commandant en Chef les Armées Françaises du Nord et du Nord-Est, cite á l'Ordre du CORPS D'ARMÉE.

39° Régiment D'Infanterie Américaine:

"Affecté à une Division Française pour tenir le secteur, a été appelé à prendre part à la bataille le 18 Juillet 1918, dès le lendemain de son arrivée. Sous le commandement du Colonel BOLLES, a fait preuve en recevant le baptême du feu, d'une vaillance admirable. A enlevé le buisson de CRESNES et le village de NOROY; s'est emparé d'une batterie ennemie, d'un grand nombre de *minenwerfer* et de mitrailleuses, et à fait plus de 100 prisonniers."

Au Grand Quartier Général, le 25 Octobre 1918.
Le Général Commandant en Chef,
Signé: PETAIN.

Pour Extrait Conformé:
Le Lieutenant-Colonel,
Chef du Bureau du Personnel.

HEADQUARTERS FOURTH DIVISION
AMERICAN E. F.

France, 28 October, 1918.

General Orders:
No. 71.

Now that the 4th Division has been withdrawn from what may be termed the first phase of the battle of the ARGONNE, it is appropriate to review its achievements.

From September 26th to October 18th, 1918, the 4th Division as a part of the III Corps, 1st American Army took part in the attack on the enemy positions between the ARGONNE FOREST and the MEUSE.

These positions were of exceptional strength, having been in existence practically since the beginning of the war, and their natural features had been strengthened with all the skill and ingenuity which the German Army, with years of experience behind it, could bring to the task.

On the first day the Division penetrated these defenses to a depth of about seven kilometers. In the days following, its lines were advanced six kilometers in the face of strong resistance on the part of the enemy. All ground won was held under the most trying and difficult circumstances, under fire of all kinds both from the front and the flank, the 4th Division being at all times in advance of the other divisions of the 1st Army.

During the period when the Division was engaged, it had opposed to it all or parts of seven German divisions, two being rated as among the best in the German Army. One of these, the 28th Division, is known as the "Kaiser's Favorite."

The Division captured 2,731 prisoners, of whom 71 were officers, 47 field guns of caliber up to 150 mm., and many minenwerfer and machine guns, as well as a great quantity of small arms and ammunition.

It is with deep pride and satisfaction that the Division Commander publishes these results. They have been gained in the face of the most determined resistance and at great cost. No men have borne themselves more gallantly than the men of this Division. No Division in France has more cause to be proud of its achievements where all have done so well. It is impossible to single out units or individuals for special commendation; infantry, artillery, police, by splendid team-work and single-hearted devotion to duty, have all contributed their share to the success we have won.

We mourn our honored dead. Those of us who remain pledge ourselves anew for effort and sacrifice, proud of our privilege of representing the American people in the struggle for world freedom.

With pride, born of past achievement, the members of the "Ivy" Division look forward to sharing in the great and final victory of our armies—a victory that cannot long be delayed.

This order will be read to each organization at the first assembly after its receipt.

By command of Brigadier General Poore:
C. A. Bach,
Colonel, General Staff,
Chief of Staff.

Official:
Lon S. Haymes,
Capt. A. G. D.
Adjutant.

HEADQUARTERS THIRD FIELD ARTILLERY BRIGADE
AMERICAN E. F.

France, 12 November, 1918.

From: Commanding General, 3rd F. A. Brigade.
To: Commanding General, 4th Div., (Thru C.G., 5th Division.)
Subject: Report on 13th and 77th F. A. Regiments.

1. The following report on the 13th and 77th F. A. Regiments of the 4th Division during their attachment to the 3rd F. A. Brigade is made for the information of all concerned.

2. The 77th F. A. Regiment.

The regiment was attached to the 3rd F. A. Brigade per Annex No. 1 to F. O. 39, 3rd Army Corps, October 28th, 1918, and moved into position the night of Oct. 29-30 astride of the Cunel-Nantillois road about ½ kilometer south of Cunel. The regiment participated in the general attack of November 1st and the subsequent pursuit of the enemy thereafter, executing harassing fire on the enemy's front lines, rear areas, and routes of retreat during the attack and during the pursuit. The regiment occupied the following successive positions: entire regiment near Cunel from October 30th to November 5th, when 2nd Battalion moved to a position northeast of Aincreville; 1st Battalion joined 2nd Battalion night of November 6-7, 1918; November 7th, 2nd Battalion crossed the Meuse at Dun-sur-Meuse and took up position about 1 km. west of Murvaux, where it remained in position until regiment was relieved; 1st Battalion moved to a position about 2½ kms. east of Murvaux November 9th, where it also remained until relieved. The entire regiment was relieved at dark November 10th, 1918, per telephone orders from 3rd Corps Artillery, and ordered to rejoin its own Brigade in the vicinity of Rampont.

3. The 13th F. A. Regiment.

The regiment was attached to the 3rd F. A. Brigade per Annex No. 1, F. O. 39, 3rd Army Corps, October 28, 1918, and moved into position in vicinity of Cunel the night of October 29-30, 1918. The regiment participated in the preparatory fire and general attack of November 1st and subsequent pursuit of the enemy across the Meuse. During this period the regiment executed gas concentration, destructive fire, and harassing fire on enemy rear areas and routes of retreat, and during the period occupied the following successive positions: from October 30th to November 5th, the entire regiment was in position in vicinity of Cunel; the night of November 5-6, 1918, the regiment, less two battalions, moved to position near Doulcon, the two rear battalions joining the regiment at Doulcon the night of November 6-7; November 8th one battalion moved across the Meuse to a position ½ km. west of Murvaux, where it remained until relieved; on November 9th the two rear battalions moved to a position in draw just west of la Sentinelle Hill about 2½ km. east of Murvaux, where it remained in position until relieved. The entire regiment was relieved at 11:00 A. M., Nov. 11, 1918, per telephone orders from Headquarters 3rd Corps Artillery, Nov. 11, 1918, and ordered to rejoin its own Brigade in vicinity of Rampont.

4. It gives me great pleasure also to report that the work of these regiments was more than satisfactory in every respect. Cheerful, loyal, and prompt obedience to orders, thoroughness and skill in execution of orders, and effective results marked their entire term of duty. Both Colonel Smith and Colonel Commiskey have excellent regiments and show that they have been carefully prepared for work in Brigade.

5. I consider Colonel Wright Smith of the 13th Regiment fully qualified for the duty of Brigadier General, and take this occasion to recommend him for promotion to that grade for the period of the war.

H. G. Bishop,
Brigadier General, U. S. A.
Commanding.

GENERAL HEADQUARTERS
AMERICAN EXPEDITIONARY FORCES

France, November 12, 1918.

General Orders:
No. 203.

The enemy has capitulated. It is fitting that I address myself in thanks directly to the officers and soldiers of the American Expeditionary Forces, who by their heroic efforts have made possible this glorious result. Our armies, hurriedly raised and hastily trained, met a veteran enemy, and by courage, discipline, and skill always defeated him. Without complaint you have endured incessant toil, privation, and danger. You have seen many of your comrades make the supreme sacrifice that freedom may live. I thank you for the patience and courage with which you have endured. I congratulate you upon the splendid fruits of victory which your heroism and the blood of our gallant dead are now presenting to our nation. Your deeds will live forever on the most glorious pages of American history.

Those things you have done. There remains now a harder task which will test your soldierly qualities to the utmost. Succeed in this and little note will be taken and few praises will be sung; fail, and the light of your glorious achievements of the past will be sadly dimmed. But you will not fail. Every natural tendency may urge towards relaxation in discipline, in conduct, in appearance, in everything that marks the soldier. Yet you will remember that each officer and each soldier is the representative in Europe of his people and that his brilliant deeds of yesterday permit no action of today to pass unnoticed by friend or foe. You will meet this test as gallantly as you have met the tests of the battlefield. Sustained by your high ideals and inspired by the heroic part you have played, you will carry back to our people the proud consciousness of the new Americanism born of sacrifice. Whether you stand on hostile territory or on the friendly soil of France, you will so bear yourself in discipline, appearance, and respect for all civil rights that you will confirm for all time the pride and love which every American feels for your uniform and for you.

JOHN J. PERSHING,
General, Commander-in-Chief.

Official:
ROBERT C. DAVIS,
Adjutant General.

HEADQUARTERS THIRD ARMY CORPS
AMERICAN EXPEDITIONARY FORCES

France, November 12, 1918.

General Orders:
No. 43.

1. With the signing of the armistice on November 11th and the enemy suing for peace, the operations of this Corps, begun on September 26th, were brought to a successful issue.

2. The Third Corps has driven the enemy from the RUISSEAU DE FORGES to the MEUSE, thence, turning east, has crossed the MEUSE, in the face of the most determined resistance, between STENAY and BRIEULLES, and, continuing its resolute advance, has forced the enemy to the line STEMAY-REMOIVILLE-PEUVILLERS.

3. In a fruitless effort to stop this victorious drive the enemy threw into the line, opposite the Third Corps, his last reserve division (192nd) on the Western Front.

4. The Corps Commander feels that his pride and gratification in the achievements of the officers and soldiers of the Third Corps are more than justified, and he desires to express to them his high appreciation of their gallant conduct and to make herewith a permanent record of the same.

JOHN L. HINES,
Major General, U. S. A.,
Commanding.

Official:
DAVID O'KEEFE,
Adjutant General.
Distribution:
4th Division.

G. H. Q.
AMERICAN EXPEDITIONARY FORCES

France, November 13, 1918.

General Orders:
No. 204.

The following communication from the Commander-in-Chief of the Allied Armies is published to the command:

G.Q.G.A., 12 November, 1918.

OFFICERS, NON-COMMISSIONED OFFICERS, SOLDIERS OF THE ALLIED ARMIES:

After having resolutely stopped the enemy, you have, during these months, with a faith and an energy unsurpassed, attacked without respite.

You have won the greatest battle of history and saved the most sacred cause; the Liberty of the World.

Be confident!

With immortal glory you have glorified your flags.

Posterity holds for you its recognition.

The MARSHAL OF FRANCE,
Commander-in-Chief of the Allied Armies:
F. FOCH.

By command of GENERAL PERSHING:
JAMES W. MCANDREW,
Chief of Staff.

Official:
ROBERT C. DAVIS,
Adjutant General.

G. H. Q.
AMERICAN EXPEDITIONARY FORCES

France, December 19, 1918.

General Orders:
No. 232.

It is with a sense of gratitude for its splendid accomplishment, which will live through all history that I record in General Orders a tribute to the victory of the First Army in the Meuse-Argonne battle.

Tested and strengthened by the reduction of the St. Mihiel Salient, for more than six weeks you battered against the pivot of the enemy line on the western front. It was a position of imposing natural strength, stretching on both sides of the Meuse River from the bitterly contested hills of Verdun to the almost impenetrable forest of the Argonne; a position, fortified by four years of labor designed to render it impregnable; a position held with the fullest resources of the enemy. That position you broke utterly, and thereby hastened the collapse of the enemy's military power.

Soldiers of all the divisions engaged under the First, Third, and Fifth Corps—the 1st, 2nd, 3rd, 4th, 5th, 7th, 26th, 28th, 29th, 32nd, 33rd, 35th, 37th, 42nd, 77th, 78th, 79th, 80th, 82nd, 89th, 90th, and 91st—you will be long remembered for the stubborn persistence of your progress, your storming of obstinately defended machine gun nests, your penetration, yard by yard, of woods and ravines, your heroic resistance in the face of counter-attacks supported by powerful artillery fire. For more than a month, from the initial attack of September 26th, you fought your way slowly through the Argonne, through the woods and over the hills west of the Meuse; you slowly enlarged your hold on the Côtes de Meuse to the east; and then on the first of November, your attack forced the enemy into flight. Pressing his retreat, you cleared the entire left bank of the Meuse south of Sedan, and then stormed the heights on the right bank and drove him into the plain beyond.

Your achievement, which is scarcely to be equalled in American history, must remain a source of proud satisfaction to the troops who participated in the last campaign of the war. The American people will remember it as the realization of the hitherto potential strength of the American contribution toward the cause to which they had sworn allegiance. There can be no greater reward for a soldier or for a soldier's memory.

This order will be read to all organizations at the first assembly formation after its receipt.

JOHN J. PERSHING,
General, Commander-in-Chief,
American Expeditionary Forces.

Official:
ROBERT C. DAVIS,
Adjutant General.

G. H. Q.
AMERICAN EXPEDITIONARY FORCES

France, December 26, 1918.

General Orders:
No. 238.

It is with soldierly pride that I record in General Orders a tribute to the taking of the St. Mihiel Salient by the First Army.

On September 12, 1918, you delivered the first concerted offensive operation of the American Expeditionary Forces upon difficult terrain against this redoubtable position, immovably held for four years, which crumpled before your ably executed advance. Within twenty-four hours of the commencement of the attack, the salient had ceased to exist, and you were threatening Metz.

Your divisions, which had never been tried in the exacting conditions of major offensive operations, worthily emulated those of more arduous experience, and earned their right to participate in the more difficult task to come. Your staff and auxiliary services, which labored so untiringly and so enthusiastically, deserve equal commendation, and we are indebted to the willing co-operation of veteran French divisions and of auxiliary units which the Allied commands put at our disposal.

Not only did you straighten a dangerous salient, capture 16,000 prisoners and 443 guns, and liberate 240 square miles of French territory, but you demonstrated the fitness for battle of a unified American Army.

We appreciate the loyal training and effort of the First Army. In the name of our country, I offer our hearty and unmeasured thanks to these splendid Americans of the 1st, and 5th Corps, and of the 1st, 2nd, 4th, 5th, 26th, 42nd, 82nd, 89th, and 90th Divisions, which were engaged, and of the 3rd, 35th, 78th, 80th, and 91st Divisions, which were in reserve.

This order will be read to all organizations at the first assembly formation after its receipt.

By command of GENERAL PERSHING:
JAMES W. MCANDREW,
Chief of Staff.

Official:
ROBERT C. DAVIS,
Adjutant General.

HEADQUARTERS THIRD ARMY CORPS
AMERICAN EXPEDITIONARY FORCES

OFFICE OF THE COMMANDING GENERAL.

January 27, 1919.

From: Commanding General.
To: Commanding General, 4th Division, A.E.F.
Subject: Soldierly Deportment of 8th Brigade.

1. Recently I had occasion to pass through the portion of the 4th Division area occupied by the 8th Brigade of your Division, and I was particularly struck by the soldierly deportment of that Command. I was moving at a rather rapid rate and I would not have been surprised if some of the members of that Command had failed to salute me during my passage through the area, but, though I observed especially, I am confident that not a single member of that Command who saw my automobile failed to salute me.

2. I desire to congratulate you upon having such an organization in your Command, and I can say that I have never seen a more punctilious command in my life. I would thank you also to convey my congratulations to the Commanding General, 8th Brigade.

J. L. HINES,
Major General, U. S. A.

G. H. Q.
AMERICAN EXPEDITIONARY FORCES

France, February 28, 1919.

General Orders:
No. 38-A.

My Fellow Soldiers:

Now that your service with the American Expeditionary Forces is about to terminate, I can not let you go without a personal word. At the call to arms, the patriotic young manhood of America eagerly responded and became the formidable army whose decisive victories testify to its efficiency and its valor. With the support of the nation firmly united to defend the cause of liberty, our army has executed the will of the people with resolute purpose. Our democracy has been tested, and the forces of autocracy have been defeated. To the glory of the citizen-soldier, our troops have faithfully fulfilled their trust, and in a succession of brilliant offensives have overcome the menace to our civilization.

As an individual, your part in the world war has been an important one in the sum total of our achievements. Whether keeping lonely vigil in the trenches, or gallantly storming the enemy's stronghold; whether enduring monotonous drudgery at the rear, or sustaining the fighting line at the front, each has bravely and efficiently played his part. By willing sacrifice of personal rights; by cheerful endurance of hardships and privation; by vigor, strength, and indomitable will, made effective by thorough organization and cordial coöperation, you inspired the war-worn Allies with new life and turned the tide of threatened defeat into overwhelming victory.

With a consecrated devotion to duty and a will to conquer, you have loyally served your country. By your exemplary conduct a standard has been established and maintained never before attained by any army. With mind and body as clean and strong as the decisive blows you delivered against the foe, you are soon to return to the pursuits of peace. In leaving the scenes of your victories, may I ask that you carry home your high ideals and continue to live as you have served—an honor to the principles for which you have fought and to the fallen comrades you leave behind.

It is with pride in our success that I extend to you my sincere thanks for your splendid service to the army and to the nation.

Faithfully,
John J. Pershing,
Commander-in-Chief.

Official:
Robert C. Davis,
Adjutant General.

HEADQUARTERS 4TH DIVISION
AMERICAN EXPEDITIONARY FORCES, GERMANY

March 11, 1919.

Memorandum No. 26.

The following memorandum from Brigadier General H. B. Fiske, General Headquarters, American E. F., dated February 17, 1919, is published for the information of all concerned:

Subject: Inspection of animal transport of the Third Army.

1. The following is a report of my recent visit of inspection to the Third Army.

2. All divisions of the Third Army were inspected, beginning on February 5th with the 3rd Division.

COMMENTS

1. The 4th Division made the best showing. The regiment designated was the 47th Infantry, commanded by Colonel T. H. Middleton. The condition of the animals, the leather, the chains, buckles, ornaments and the cleanliness and upkeep of the transportation of all kinds was magnificent.

By command of MAJOR GENERAL HERSEY:
C. A. BACH,
Colonel, General Staff,
Chief of Staff.

Official:
MAX. B. GARBER,
Lt. Col. 58th Inf.,
Acting Adjutant.

ARMY OF OCCUPATION
THIRD U. S. ARMY

OFFICE OF COMMANDING GENERAL

Coblenz, Germany,
March, 19, 1919.

MAJOR GENERAL M. L. HERSEY,
Commanding 4th Division, A.E.F.,
MY DEAR GENERAL HERSEY:

The appearance of your Division was a source of great pleasure to me yesterday. I think your transportation is beyond the reach of competition. The set-up, clothing, equipment, and bearing of the men of your command was very fine and the subject of universal comment by officers and civilians alike. I am sure that the Commander-in-Chief was greatly pleased and that the Army of Occupation has established a reputation which will extend over the civilized world.

Please convey my appreciation to the officers and men of your command.

With best wishes,

Sincerely yours,
J. T. DICKMAN.

ARMY OF OCCUPATION
THIRD U. S. ARMY

OFFICE OF THE CHIEF OF STAFF

Coblenz, Germany,
March 19, 1919.

From: Chief of Staff, Third Army, American E. F.
To: Commanding General, Fourth Corps, American E. F.
Subject: Commendation.

1. The Army Commander directs me to congratulate the Commanding General of the Fourth Corps on the condition and appearance of the Fourth Corps troops and the Third and Fourth Divisions during the recent inspections by the Commander-in-Chief.

2. The condition of these troops is the direct reflection of the intelligent work they have done in training and instruction in spite of adverse conditions since the arrival in their present areas.

By command of MAJOR GENERAL DICKMAN:
MALIN CRAIG,
Brigadier General, U. S. A.
Chief of Staff.

1st Ind.

Headquarters IV Army Corps, American E. F., March 23, 1919.
To—Commanding General, 4th Division.

1. Official copy furnished to the C. G., 4th Division.
2. The Corps Commander directs me to inform you that it gives him great pleasure to transmit to you the congratulations of the Army Commander and to say that he appreciates the fact that nothing but the continuous and energetic efforts of the Division Commander and the loyal support of his subordinates could have brought your division to its present high standard of efficiency.

B. H. Wells,
Chief of Staff.

AMERICAN EXPEDITIONARY FORCES
Office of The Commander-in-Chief

France, March 25, 1919.

Major General Mark Hersey,
Commanding 4th Division, A.E.F.,
Germany.

My dear General Hersey:

It was with deep gratification that I observed the excellent condition of the 4th Division on the occasion of my inspection on March 18th. The transportation and the artillery of the Division were in splendid shape, and the general appearance of the men was equal to the highest standards. Throughout the inspection and the review, the high morale existing in all ranks was evident.

Arriving in France in May, the 4th Division was first engaged in the Marne counter-offensive on July 18th as a part of the French VIth Army. Detachments aided in the crossing of the Ourcq and on August 3rd-4th the Division advanced to the Vesle. In the reduction of the St. Mihiel Salient, it carried its objectives with effectiveness and precision. For the opening attack of the Meuse-Argonne offensive, the 4th Division was put into line as the center unit of the 3rd Corps and by its aggressiveness made a total advance of 13 kilometers despite continued and heavy resistance.

As a part of the 3rd Army, the Division participated in the march into Germany and the subsequent occupation of enemy territory. I am pleased to mention the excellent conduct of the men in these difficult circumstances, for which, as well as for their services in battle, they are due the gratitude of the nation.

I wish to express to each man my own appreciation of the splendid work that has been done and the assurance of my continued interest in his welfare.

Most sincerely yours,
John J. Pershing.

HEADQUARTERS THIRD ARMY
AMERICAN EXPEDITIONARY FORCES
GERMANY

16 May, 1919.

From: The Chief of Staff, Third Army.
To: The Commanding General, 4th Division.
Subject: Appreciation of the Services of the Division.

Upon the eve of the departure of the 4th Division from the Army of Occupation, the Commanding General takes pleasure in expressing to you and to the officers and men of your Division his appreciation of the services rendered by them to the Nation.

During the operations on the Marne and the Vesle, in the St. Mihiel Salient and the Meuse-Argonne Offensive, the fighting power of your Division greatly enhanced the success of the American arms, and contributed largely to the defeat of a stubborn enemy.

As one of the Divisions forming the Army of Occupation, the bearing and conduct of your officers and men has been exemplary and your animals and transportation have been maintained in such a high state of excellency as to evoke much favorable comment.

The splendid achievements of your Division while in the American Expeditionary Forces are symbolic of the best traditions of the service, and its worth has been noted with pride by the Commander-in-Chief, the Army Commander, and others who have come in contact with it.

By command of LIEUTENANT GENERAL LIGGETT:

GEORGE GRUNERT,
Acting Chief of Staff.

HEADQUARTERS IV ARMY CORPS
AMERICAN EXPEDITIONARY FORCES
GERMANY

17 May, 1919.

From: Commanding General, IV Army Corps.
To: Commanding General, 4th Division, American E. F.
Subject: Commendation.

1. The 4th Division having been relieved from duty with the IV Corps, of which it formed a part from October 24th, 1918, to October 26th, 1918, and from November 10th, 1918, to April 30th, 1919, I take this opportunity to express to its commander and to its entire personnel my sincere appreciation of their loyal, faithful, and efficient service.

2. With an excellent record of achievement in battle behind it, the Division became a part of the Army of Occupation. It joined the IV Corps in the Woevre just before the Third Army began its march to the Rhine. Throughout that march and during the trying days of the Military Occupation of the Rhineland, the Division not only always maintained its high standards of efficiency, devotion to duty, discipline and esprit, but constantly improved those standards until to-day it is rated a Division second to none in the A. E. F.

3. The officers of your staff have always coöperated in the most loyal way with the officers of the Corps Staff. I congratulate you on the record your Division has made in all its activities. My best wishes follow your officers and men in their future careers.

C. P. SUMMERALL,
Major General, U. S. Army.

THE ROLL OF HONOR

APPENDIX D

KILLED IN ACTION

* Indicates died from wounds received in action.

7TH INFANTRY BRIGADE

BRIGADE HEADQUARTERS

Officers

*Captain Jean Renard, French Army

39TH INFANTRY REGIMENT

Officers

Lieutenant Colonel	William E. Holliday
1st. Lieutenant	Philip J. Davidson
*1st. Lieutenant	George D. Parnell
*1st. Lieutenant	Theodore D. Schmidt
1st. Lieutenant	Andrew V. Seipel
1st. Lieutenant	James A. Edmond
2nd. Lieutenant	Clyde D. Funderberg
2nd. Lieutenant	William S. Graham
*2nd. Lieutenant	David S. Grant.
*2nd. Lieutenant	William G. Hodge
2nd. Lieutenant	Gerald F. Kelly
2nd. Lieutenant	James A. Kelly
2nd. Lieutenant	George D. Montgomery
2nd. Lieutenant	Edward L. Moore
2nd. Lieutenant	Randolph C. Stocker
*2nd. Lieutenant	Paul S. Strickland
*2nd. Lieutenant	Elliott C. Wellar

Enlisted Men

COMPANY A

1st. Sgt.	John C. Keller	Sgt.	John Greenfield
Sgt.	Robert D. Winters	Cpl.	Milford E. Alkire
Cpl.	Einer Brateng	Cpl.	Glen Croew
Cpl.	Patrick J. Farrell	*Cpl.	Harry Halverson

Cpl.	Ralph Milburn	Cpl.	Angelo Massera
Cpl.	Daniel W. Rabun	Cpl.	John F. Urys
Cpl.	Lewis Wiener	Cpl.	William A. Welch
Cpl.	Jacob H. Yenner	Bug.	Charles C. Kreider
Bug.	Pasquale Maxello	*Pvt. 1 cl.	James Davis
Pvt. 1 cl.	Cecil Fosnaugh	*Pvt. 1 cl.	Frank Wojciechowski
Pvt.	Roland S. Andreas	Pvt.	Alfred D. Antoni
Pvt.	Emanuel Abel	Pvt.	Eddie A. Anderson
Pvt.	Elmer E. Dobson	Pvt.	Richard R. Chezum
*Pvt.	Frank Domask	Pvt.	Joseph Dilworth
Pvt.	Frank Dipesa	Pvt.	Roman P. Friesenhahn
Pvt.	Floyd E. Ferguson	Pvt.	Rimothy A. Harrington
Pvt.	Grover C. Hanna	Pvt.	Joseph Jurewiecz
Pvt.	August A. Gross	*Pvt.	Edward Kraemer
Pvt.	Alfred A. Olson	Pvt.	Henry A. Peters
*Pvt.	Charlie G. Stolz	Pvt.	Ralph L. Smith
Pvt.	Joseph Swinski	Pvt.	Lebora Scarantino
Pvt.	Robert W. Thompson	Pvt.	Farina Savatore
Pvt.	August Trhlik	Pvt.	Walter G. Thompson
*Pvt.	James M. Sullivan	Pvt.	William T. Wren

Company B

Sgt.	Hyman L. Ambos	Sgt.	William E. Mooreheart
Cpl.	Simon E. Harrel	Cpl.	Charles J. Tipil
Pvt. 1 cl.	Roman Profoskie	Pvt. 1 cl.	Emile A. Toudreau
Pvt. 1 cl.	Hermen Jerrells	Pvt. 1 cl.	John Kane
Pvt. 1 cl.	Luther Warren	Pvt.	Peter Anderson
Bug.	Phillip Mazzelo	Pvt.	George Buchmeier
Pvt.	Ferdinand Basel	Pvt.	Walter F. Bauman
Pvt.	Geremia Di Bernardino	Pvt.	Cornelius E. Chatterton
*Pvt.	George H. Clatworthy	Pvt.	Sam Crumpley
Pvt.	Francesco Cassano	Pvt.	Thomas J. Fitzgerald
*Pvt.	David Eisenberg	Pvt.	Frank Gonyea
Pvt.	Thomas H. Franks	Pvt.	Thomas Huntley
Pvt.	Frank N. Guenther	Pvt.	John Murray
Pvt.	Henry F. Holtz	Pvt.	Edward Magnison
Pvt.	Samuel M. McKinney	Pvt.	James E. Nowlin
Pvt.	Thomas E. Mangion	*Pvt.	Mike Podlesny
Pvt.	Louis J. Paetz	Pvt.	Edward M. Sannes
Pvt.	James Reid	Pvt.	Lyle W. Stockton
Pvt.	Joseph Smith	Pvt.	Arthur D. Wohlt
Pvt.	John W. Swackhamer		

Company C

Sup. Sgt.	Walter T. Mann	Sgt.	Thomas L. Whalen
Sgt.	Lottie A. Zell	Cpl.	Alex Jameson
Cpl.	George W. Reber	Cpl.	Frank Raszeja
Cpl.	Patrick J. Kelly	Cpl.	Aubrey E. Robinson
Cpl.	Valentine Volpe	Mech.	Harvey Akes

Pvt. 1 cl. Alvin Baker
Pvt. 1 cl. John F. Borski
Pvt. 1 cl. Ned Hoskins
Pvt. 1 cl. Charles F. Smith
Pvt. 1 cl. William Zalabek
Pvt. Grover Adams
Pvt. Elmer Crowe
Pvt. Grover C. Dennis
Pvt. William H. Dennis
Pvt. Charles O. Gourley
Pvt. Daniel A. Kautz
Pvt. Edward Lippman
Pvt. Barton W. Masterson
Pvt. Erric J. Oslin
Pvt. William F. Passow
Pvt. Charles A. Stonestreet
Pvt. 1 cl. Clarence Butler
Pvt. 1 cl. Errico De Loreto
Pvt. 1 cl. Harry Rattenbury
Pvt. 1 cl. John C. Wackowiak
Pvt. Orajio Alasce
Pvt. Andrew M. Anderson
Pvt. James Conty
Pvt. Amy A. Davis
*Pvt. Elzear De Schenes
Pvt. Paul Herington
Pvt. Daniel Lettiere
Pvt. Edward T. Martin
Pvt. Trueman Melvin
Pvt. Frank M. Popa
Pvt. Santiago Satillo
*Pvt. Albert Beack

Company D

1st. Sgt. Orion E. Collier
Cpl. William Kosakiewicz
Cpl. Michael Riley
Cpl. Alex D. Weinberg
Bug. Marshall C. Young
Pvt. 1 cl. Matthew W. Kirzinger
Pvt. 1 cl. Dan Cohen
Pvt. Louis W. Ackerman
Pvt. Walter Baltus
Pvt. Joseph Czerna
Pvt. Murl W. Crocker
Pvt. John W. Crowder
Pvt. Bernie Edman
Pvt. Ralph Ferguson
Pvt. Goerge H. Kennedy
Pvt. Harold G. Kennedy
Pvt. Harry Mondress.
Pvt. Basil P. Palmer
Pvt. James W. Roebuck
Pvt. Joseph J. Wetzel
*Pvt. Harry A. Cook
Cpl. Peter Gedda
Cpl. Frank Marchec
*Cpl. Thomas J. Siaflis
Cpl. Thomas B. Yates
Pvt. 1 cl. Edward Campbell
*Pvt. 1 cl. Derwin N. Schroeder
Pvt. Henry C. Apgenhaffen
Pvt. Felikas Baserewski
*Pvt. Fred Bug
Pvt. Charles L. Crawford
Pvt. Luigi Corccorello
Pvt. Leo Downey
Pvt. Sidney Ferris
Pvt. Charles J. Herbert
Pvt. William R. Kent
Pvt. Edd Lowry
Pvt. Ernest B. Presse.
Pvt. Kolebrus Potoski
Pvt. Robert E. Weekly
Pvt. Otto Wiebell

Company E

Sgt. Frank Gaynor
Cpl. Seferino Arruppe
Cpl. Bernard L. Buente
Cpl. Frank E. Drew
Cpl. Frank Jones
Cpl. Harry Luby
*Pvt. Howard A. Bolin
Pvt. Ceasare Constantino
Pvt. Louis Farrell
Sgt. Harry Hobart
*Sgt. Mark S. Reed
Cpl. Edmund Blaine
*Cpl. Almon N. Fowler
Cpl. Harry Lawton
Pvt. Allen Atwater
Pvt. Joseph C. Brown
Pvt. Joseph Erickson
Pvt. Avon D. Flanders

Pvt.	Gabriel Ferri	Pvt.	Arthur Gauthier
*Pvt.	Antonio Gianette	Pvt.	Eugene Gregoire
Pvt.	Anton Gustas	Pvt.	Francesco Gregorio
Pvt.	James E. Howell	Pvt.	Frank Hegner
Pvt.	Matto Ifallo	Pvt.	Samuel Ingber
*Pvt.	Theodore T. Jensen	Pvt.	Ole Johnson
Pvt.	Snowden M. Kessell	Pvt.	Morris Kulfsky
Pvt.	Morris Kurloff	*Pvt.	Edmond Leblond
Pvt.	Howard E. Laughlin	Pvt.	Willie Myhre
Pvt.	John P. Morrissey	Pvt.	Oliver E. Morlock
*Pvt.	Floyd McNew	Pvt.	Ralph Nebbelin
Pvt.	Alfred A. Olson	Pvt.	Michael Przybysz
Pvt.	Fred G. Peetz	Pvt.	Samuel Siegel
Pvt.	Harvey P. Street	Pvt.	Ernest L. Simmons
*Pvt.	Norris Shaff	Pvt.	William J. Schneider
Pvt.	Gregory Schussman	Pvt.	Floyd Thompson
Pvt.	Edgar O. Wright	*Pvt.	Herman Zunker
Pvt.	Joseph P. Tamkevitch		

Company F

Sgt.	Benjamin Clabots	Sgt.	Clyde Hickman
Sgt.	Paul B. Halajian	Sgt.	Raymond Jones
Sgt.	Peter W. Larsen	Cpl.	Le Roy Aspden
Cpl.	Thomas Lionello	Cpl.	Mike Singer
Mech.	Peter A. Kosso	Pvt. 1 cl.	John M. Morrison
Pvt. 1 cl.	William F. Brinkman	Pvt.	Nelson R. Burner
Pvt. 1 cl.	Charles F. Smith	*Pvt.	Loyd C. Bute
Pvt.	George Boness	Pvt.	Sidney T. Foye
Pvt.	Charles Doby	*Pvt.	Frank B. Ewell
Pvt.	Claude W. Eddleman	Pvt.	Louis Hughes
Pvt.	Eugene Gignac	Pvt.	Charles Johnson
Pvt.	Alfred M. Hodge	Pvt.	Emanuel Koellen
Pvt.	William M. Kapschull	Pvt.	Luke J. Mayette
Pvt.	Carl B. Lewis	Pvt.	Theodore C. Moss
Pvt.	Obert J. Mjelde	Pvt. 1 cl.	Floyd D. Murphy
*Pvt.	Milton J. Newmeyer	Pvt.	Michael Olskeoich
Pvt.	Zasmo Priejeff	Pvt.	John Petroskie
Pvt.	Ernest Plant	Pvt.	George Poulas
Pvt.	George L. Putman	Pvt.	Ernest Swan
Pvt.	Henry Schutt.	Pvt.	Harry C. Surface
Pvt.	Charles J. Sobotta	Pvt.	Theron M. Spencer
Pvt.	Perry L. Toombs	Pvt.	Gaetano Vindigno
Pvt.	Jennings B. Wilson	*Pvt.	William M. Wright

Company G

Sgt.	John A. Murphy	Sgt.	Mark A. Pooly
Cpl.	James A. Gere	Cpl.	Clarence Hill
Cpl.	William Maly	Cpl.	Jacob F. Ouhl
Cpl.	Raymond J. White	Mech.	Harry Roy
*Pvt. 1 cl.	Pierre Barbe	Pvt. 1 cl.	Frank King

Pvt. 1 cl. George W. Hill
Pvt. 1 cl. Henry Murray
Pvt. Carl E. Burrill
Pvt. Jesse N. Cherrington
Pvt. Antonio Delgiomo
Pvt. Ralph Freer
Pvt. Walter A. Hake
*Pvt. John Heustis
Pvt. Andrew Polash
Pvt. Walter L. Pinnecker
*Pvt. Morris E. Rosenfield
Pvt. Robert Strong
Pvt. Harry Thoman
Pvt. 1 cl. Alexander Kaminski
Pvt. Joseph Antrillo
Pvt. Daniel P. Carey
Pvt. Adolph Demel
Pvt. Prentis S. Edwards
Pvt. Augusto Fornacirai
Pvt. Jocob C. Harrall
Pvt. Peter A. Mallot
Pvt. Arthur Paepke
Pvt. Michael F. Pruska
Pvt. George E. Smith
Pvt. Earl D. Umbenhouer

Company H

Sgt. Emil Dallman
Cpl. Charles Baburek
Cpl. Jesse C. Harding
Cpl. George M. Shannon
Pvt. 1 cl. Albert Schneider
Pvt. Wiliam Barnes
Pvt. Harvey W. Dodds
Pvt. Harry W. Fair
Pvt. Dexter L. Henderson
*Pvt. William M. Hutchinson
Pvt. Christ Koenig
Pvt. Francesco Lopario
Pvt. Mike Manos
*Pvt. Vithia I. Perry
Pvt. Abram Van Pelt
*Pvt. Raymond T. Reichard
Pvt. Arthur Schultz
Pvt. William Shaefer
Pvt. Walter Szyjka
Pvt. Joe Zurfluh
*Sgt. Israel Wagstaff
Cpl. Walter A. Engleman
Cpl. Antone Parades
Pvt. 1 cl. Jesse O. Burns
*Pvt. 1 cl. Roy Gill
Pvt. Rome Antrobus
Pvt. Ralph Burkey
*Pvt. George J. Elbus
Pvt. Waclowas Gecas
*Pvt. Harry J. Heartter
*Pvt. Adolph Haug
*Pvt. Alexander Loeser
*Pvt. Elbert Marvin
*Pvt. Howard A. Michel
Pvt. William H. Phillips
Pvt. John R. Rowe
*Pvt. Joseph P. Rooney
Pvt. Leon Sayler
Pvt. Alex Stern
Pvt. Fred L. Wild

Company I

Sgt. Abraham Friedman
*Sgt. Walter H. Garnett
Cpl. Samuel Clifton
Cpl. Edward A. Gilbert
Cpl. Herman Goltz
*Cpl. Clarence E. Brown
Mech. Kohn Patarino
Pvt. 1 cl. Thomas P. Mann
Pvt. 1 cl. John Proszyz
Pvt. 1 cl. John Nagle
Pvt. 1 cl. Arnold Schwieger
Pvt. Forrest F. Blough
Pvt. Thomas A. Birnes
Sgt. John M. Grier
Cpl. Robert R. Clemmons
Cpl. William Coltz
Cpl. Edward Gill
Cpl. Joe O'Brien
Cook Albert W. Bergonzi
Pvt. 1 cl. Eli Blaksley
Pvt. 1 cl. Alfred Mason
*Pvt. 1 cl. Tull Newton
Pvt. George L. Atwood
Pvt. Simon F. Black
Pvt. James J. Birne
Pvt. Adelard Bibeau

*Pvt.	Clyde E. Buchanan	Pvt.	Charles E. Canavan
Pvt.	Antonio Cannon	Pvt.	Mood Craig
*Pvt.	Stonie Denton	*Pvt.	Michele Di Dionigi
*Pvt.	August Donner	Pvt.	Elbridge Hanson
Pvt.	Leroy Hanson	Pvt.	Joseph Hendricks
Pvt.	Frank F. Indehar	Pvt.	Willie A. Krueger
Pvt.	Arthur A. Mellin	*Pvt.	Stephen H. Oats
Pvt.	Mariano Palambi	Pvt.	Alexandro Patete
*Pvt.	Harry Peterson	Pvt.	Joseph C. Racine
Pvt.	Fred C. Schreiber	Pvt.	Walter A. Smith
Pvt.	Andreas Steenkiste	Pvt.	John W. Temple
Pvt.	John G. Thordson		

Company K

Sgt.	Joseph B. Roberts	Cpl.	John Jansen
Cpl.	Joseph E. Wazny	Mech.	Ralph W. Wright
Pvt. 1 cl.	Raffalle Bolea	Pvt. 1 cl.	Walter T. Brown
Pvt. 1 cl.	Philil Fedoruk	Pvt. 1 cl.	Zigmund Lipchick
Pvt. 1 cl.	Hugh Murmane	*Pvt.	Henry O. Brendon
*Pvt.	Robert Cage	Pvt.	Harry A. Cook
Pvt.	Thomas H. Crosby	Pvt.	Harold H. Emerson
Pvt.	Herbert Gowan	*Pvt.	Carl J. Jansen
*Pvt.	Austin Lewis	*Pvt.	Charles C. Lilley
Pvt.	Selvas Monas	Pvt.	Constantine Pedranti
Pvt.	Edward E. Ramsdell	Pvt.	Jasper Skeeter
Pvt.	Gustave H. Schulze	*Pvt.	Emil Schwertfeger
Pvt.	Giovanni Toto	Pvt.	Frank Wylie
Pvt.	Lester Wells	Pvt.	Paul Zimmerman

Company L

Sgt.	Daniel Healy	Sgt.	John V. L. McKee
Sgt.	Ora E. Paul	Cpl.	Joseph Hendricks, Jr.
Cpl.	Ulysses LaVoie	Cpl.	Francis Magrane
Cpl.	Walter A. Looman	Cpl.	Joseph Murray
Cpl.	Ralph Oepen	Cpl.	James F. Taylor
Cpl.	Paul L. Stewart	Pvt. 1 cl.	John J. Briganski
Cpl.	Leroy W. Upton	Pvt. 1 cl.	John C. Frederick
Pvt. 1 cl.	Howard S. Graham	Pvt.	Christ P. Bogstead
Pvt. 1 cl.	George L. Schroeder	Pvt.	John E. Brown
Pvt.	Carl L. Bowman	Pvt.	Patrick J. Gallagher
Pvt.	James E. Dooley	Pvt.	Ross M. Heard
Pvt.	Sylvanus H. Hoxie	*Pvt.	John T. Hutton
Pvt.	William C. Kiser	Pvt.	Antonio Manfredi
Pvt.	Morris L. Miller	Pvt.	Allen E. Moon
Pvt.	Luther A. Pettigrew	Pvt.	Henry Peterson
Pvt.	Charles A. Pierce	*Pvt.	Charles Runge
Pvt.	Bertel G. Stadig	Pvt.	James F. Stanley
Pvt.	Mardi E. Whelar		

Company M

1st. Sgt.	Walter H. Larman	*Cpl.	Austin Draper
*Cpl.	Clinton Bressett	Cpl.	Owen Davis
Cpl.	Walter L. Byrne	Cpl.	Dennis Sullivan
Cpl.	Corbit Mitchell	Cpl.	John L. Weathers
Cpl.	Guy Taylor	Bug.	Ralph J. Goddard
Cpl.	Charles F. Wood	Pvt. 1 cl.	Jesse B. Jones
Mech.	Walter S. Gray	Pvt.	Alexander Abramowitz
Pvt. 1 cl.	William Ledergerber, Jr.	Pvt.	George Beck
*Pvt.	Sedor P. Bayko	Pvt.	Julius F. Collins
Pvt.	Andrew Beyer	Pvt.	Haurile Cozachuck
Pvt.	Travis D. Cook	Pvt.	John D. Duncan
Pvt.	Joseph A. Dion	Pvt.	John F. Funkhouser
Pvt.	Clifton R. Foster	Pvt.	Vern R. Hawbaker
Pvt.	Arthur G. Fish	Pvt.	Robert L. Millian
Pvt.	Samuel A. Miller	Pvt.	George H. Moore
Pvt.	Robert H. Mulligan	Pvt.	William Parry
*Pvt.	Henry Peterson	Pvt.	Claude Puplinskie
Pvt.	William Poly	Pvt.	Hyram D. Rains
Pvt.	Daniel F. Reagan	Pvt.	Joseph Reano
Pvt.	Anton Schoenberger	Pvt.	Stephen F. Shager
Pvt.	James Stedry	Pvt.	Walter Taylor
*Pvt.	Parley C. Turner	Pvt.	Walter Thompson

Headquarters Company

Sgt.	Joseph A. Grace	*Sgt.	Feve L. Hunt, Jr.
Sgt.	Maurice Kinsey	Sgt.	Harold O. Young
Band Cpl	Isaac E. Holloway	Cpl.	John J. Gill
*Mus. 3rd cl.	Edwin R. Bennett	*Cpl.	John C. Schoenbauer
Pvt. 1 cl.	Claude A. Huff	*Mus. 3rd cl.	Oscar W. Lindsey
Pvt. 1 cl.	William Lawson	Cook	James C. Kahl
*Pvt.	Gilbert McNutt	Pvt. 1 cl.	William R. Lagragren
*Pvt.	Peter Neimczyk	*Pvt. 1 cl.	Arthur W. Lyford
Pvt.	Elmer G Stevenson	Pvt.	Nicholas Malooly
Pvt.	Baleslow Potosky	Pvt.	William R. Thorson

Machine Gun Company

Sgt.	William F. Curvow	Cpl.	Raymond Buma
Cpl.	Thomas O. Cyr	Cpl.	Charles R. Hogan
Cpl.	Alfred A. Pennings	Mech.	Peter Gorgen
Pvt. 1 cl.	Stephen Crowl	Pvt. 1 cl.	Thomas J. Hassett
*Pvt. 1 cl.	Don M. Huffman	Pvt. 1 cl.	Mack M. Jenkins
Pvt. 1 cl.	Russell McElfresh	Pvt. 1 cl.	Leo Sanford
Pvt. 1 cl.	Adolph J. Wotruba	Pvt.	William Allen
*Pvt.	Waldo R. Brown	Pvt.	John Brown
Pvt.	John Brekke	Pvt.	William D. Call
Pvt.	John L. Evans	Pvt.	Charles D. Fontanella
Pvt.	Albert Gagnon	Pvt.	Clarence H. Huling

Pvt.	Charles C. Jones	Pvt.	William T. Hyder
Pvt.	George B. Johnson	Pvt.	Raymond J. Kuhlmeier
*Pvt.	Don A. McKie	*Pvt.	William Sullivan

Supply Company

*Cpl.	Stephen Mercer	*Wag.	Alexander J. Stricker
Pvt.	Nevin Fisher	Pvt.	Wm. C. Foreaker
*Pvt.	Omer Loop		

Medical Detachment

Sgt. 1 cl.	Francis P. Baker	Pvt. 1 cl.	Phillip J. O'Connell
Pvt.	Paul G. Kimball	Pvt.	John Wendland

47TH INFANTRY REGIMENT

Officers

Major..................Harrison B. Webster, M.C.
Captain..................Ross Snyder
1st. Lieutenant...........Gail H. Alexander
1st. Lieutenant...........Oliver W. Bailey
*1st. Lieutenant...........Julius E. Bell
1st. Lieutenant...........Raymond F. Houston
*1st. Lieutenant...........Lee C. Lewis
1st. Lieutenant...........Robert H. Murdock, M.C.
1st. Lieutenant...........James P. Over
1st. Lieutenant...........Thomas L. Russell
1st. Lieutenant...........H. C. Spengler
2nd. Lieutenant ....Conrad Crawford
2nd. Lieutenant...........James G. Elder
*2nd. Lieutenant...........Thomas M. Hamer
2nd. Lieutenant...........William F. Hanlin
2nd. Lieutenant...........Earl Knecht
*2nd. Lieutenant...........Alexander G. MacKay
*2nd. Lieutenant...........Joseph G. MacDonough
*2nd. Lieutenant...........Richard D. Robinson
2nd. Lieutenant...........Egbert F. Tetley
2nd. Lieutenant...........William H. Thomas
2nd. Lieutenant...........Uriel G. Utley

Enlisted Men

Company A

Sgt.	Richard Blankenship	Sgt.	Zygmond Dluzak
Cpl.	William Rizos	Bug.	Ernest C. Richardson
Pvt. 1 cl.	Bernard J. Behringer	Pvt. 1 cl.	George Cottam
Pvt. 1 cl.	Nelson Dube	Pvt. 1 cl.	Henry M. Gerald
Pvt. 1 cl.	Arthur J. Higby	Pvt. 1 cl.	Alfred R. Hinkle
Pvt. 1 cl.	Leon A. Jacquemet	*Pvt. 1 cl.	Adam Klein

Pvt. 1 cl.	Sam G. Lindsey	Pvt. 1 cl.	Carl T. Medford
*Pvt. 1 cl.	Michael E. Parchimovitch	*Pvt. 1 cl.	William W. Seim
Pvt.	Harold Berg	Pvt.	Victor Bershinsky
Pvt.	James Baker	*Pvt.	Tommie Bell
*Pvt.	Axel C. Carlson	Pvt.	William A. Davis
Pvt.	Edward Emmons	Pvt.	James G. Edwards
Pvt.	Earl S. Gatlin	Pvt.	Louis Generous
Pvt.	John M. Gentry	Pvt.	Oliver L. Harrell
Pvt.	William H. Horton	Pvt.	Philip Kirk
Pvt.	Marion J. Losco	Pvt.	William W. Leonard
Pvt.	Floy Martin	*Pvt.	John W. Montague
*Pvt.	Manuel Olivares, Jr.	Pvt.	Charles J. Rice
Pvt.	George E. Schmidt	Pvt.	Samuel Stein

COMPANY B

Sgt.	Walter H. Detrow	Sgt.	Patrick Neville
Cpl.	Emil Behling	Cpl.	John F. Holland
Cpl.	Glenwood G. Hopkins	*Cpl.	John F. Moriarity
Cpl.	Virgil C. Wommack	Pvt. 1 cl.	George R. Ayers
Pvt. 1 cl.	Emil F. Jackson	Pvt. 1 cl.	Arthur M. Miller
Pvt.	Thomas J. Ayers	Pvt.	John P. Baroutsas
Pvt.	Reuben Baskin	Pvt.	Joseph Benhoff
Pvt.	Reinhold Benson	*Pvt.	John J. Blaser
*Pvt.	Albert H. Briegel	*Pvt.	Byrd W. Boggs
Pvt.	James Catalano	Pvt.	John A. Cliff
*Pvt.	Charles S. Cratt	*Pvt.	Lester M. Crisp
Pvt.	Frank Curtis	Pvt.	Carl W. Escover
Pvt.	Solomon Friedman	Pvt.	Frank J. Gabriell
Pvt.	William J. Gentholtz	Pvt.	William F. Goodrich
Pvt.	Donaciano Gonzales	Pvt.	Eric J. Johnson
Pvt.	Albin R. Johnson	Pvt.	Frank B. King
Pvt.	Harry Leach	Pvt.	Frank Lulay
Pvt.	Emanuel M. Lucas	Pvt.	Jerry McCoy
Pvt.	Cyrus F. McCartney	Pvt.	Frank E. Niemiec
Pvt.	Joe Mrva	*Pvt.	John Pacquer
Pvt.	Owen O'Donnell	Pvt.	Grover C. Reid
Pvt.	John C. Reeves	Pvt.	Amal Schwartz
Pvt.	Charles H. Schaefer	Pvt.	Paul E. Sparrow
Pvt.	Seymour F. Skutt	*Pvt.	Ray W. Stonebrook
Pvt.	Raymond G. Swanboro	*Pvt.	Fred Vishnifsky
Pvt.	Verdie L. Venters	Pvt.	George C. Winkler
Pvt.	John S. Walker		

COMPANY C

*Sgt.	Arthur C. Finkbeiner	Sgt.	Antonio Louis
Sgt.	Marlo Menzie	Sgt.	Franklin S. Robison
Sgt.	Alvia Rogers	Sgt.	Edmund J. Welton
Cpl.	Earl W. Andrews	Cpl.	Arthur Dean

Cpl. Herman Korringa
Pvt. 1 cl. Burchard C. Bargy
Pvt. 1 cl. Frank Bennett
*Pvt. 1 cl. Harry Cornwell
Pvt. 1 cl. William Goetzinger
Pvt. 1 cl. James M. Miller
*Pvt. 1 cl. Edwin Peterson
Pvt. 1 cl. Joseph M. Walloch
*Pvt. Albert Anderson
Pvt. Nels Anderson
*Pvt. Ewald H. Brandau
Pvt. Harry L. Clinton
Pvt. Frank C. Dorn
Pvt. Magnus Grondal
Pvt. Peter G. Higgins
Pvt. Robert E. Hooper
Pvt. Samuel E. Johnson
Pvt. Adam Marach
*Pvt. Orian A. Miller
Pvt. Karl A. Morandi
Pvt. Earl H. Opel
*Pvt. Henry Rider
Pvt. Karl Sawicky
Pvt. Irwin T. Smith
Pvt. Thomas F. Stanek
Pvt. Myles Swain
Pvt. Frank Troia
Pvt. Jesse J. Wilson
Pvt. Dan. H. Yaple
Pvt. William E. Zierke
Cpl. John J. Morissey
Pvt. 1 cl. Asa Beard
Pvt. 1 cl. Reuben A. Clevenger
*Pvt. 1 cl. George A. Goodney
Pvt. 1 cl. Christen V. Jensen
Pvt. 1 cl. Michael P. Pawlowski
Pvt. 1 cl. Wincenty Pietreniuk
*Pvt. 1 cl. Edward Zachowski
Pvt. John E. Anderson
Pvt. Arthur E. Bush
Pvt. Joseph J. Cawley
Pvt. Nelson Conant
Pvt. Earl D. Findlay
*Pvt. John S. Haug
*Pvt. Charles Hovey
*Pvt. William C. Hutchinson
Pvt William A. Jackson
Pvt. Thomas R. Lynch
*Pvt. Robert L. Marolf
Pvt. John H. Mitchell
Pvt. Samuel E. Murray
*Pvt. John W. Reid, Jr.
Pvt. Albion E. Robison
Pvt. Peter R. Schmidt
*Pvt. William E. Stout
*Pvt. Joseph Timmerman
*Pvt. Antonio Teixeira
Pvt. Robert Wedgewood
Pvt. Lonnie Yancy

Company D

Sgt. Clarence W. Bailar
Sgt. Eugene P. Canty
Cpl. Grover Whitton
Pvt. 1 cl. Charlie T. Gaines
Pvt. 1 cl. James W. Fields
Pvt. William Adams
Pvt. Luther Browning
Pvt. Shade Carlton
Pvt. John L. Cowen
Pvt. William P. Darnall
Pvt. Conrad M. Ellingson
Pvt. William C. Foreman
Pvt. Philip Ghelfi
Pvt. Florian Gosinski
Pvt. Henry Hanson
Pvt. Lewis Heller
*Pvt. Stanley F. Husak
Sgt. Frank G. Burt
Cpl. George I. Furo
*Pvt. 1 cl. John F. Connelly
Pvt. 1 cl. William G. Geno
*Pvt. 1 cl. Cyril Whitman
Pvt. Alex Blinder
Pvt. Harry H. Brueggenjohaan
Pvt. Larkin J. Clay
*Pvt. Louis Cooper
Pvt. George A. Eden
Pvt. Abraham Fine
Pvt. Ira L. Freeman
*Pvt. Stani Gimbts
Pvt. Ole Hamilton
Pvt. Dave C. Harry
Pvt. William A. Herren
Pvt. Emil K. Johnson

*Pvt.	Olaf M. Johnson	Pvt.	Gilbert M. Koontz
Pvt.	Ralph Lamb	*Pvt.	John E. Luibel
Pvt.	William C. Lynch	*Pvt.	Joseph McFadden
Pvt.	Charles H. Olesch	Pvt.	Raymond Pond
Pvt.	Albert L. Roberts	*Pvt.	William Stock
Pvt.	Joseph J. Taggart	Pvt.	Dee Wilden
Pvt.	Beachem W. Willoughby		

Company E

Sgt.	John F. Donovan	Sgt.	Joe F. Koszeuski
Sgt.	Edmund Miodynski	Sgt.	Harold B. Taft
Cpl.	Walter E. Cassens	Cpl.	Warren Cooley
Cpl.	Charles Delling	Cpl.	Hez. G. DeLong
Cpl.	Joe A. De Sousa	*Cpl.	Charles L. Dunn
Cpl.	Frank Elrod	Cpl.	James B. Fellenzer
Cpl.	Otto D. Gottsch	Cpl.	Caesar A. Hollberg
Cpl.	Henry E. Hopper	Cpl.	David G. Jones
Cpl.	John Jurenos	Cpl.	Louis K. Wielock
*Mech.	William L. Spencer	Pvt.	Frank E. Anderson
Pvt.	Oscar Anderson	Pvt.	Philip G. Arnet
Pvt.	Herman S. Artenoff	Pvt.	Hugp Asmus
Pvt.	Hugo Baessler	Pvt.	Harvey H. Barlow
Pvt.	Clawerence Bartram	Pvt.	Adolph C. Bauer
*Pvt.	Celestino Bechelli	Pvt.	James C. Bell
Pvt.	Charles Biaselle	Pvt.	Fred. W. Boerger
*Pvt.	Mitchell Brennen	Pvt.	John H. Brown
*Pvt.	Frank Budnik	Pvt.	Joseph G. Campodonico
Pvt.	Leon Carollo	Pvt.	Walter Carpenter
Pvt.	Oreste Carrai	Pvt.	Colombo Cecchi
Pvt.	Ignacy Chojnacki	Pvt.	Nick Costas
Pvt.	Seth A. Crawford	Pvt.	Leland S. Cunningham
Pvt.	Joseph C. Damico	Pvt.	Henry D. Dardenne
Pvt.	Milyo Denato	Pvt.	George Dodenhoff
*Pvt.	George W. Draeger	Pvt.	William Dreps
Pvt.	Edward T. Edwards	Pvt.	Fred H. Ellis
Pvt.	John A. Ehteridge	Pvt.	Park W. Etter
Pvt.	Arthur R. Felk	Pvt.	Charles Fisher
Pvt.	James Flinn	Pvt.	Luther Flora
Pvt.	Arthur R. Folke	Pvt.	Guisseppe Ferraro
*Pvt.	William E. Flynn	*Pvt.	Tony F. Harmer
Pvt.	Elvin C. Hazelbaker	Pvt.	Chris M. Herbig
Pvt.	Robert J. Hillman	Pvt.	Earl L. Hovey
Pvt.	Lonnie Hundley	Pvt.	Stanley Jablonski
Pvt.	Stanley J. Jarecki	Pvt.	Yames N. Johnson
Pvt.	Gustave A. Johnson	Pvt.	Ernest Keeling, Jr.
Pvt.	Thorwald Kilen	Pvt.	Emanuel King
Pvt.	Julius A. Kopple	Pvt.	Albert J. Kramer
Pvt.	Elmer Lindberg	Pvt.	Peter LaMaack
Pvt.	Thomas M. McMillan	*Pvt.	William F. Marohn
Pvt.	Alvin L. Mattson	Pvt.	John L. Meyer

Pvt.	John Murphy	*Pvt.	John L. Nacke
*Pvt.	Grover Neal	Pvt.	Jim O'Connor
Pvt.	Carl Olson	Pvt.	Frank O. Parsons
Pvt.	George W. Pfitsch	Pvt.	Fernie Plumlee
*Pvt.	Auis Porter	Pvt.	James J. Roach
Pvt.	Biagio B. Saccone	Pvt.	Daniel W. Shaulis
*Pvt.	Posie E. Suthpin	Pvt.	Antonio Testani
Pvt.	Harold H. Tyrell	Pvt.	Harvey F. Welborn
Pvt.	Solon Whitehead		

Company F

*Sgt.	Marion Dale	*Sgt.	Oscar Goyea
Cpl.	Hymen Feldman	*Cpl.	George S. Koonce
*Cpl.	Emerson Lee	Cpl.	Albert G. Mason
*Cpl.	Daniel G. Norton	Cpl.	Bruno Smith
Mech.	Raymond J. Broghamer	Bug.	Grover C. Jenkins
Pvt. 1 cl.	William E. Boetel	*Pvt. 1 cl.	Charles Eissens
Pvt. 1 cl.	Harry A. Layfield	Pvt. 1 cl.	John J. Zankowich
*Pvt.	Otto F. Beitz	Pvt.	James H. Bicknell
*Pvt.	John Bono	*Pvt.	Clair Demott
*Pvt.	Walter Dyrland	Pvt.	Alfred E. Fuerst
Pvt.	Ray E. George	Pvt.	Everett M. Guynn
Pvt.	Ervin Hatfield	Pvt.	Henry G. Hodge
Pvt.	Chris Kowalski	Pvt.	Emanuel Leberer
Pvt.	Robert P. Lyle	Pvt.	Floyd McGee
Pvt.	Guglielmo Pasetto	Pvt.	August Tordrup
Pvt.	Frederick Von Behren	*Pvt.	Angelo Vitro

Company G

*1st. Sgt.	Charles Macken	Sgt.	Louis L. Addonizio
Sgt.	Patrick Conway	Sgt.	Thomas E. Harkins
Sgt.	Argyrios Koyvaras	Cpl.	Francis Donaghue
Cpl.	James A. Dunn	*Cpl.	Warren Flansburg
*Cpl.	Harry E. Glass	*Cpl.	Robert L. Heighton
*Cpl.	Verner B. Parker	Bug.	Edward Wittner
Pvt. 1 cl.	Joseph L. Murray	Pvt.	John D'Amore
*Pvt.	Frank R. Anderson	Pvt.	Spiros Collian
Pvt.	Charles H. Crocker	Pvt.	Frank Dupre
Pvt.	William J. Eagleson	Pvt.	George D. Eliopulos
*Pvt.	Walter H. Franklin	*Pvt.	Roscoe R. Groff
Pvt.	George C. Grohens	*Pvt.	Andrew C. Hansen
Pvt.	Albert Hoium	Pvt.	Theodore Hoes
*Pvt.	James G. John	Pvt.	Ernest J. Kies
Pvt.	Joseph V. Leech	*Pvt.	William G. Laplante
Pvt.	John Libecki	*Pvt.	Edgar V. Le Claire
*Pvt.	Joseph S. Mallak	Pvt.	Nicola Maido
Pvt.	Louis J. Morris	Pvt.	Sidney E. Metz
Pvt.	Stanislaw Masynski	*Pvt.	Emil L. Mullaert
Pvt.	Jim Papavasil	Pvt.	Newell Pancoast

Pvt.	Rudolph Rittar	Pvt.	Svend P. Rasmussen
Pvt.	Louis T. Savage	*Pvt.	Carroll T. Rankin
*Pvt.	Edward A. Sizler	*Pvt.	Pancrazio Simone
*Pvt.	Clarence A. Wiles		

Company H

Mess Sgt.	Clarence R. Kimberlin	*Cpl.	Thomas J. Burke
Cpl.	Royal O. Gray	Cpl.	Oliver P. Mock
Cpl.	Charles Stoffel	Bug.	Earnest C. Richardson
Pvt. 1 cl.	Edward C. Anderson	*Pvt. 1 cl.	Winston Arnett
*Pvt. 1 cl.	Eldo Buzzard	Pvt. 1 cl.	Herman C. H. Gaulke
Pvt. 1 cl.	James M. Ollis	Pvt. 1 cl.	Clarence E. Orr
Pvt. 1 cl.	Robert L. Pentecost	Pvt. 1 cl.	Joseph J. Sullivan
Pvt.	James B. Carpenter	Pvt.	Charles H. Crane
*Pvt.	Harry J. Davis	*Pvt.	Lenigi G. Gogna
Pvt.	Arthur M. Haring	Pvt.	William O. Isham
Pvt.	Benjamin H. Mahaffey	Pvt.	Bennie A. Meyer
*Pvt.	Harry L. Mau	*Pvt.	Victor H. Nelson
Pvt.	Ira R. Palmer	Pvt.	George F. Robertson
Pvt.	Grover L. Roland	Pvt.	Anastacio Trujillo
*Pvt.	Doak Vandyke		

Company I

Sgt.	Ernest Blantin	Sgt.	James R. Patterson
*Cpl.	John W. Houts	*Cpl.	Jacob Kreis
*Cpl.	Arthur E. Wallin	*Pvt. 1 cl.	George A. Buzacott
Pvt. 1 cl.	Royal L. Dahrens	Pvt. 1 cl.	Cecil M. Martin
*Pvt. 1 cl.	Clarence L. Reinig	Pvt.	Albert B. Alexander
Pvt.	Irvan A. Bennett	Pvt.	Ross E. Bankson
*Pvt.	Fred Boss	Pvt.	Jim W. Caliborne
Pvt.	Harry B. Cooper	Pvt.	John S. Hobaugh
Pvt.	Monroe R. Hunter	Pvt.	John Hosdough
*Pvt.	Emanuel Kjelson	Pvt.	Adolph A. Lekanger
Pvt.	Alexander Linde	*Pvt.	Robert A. Madden
Pvt.	William H. McDaniel	*Pvt	Get Micklow
Pvt.	Henry A. Phippin	Pvt.	Thomas J. Prendergast
*Pvt.	Oliver E. Pumphrey	*Pvt.	John Riva
Pvt.	Conford C. Robinson	Pvt.	Frank S. Roethler
Pvt.	Harry Siegel	Pvt.	Arthur F. Scheider
Pvt.	Parks H. Simpson	*Pvt.	John C. Schumacher
Pvt.	Benjamin Thompson	Pvt.	Thomas R. Thompson
*Pvt.	Albert Trebesch	Pvt.	William J. Wilson

Company K

Sgt.	Ross Boyer	*Sgt.	William Wood
Cpl.	Nicholas Alicaris	*Cpl.	De Witt T. Logsdon
Pvt.	Ernest Bennett	Pvt.	Edward H. Brandenberg

Pvt.	Maurice Davis	Pvt.	George Evans
Pvt.	Bert Fox	*Pvt.	George A. Harper
Pvt.	George T. Harper	Pvt.	Forrest Harry
Pvt.	Posey Lamkin	*Pvt.	William Lee
Pvt.	Clyde W. McClelland	Pvt.	Willis McKinnis
*Pvt.	Arthur Millsted	Pvt.	Edward Nathan
*Pvt.	Cyril J. O'Connell	*Pvt.	Anthony J. Parillo
Pvt.	Everett K. Reedy	Pvt.	John Scanlon
Pvt.	Bertrum C. Sherwood	Pvt.	Carl Simon
Pvt.	John Slubowski	Pvt.	Irving H. Sten
Pvt.	Grover C. Stroberger	Pvt.	Earl R. Swanson
*Pvt.	Halvard H. Shogen	*Pvt.	John Short
*Pvt.	Louis J. Shutsa		

COMPANY L

1st Sgt.	Maurice J. Lynch	Cpl.	Pasquale Frezza
Cpl.	Reuben Hanson	Cpl.	Edward L. Rosenthal
*Cpl.	Jacob J. Statz	Pvt. 1 cl.	William R. Hunt
*Pvt. 1 cl.	Sherman Olsen	Pvt. 1 cl.	Lawrence Rehme
*Pvt. 1 cl.	Anton Simiatis	Pvt.	John H. Gardella
Pvt.	Charles C. Giffhorn	*Pvt.	Edward G. Hackman
Pvt.	Claude L. Halsey	Pvt.	Michael Harvey
Pvt.	George E. Hearn	Pvt.	Charles L. Hilsabeck
Pvt.	Louis H. Hosler	Pvt.	Thomas W. Hughes
Pvt.	Ray J. Hulett	Pvt.	Arthur J. Jeffree
Pvt.	Frank Jirous	*Pvt.	Severt Johnson
Pvt.	George P. Kalknarf	Pvt.	Vincent Kirvin
Pvt.	George W. Knight	Pvt.	William Libao
Pvt.	Lawrence McCarthy	Pvt.	Stefan Milton
Pvt.	John J. Molyneaux	*Pvt.	Roy A. Musser
Pvt.	Fred L. Nichols	Pvt.	Nick Palico
*Pvt.	John L. Parks	Pvt.	Walter A. Rautkis
Pvt.	William L. Robideau	*Pvt.	Thomas Robson
Pvt.	Otto Schultz	Pvt.	Leon N. Smith
Pvt.	William W. Spragg	*Pvt.	Fred C. Splittgerber
*Pvt.	Stevens N. Tate	Pvt.	Evart Thomas
Pvt.	Arthur H. Vedelius	Pvt.	Homer Weiss

COMPANY M

*1st. Sgt.	Paul Von Krebs	Sgt.	Forest N. Burns
*Sgt.	George N. Sleezer	Sgt.	Thomas Vitovec
Sgt.	Francis P. Williams	Cpl.	Ira E. Bonnell
*Cpl.	John Lewis	Cpl.	Hamlet Lombardy
Cpl.	Sylvester Tomberlain	Cpl.	Jefferson L. Winn
Pvt. 1 cl.	James L. Allen	*Pvt. 1 cl.	Henry H. Barrows
Pvt. 1 cl.	Richard B. Phillips	Pvt. 1 cl.	Martin Troy
*Pvt. 1 cl.	William Tech	Pvt. 1 cl.	David D. Sloan
*Pvt. 1 cl.	Joseph Wroble	*Pvt.	Fred Arighi
Pvt.	Hans T. E. Anderson	*Pvt.	Claude C. Courteney
Pvt.	Martin Biesterfield	Pvt.	Andrew J. James

Pvt.	Charles C. Curry	*Pvt.	Philip Link
Pvt.	Joe Limon	*Pvt.	William R. Patzer
Pvt.	John C. Meeker	Pvt.	Grover C. Sheets
*Pvt.	William Phillips	Pvt.	Jesse C. Williamson
Pvt.	Lee H. Whitt	Pvt.	Thomas L. Wray
Pvt.	Eugene Wilson	Pvt.	Frank M. Wright
Pvt.	Clarence O. Wright	Pvt.	Charles D. Allen

Headquarters Company

1st. Sgt.	Erdroy E. Peterson	Sgt.	Leo R. Farmer
*Cpl.	Howard W. Bitner	*Cpl.	Forrest L. Cooney
*Cpl.	Albert S. Weeks	*Mus.	Fred J. Slager
Pvt. 1 cl.	Paolo La Bruna	Pvt. 1 cl.	Donald Lewis
Pvt.	Adolph Jacowicz	*Pvt.	Willie E. Little
*Pvt.	Anthony Marfaro	*Pvt.	Steve Meyers
Pvt.	Frank C. Mitchell	Pvt.	Joseph G. Mitchell
Pvt.	James Murray	Pvt.	John M. Murray
Pvt.	Earl M. Scherrer	*Pvt.	Frank W. Spafford
*Pvt.	Otis Turbaville	Pvt.	Edmund H. Wagner

Machine Gun Company

Sgt.	Alfred Carter	Sgt.	Frank G. Flood
*Pvt. 1 cl.	Alexander B. Linnett	Pvt. 1 cl.	Clinton H. Morris
Pvt. 1 cl.	Louis Roberson	Pvt. 1 cl.	James P. Walker
Pvt.	Thomas Allen	Pvt.	Patrick H. Buckley
Pvt.	Irvin E. Davis	Pvt.	Dennis J. Ford
Pvt.	Warner C. Johnson	Pvt.	Ira F. Sidell
Pvt.	Henry Thomas	Pvt.	James K. Ward

Supply Company

Wgr.	Henry Kapperman	*Pvt.	Paul Blumke

Medical Detachment

Pvt. 1 cl.	Jesse J. Fairbrother	*Pvt.	Richard Jasper
*Pvt.	Maurice H. Johnson		

11TH MACHINE GUN BATTALION

Officers

*1st. Lieutenant .... Albert E. Simpson

Enlisted Men

Company A

Cpl.	Raymond Cordi	Cpl.	Leam Shelsdon
*Pvt.	Lacey T. Carey	Pvt.	James A. Merchand
Pvt.	Jack W. Montague	Pvt.	Joseph Lacamera

Pvt.	Vernon C. Cameron	Pvt.	Jacob L. Christ
*Pvt.	William Horsepool	Cpl.	William Evans

Company B

Sgt.	Earl Lafferty	Sgt.	Claude E. Cherry
Cpl.	Charles H. Bender	Pvt. 1 cl.	John P. Osborn
Pvt.	Ralph E. Williams	Pvt.	Stanley Savitski
Pvt.	Harry E. Murphy	Pvt.	Elmer D. Bloomingdale
*Pvt.	Harvey Lawaster	Pvt.	Harry J. Waywacker

Company C

Sgt.	Richard W. Stonestreet	Sgt.	Fred C. Gwozdg
*Sgt.	George Payne	Pvt.	Edward C. Floding
Pvt.	Harry J. Waywacker	Pvt.	Huge Basseler
*Pvt.	John W. Eardley	Pvt.	William H. Bogard

Company D

Sgt.	Sam T. Turner	Cpl.	Harry C. Kratzer
*Pvt.	John Smore	Pvt.	Edward J. Brady

Medical Detachment

Pvt. Harold J. Martin

8TH INFANTRY BRIGADE

BRIGADE HEADQUARTERS

Officers

*1st. Lieutenant................Barrington W. Sellers
*2nd Lieutenant................Tom D. Halliday

Enlisted Men

Sgt.	Floyd E. Nevins	Pvt. 1 cl.	Lowrey G. Warren
Pvt.	Paschel Ashmore		

58TH INFANTRY REGIMENT

Officers

Major.......................Samuel H. Houston
Captain.....................William H. Humphreys
Captain.....................Jeremiah E. Murphy
*Captain....................Harry E. Pond
1st. Lieutenant..............William E. G. Cooper
1st. Lieutenant..............Harold D. Cupitt
1st. Lieutenant..............Peter W. Ebbert
1st. Lieutenant..............Neal B. Finley

*1st. Lieutenant ....Harry Green
1st. Lieutenant ....Robert E. Kennington
1st. Lieutenant ....Goerge W. Long, Jr.
1st. Lieutenant ....Edgar E. Roberts
2nd. Lieutenant..............Frederick M. Atwood
2nd. Lieutenant..............Eldon F. Brewster
2nd. Lieutenant..............Burton N. Chase
*2nd. Lieutenant..............Kenneth V. Elliott
*2nd. Lieutenant..............Thurman G. Flannigan
2nd. Lieutenant..............John W. Hassell
2nd. Lieutenant..............Harry E. Hernandez
2nd. Lieutenant..............Harry A. Kroeger
2nd. Lieutenant..............John A. Long
2nd. Lieutenant..............Lowell H. Riley
2nd. Lieutenant..............Charles T. Roby
*2nd. Lieutenant..............Charles N. Bajart

Enlisted Men

Company A

*Sgt.	William H. Hartung	Cpl.	Theodore W. Peterson
Cpl.	Otto Lee	Cpl.	Joseph P. Voliman
*Cpl.	Mike Ofsovich	Cpl.	Emerson G. Woelfel
Pvt. 1 cl.	John Shubick	Pvt.	John Draper
*Pvt.	Jess Adams	*Pvt.	Edward Dwyer
*Pvt.	Robert L. Adamson	Pvt.	Willis Foster
Pvt.	Henry Apel	Pvt.	Floyd Gallagher
Pvt.	Peter Arthurs	Pvt.	Delbert Gass
Pvt.	James Aupperle	Pvt.	Hugh A. Golden
Pvt.	Henry C. Bernhart	Pvt.	Louis Gordon
Pvt.	Louis E. Berry	Pvt.	Northern C. Green
Pvt.	John Bohl	*Pvt.	Dan F. Kaufman
Pvt.	Harry Betts	Pvt.	Domenico Leo
Pvt.	Clarence J. Brandley	Pvt.	Fred Livermore
*Pvt.	Leo Braun	Pvt.	John Pichelmeyer
Pvt.	Thomas E. Callahan	Pvt.	William C. Schleiger
Pvt.	George E. Chaddick	*Pvt.	Harry Sheenan
Pvt.	Thomas Chimielowjnz	Pvt.	Charley H. Vetter
Pvt.	John Damas	Pvt.	Dewey Woody
Pvt.	John Doles		

Company B

1st. Sgt.	John L. Billing	Sgt.	Theodore Patnaude
Sgt.	John Millsop	Sgt.	David M. Smith
Cpl.	George M. Bissett	*Cpl.	John T. Markley
Cpl.	Fred Chappell	Cpl.	Castos P. Rados
*Cpl.	Gilbert Irvin	Cpl.	Roy E. Shenk
Cpl.	Carl J. Jensen	*Mech.	James Peightal
Pvt. 1 cl.	Nicholas E. Blanc	Pvt. 1 cl.	Claude B. Vaughn

Pvt. 1 cl. Louis Caneto
Pvt. Henry Anderson
Pvt. Clinton O. Allard
Pvt. Oscar C. Armstrong
Pvt. Brunon Bichniewicz
Pvt. Luciano Billanuego
Pvt. Elmer E. Bowbey
Pvt. Leslie C. Bracken
Pvt. Walter G. Bracken
Pvt. George N. Buchanan
Pvt. Emile Bucher
Pvt. William A. Brown
Pvt. Emil J. Bosh
Pvt. George D. Booselis
Pvt. Erwin W. Bosley
Pvt. Clyde Ellison Bosley
Pvt. Andrew Blackwell
Pvt. Henry Campbell
Pvt. Andrew A. Christiansen
Pvt. William J. Croatt
Pvt. Louis V. Castro
Pvt. Joseph P. Callan
Pvt. Edwin L. Clausing
Pvt. Fred. D. Canwell
Pvt. Virgil C. Cook
Pvt. Thomas Courtney
Pvt. Charles Cushman
Pvt. Edward C. Delray
Pvt. Domenico Decao
Pvt. Herman Diehl
*Pvt. Herman W. Dierks
Pvt. Conrad Eckel
Pvt. Francis W. Fagan
Pvt. Joseph Feest
Pvt. William B. Fowles
Pvt. John Gianettoni
Pvt. Edward L. Gerl
Pvt. Redwald Gottenburg
Pvt. Guiseppi Graci
Pvt. Fred Gerhardt
Pvt. Thorvald Granby
*Pvt. Herman F. Gustafson
Pvt. Armilo Gliottoni
Pvt. Peter Graham
Pvt. Cosmo Griego
Pvt. Eliseo Griego
Pvt. Mike Gryncewicz
Pvt. Henry M. Guttormson
Pvt. Charles F. Hackler
Pvt. Thadius Hadges
Pvt. Cleveland Head
*Pvt. Frank V. Harpur
Pvt. John Jeselskis
Pvt. Clim Johnson
Pvt. Anthony C. Klein
Pvt. John Kobus
Pvt. Isidore M. Kneip
Pvt. Diamond L. Larson
Pvt. Carl Latak
Pvt. John S. Larson
Pvt. Anton W. Lundell
Pvt. Frank Lewandowski
Pvt. Henry C. Lading
Pvt. Clyde B. Lindsey
Pvt. Henry Maertz
Pvt. Jesse Mars
Pvt. Rudolph Mikle
Pvt. Frank McKinney
Pvt. Jesse Milone
Pvt. James G. McCarthy
Pvt. Frank O'Dell
Pvt. Peter Rosle
Pvt. Albert H. Rozell
Pvt. Lee Reaser
Pvt. Frank Reaux
Pvt. Frank J. Slovich
Pvt. Fred E. Sober
Pvt. Walter G. Sautter
Pvt. Louis P. Spies
Pvt. Ray Swartz
Pvt. John J. Schuh
Pvt. Joseph Sherman
Pvt. Maurice C. Sweetland
*Pvt. Elbert Stone
Pvt. John J. Springer
Pvt. Founth Thetford
Pvt. Anton Tomek, Jr.
Pvt. Willow Trapp
Pvt. John M. Trenda
Pvt. Barney B. Williams
Pvt. Edward N. Weber

Company C

*Sgt. Frank Fagan
*Cpl. Walter F. Brostrom
Cpl. Max R. Grundkovski
Cpl. Stanley Morozik
Cpl. Michael J. Oberst
*Cpl. Robert Race

Cpl. John D. Kirkpatrick
Mech. Jacob Honig
Pvt. 1 cl. Henry Das
Pvt. 1 cl. Patrick A. Robarge
Pvt. 1 cl. John Schwatz
Pvt. Joseph F. Andrade
*Pvt. David L. Berg
*Pvt. Philip L. Brevold
Pvt. Philip S. Boldridge
Pvt. Giorgi Bonomo
Pvt. Andre Burwan
Pvt. Arthur E. Bush
Pvt. Lester G. Carter
Pvt. William McK. Crites
*Pvt. James A. Crosland
Pvt. Martin J. Dahl
*Pvt. Fred Dube
Pvt. Lorenzo D'Amico
Pvt. Frank P. DiPesa
Pvt. Ralph L. Felix
Pvt. Paul F. Fountain
*Pvt. Matthew W. Gerra
*Pvt. Charlie C. Gourd
*Pvt. Herman Gulbranson
Pvt. Frederick Hammel
*Cpl. Edward P. Santaw
Mech. Theodore E. Jacquet
*Pvt. 1 cl. Berthold C. Schwartz
Pvt. 1 cl. Robert Shipbaugh
*Pvt. 1 cl. Edward H. Uleary
Pvt. 1 cl. Jay C. Williams
Pvt. Bendik Hauan
Pvt. Paul Herrera
*Pvt. Henry M. Jones
Pvt. John Luczyk
Pvt. Kurth Meyer
Pvt. William J. Moyer
Pvt. Alfred Pagers
Pvt. Manuel J. Pice
Pvt. John A. Pirtie
*Pvt. Clarence L. Perry
*Pvt. Earl F. Philips
Pvt. Roy E. Power
Pvt. Clarence R. Roberts
Pvt. Edwin Roser
Pvt. Wait F. Worrell
Pvt. William C. Wolters
*Pvt. William L. Whidden
Pvt. Henry Winters, Jr.

Company D

Sgt. Arthur R. Lemke
Sgt. Joseph Messina
Cpl. Robert Kellner
Pvt. Stanislaw Andrych
Pvt. John Ayler
*Pvt. Stephen Bogus
Pvt. Ralph Copola
Pvt. Joe Dennietio
*Pvt. Albert L. Dolven
Pvt. Fred Ekelman
*Pvt. Stanley L. Elliott
Pvt. Ariel N. Facey
Pvt. Adam Felter
Pvt. John Fisher
*Pvt. Okey R. Ferrell
*Pvt. Charley Forsberg
*Pvt. Gilmore L. Foss
Pvt. Tom Findley
Pvt. Stephen Gust
Pvt. Grover Harter
*Pvt. Arthur Hyatt
Pvt. Evald Helstad
Pvt. Elbert C. Ivy
Sgt. Ennis R. Patton
Cpl. John W. Wilson
Cook Frank Coleman
*Pvt. Walter W. Jackson
Pvt. George L. Jackson
Pvt. Arndt T. Johnson
Pvt. Carl M. Johnson
Pvt. Olaf Knutson
Pvt. Otto Kostelak
Pvt. Steward Kroll
Pvt. Thomas V. Lynch
Pvt. Hans P. Marcher
Pvt. John Mares
Pvt. Stanley Marzee
Pvt. John W. McLoughlin
Pvt. Alf Meyer
Pvt. August Nelson
Pvt. Thomas Orgorodnik
*Pvt. John L. Paulson
Pvt. Walter Pluecker
*Pvt. Carrey Redkey
*Pvt. Raymond Roberts
Pvt. Robert C. Richardson

Pvt.	Willie Schlie	Pvt.	Ralph J. Thibodeau
Pvt.	Stanley Semrau	Pvt.	Joseph Wesoloski
*Pvt.	John Sulzicki	Pvt.	Albert Willman

Company E

1st. Sgt.	Frank J. Sarley	Sgt.	Rush C. Davis
Sgt.	Lloyd Ackerman	*Cpl.	Louis G. Peters
Cpl.	Frank J. Bowman	Cpl.	Phillip Postier
Cpl.	Vincent L. Giantvalley	Cpl.	William H. Patton
*Cpl.	Harold F. Mason	Cpl.	Charles S. Ross
Cpl.	Anton Maleski	*Cpl.	Andrew C. Tallman
Cpl.	John F. O'Reilly	Cpl.	Arthur L. Yost
Mech.	Patrick J. Eustace	Mech	Patrick J. Rodgers
Pvt. 1 cl.	Vito Gerondo	Pvt.	William Myers
Pvt.	Clarence L. Armstrong	Pvt.	Ray A. Myus
Pvt.	Herman Baker	Pvt.	Earl Nelson
Pvt.	Henry E. Binet	Pvt.	Frank Nelson
Pvt.	Edward S. Canavan	Pvt.	John Nilson
Pvt.	Raphaele Capobianco	*Pvt.	Alfred Oppegaard
Pvt.	Owen C. Channel	Pvt.	Andrew Ottum
Pvt.	Marcel Debaight	Pvt.	Fred Pagano
Pvt.	Peter Della Rocca	Pvt.	Spero A. Pappandrikopoulos
Pvt.	Simon E. Frederickson		
Pvt.	Casper A. Gilson	Pvt.	John Pavel
Pvt.	Walter Haynes	Pvt.	Harrison F. Pederson
Pvt.	Thomas H. Irwin	Pvt.	Andrew Peterson
Pvt.	Fred Jones	Pvt.	Arthur L. Peterson
Pvt.	Lester King	Pvt.	Nick Philipchuck
Pvt.	Harry Lantz	Pvt.	Howard M. Polen
Pvt.	Herman L. Lee	Pvt.	Herman Reis
Pvt.	Robert R. Logan	Pvt.	William Richards
Pvt.	Pete Lucak	Pvt.	Harry Rock
Pvt.	Oscar Lura	Pvt.	Pete Rodovanovich
Pvt.	Tony Marzoville	Pvt.	Ethel E. Roberts
Pvt.	Emil Mase	Pvt.	Viktor Schberinski
Pvt.	Anthony R. Matos	Pvt.	Frank P. Trester
Pvt.	William M. Mitchell	Pvt.	Alonzo L. Williams
Pvt.	Frank E. Morton	Pvt.	Herbert L. Wilson

Company F

Sgt.	Edward Ailes	Sgt.	Lewis A. Pye
Sgt.	Roy H. Chandler	Sgt.	Mark D. Singross
Cpl.	Clifford P. Lewis	Cpl.	William O. Louden
Cpl.	Daniel P. McGrath	Cpl.	Donald E. Turner
Pvt. 1 cl.	Harold F. Canfield	Pvt. 1 cl.	James Nogulic
Pvt. 1 cl.	Clyde Comer	Pvt. 1 cl.	Daniel Pietrogiacomo
Pvt. 1 cl.	Bruno Cosmond	Pvt. 1 cl.	Niels Skeem
*Pvt. 1 cl.	William F. Gavin	Pvt. 1 cl.	Gaetano Sorforte
Pvt. 1 cl.	Frank Mirachi	Pvt. 1 cl.	George W. Stockwell
*Pvt. 1 cl.	James C. McNeil	Pvt. 1 cl.	Alexander Udin

Pvt. Charles F. Batson
Pvt. William A. Billbe
Pvt. Giovanni Cannone
*Pvt. Aaron Caplan
Pvt. Raymond H. Dailey
*Pvt. John Deusswicz
Pvt. Carl J. Eckburg
Pvt. John E. Ellingsberg
Pvt. George H. Fritz
*Pvt. Luigi Gavazi
Pvt. John Gear
Pvt. Guiseppe Giardino
Pvt. Isidor Gingold
Pvt. John M. Ginney
Pvt. James E. Glenn
Pvt. Ben O. Hill
Pvt. Walter Herkowski
Pvt. Finis E. Johnson
Pvt. Gustaf E. Lindberg
Pvt. Stanley Michalski, Jr.
Pvt. Neal Morgan
Pvt. Fred W. Neagle
Pvt. William C. Nippert
Pvt. Elmer A. Oakley
Pvt. Piotr Oleksisk
Pvt. Viktor Ostrowski
Pvt. Leopoldo Pasolini
Pvt. Henry Rivers
*Pvt. Joseph Schulte
Pvt. Alexander Silva
Pvt. William E. Silvers
Pvt. Thorolf Simmonson
Pvt. William Thompson
Pvt. Lawrence G. Welch
Pvt. Morgan E. Welker
Pvt. William A. Weibel
Pvt. Walter W. Weiher
Pvt. Chris S. F. Willardson
Pvt. Linus L. Walters

Company G

*Sgt. Martin D. Foy
Cpl. Charles N. Franks
Cpl. James W. McCoy
Pvt. 1 cl. Eddie O. Anderson
Pvt. 1 cl. Charlie A. Feldman
Pvt. 1 cl. Alvin L. Henry
Pvt. James Z. Breeland
Pvt. Frank E. Caesar
*Pvt. James Chio
Pvt. James C. Coarsey
Pvt. William Corcoran
*Pvt Thurman Davis
Pvt. Charles Denaro
Pvt. Thomas G. Doris
Pvt. John Enno
*Pvt. William D. Eubanks
Pvt. Albert Fines
*Pvt. Harry M. Fitzgerald
*Pvt. Edward Flaherty
Pvt. Henry W. Flessner
Pvt. John J. Frahm
Pvt. Luther M. Gardner
Pvt. Louis R. Greco
Sgt. Harry Rubenstein
Cpl. Glen H. Shipley
Cpl. Clarence C. Venuti
Pvt. 1 cl. Felix Kuzbarkis
Pvt. 1 cl. Bollos Stathakis
*Pvt. 1 cl. Hall W. Watts
Pvt. Harry Haddix
Pvt. Thomas T. Hammer
Pvt. Homer W. Harrouf
Pvt. William M. Helms
Pvt. Edward E. Hayes
Pvt. George W. Hutchinson
Pvt. Frank Lada
Pvt. Andrew Lillico
Pvt. Irvin E. Larson
*Pvt. Roland F. McLaughlin
Pvt. Henry Peas
Pvt. Henry F. Peterson
*Pvt. Herschel Rongui
*Pvt. Harry Sullins
Pvt. Lee Trammell
Pvt. Charles E. Voris

Company H

1st. Sgt. Charles Stephens
Sgt. Robert E. Conway
Cpl. Charles C. Cook
Sgt. Herbert Miller
*Sgt. Fred Strassburg
*Cpl. Charles H. Miller

*Cpl.	William Hank
Cpl.	Louis S. Hartshorn
Cpl.	Jacob J. Jockett
Cpl.	Albert O. Kjelde
Cpl.	William R. Marley
Pvt. 1 cl.	Henry G. Knox
Pvt. 1 cl.	Edward Roche
Pvt.	William C. Coleman
Pvt.	John G. Dietz
Pvt.	George Dubie
Pvt.	Jim Galone
Pvt.	Jacob G. Heppner
Pvt.	Swen M. Johnson
Pvt.	Victor Johnson
Pvt.	Otto H. Klein
Pvt.	Harry A. Lee
Pvt.	Ed Martin
Pvt.	John F. Ming
Pvt.	Joseph Monkowski
Pvt.	Tommy H. Moore
Cpl.	Frank L. McTighe
Cpl.	Stacey J. Pine
Cpl.	Edward Pomarnky
Cpl.	Osco Tucker
Mech.	Lorenzo Martell
Pvt.	Edmund C. Naber
Pvt.	Miles T. Nix
*Pvt.	Michael Palese
Pvt.	Caleb W. Pearson
Pvt.	Joe B. Philapy
Pvt.	Albert E. Rainville
Pvt.	Earl J. Roggensees
Pvt.	Max Schaff
Pvt.	Carl C. Schmidt
Pvt.	Manuel Soto
Pvt.	Harry Tighe
*Pvt.	Ira Welliver
*Pvt.	Thomas D. Wells
Pvt.	Nugent U. Whitton
Pvt.	Emil Wohlgemuth
Pvt.	John J. Wyss

Company I

Sgt.	Charles J. Kral
Pvt. 1 cl.	William C. Baumgart
Pvt. 1 cl.	George M. Bodde
Pvt.	Toney Alasky
Pvt.	William E. Bredendick
Pvt.	Florianno Didenato
Pvt.	Joseph Fortin
Pvt.	Frank A. Griffith
Pvt.	Arthur Hulstrand
Pvt.	Adam Kweder
Pvt.	Gurdon E. May
Pvt. 1 cl.	William Lesher
Mech.	Homer Reed
Pvt.	Frank Meconi
*Pvt.	James Niel
Pvt.	Elmer L. Parker
Pvt.	Storr Persinski
Pvt.	Voncenzo Russo
Pvt.	Jacob R. Smolsvik
Pvt.	Albert J. Supernor
Pvt.	Guy R. Vaughn
Pvt.	Mathias Webber
Pvt.	Loreto Zuccoro

Company K

*1st. Sgt.	William G. Jackson
Sgt.	Frederick A. Dockendorf
Cpl.	William Callahan
Pvt. 1 cl.	Harry B. Adelsbach
Pvt. 1 cl.	Conrad F. Ekblom
Bug.	Frank C. Fergus
*Pvt.	William Bertell
Pvt.	John C. Bridgham
*Pvt.	Peter K. Hapes
Pvt.	Anthony N. Kling
Pvt.	Paul Kowalski
Sgt.	Frank J. Fisher
Sgt.	Donald Forsyth
Cpl.	John A. Yanoscak
*Pvt. 1 cl.	John Lankheet
*Pvt. 1 cl.	Thomas Robison
Pvt.	Philip LaBona
Pvt.	Gerhard K. Larson
Pvt.	William Larson
Pvt.	Adel L. Peterson
Pvt.	Louis A. Premo
*Pvt.	James A. Quick
Pvt.	Joe Siminowsky

Company L

*1st. Sgt.	Kenneth H. Thompson
Sgt.	Elwood T. Abbott
Cpl.	John Allen
Cpl.	William Frey
Pvt. 1 cl.	Clarence A. Stroud
Mech.	Ira Phillips
Pvt.	Alphonso Boissenau
Pvt.	Fedor Borovik
Pvt.	Clyde T. Carson
Pvt.	John J. Carter
Pvt.	Merlin Clark
Pvt.	Linas W. Duffany
Pvt.	Richard Flotre
Pvt.	Gale Hamrick
Pvt.	Alfred A. Hidde
Pvt.	Vernon Hill
*Pvt.	Michael T. Jablonski
Pvt.	Uroan Karcher
Sgt.	Cecil C. Martin
Cpl.	George F. Jones
Cpl.	Earl A. Tyrer
Cpl.	Elven W. Young
Pvt. 1 cl.	Clarence Sween
Pvt.	Ira C. Lewis
Pvt.	Raymond E. Lowe
Pvt.	Lewis Milyaro
Pvt.	Bart C. Okerson
Pvt.	Andrew J. Ottinger
*Pvt.	Ray D. Stubbs
Pvt.	Howard E. Swanson
*Pvt.	Edward Vargo
Pvt.	Paul Wellens
Pvt.	Englebert Wenzler
*Pvt.	John S. Wysocki
Pvt.	Charles R. Wilhite
Pvt.	John C. Young

Company M

Cpl.	Lewis Daniel Clark
*Pvt. 1 cl.	Raymond Gilger
Pvt.	Frederick W. Cochrane
Pvt.	William R. Cottrell
Pvt.	Frederick Harter
*Pvt.	Louis Heacock
Cpl.	Frank H. Reese
Pvt.	Alfred Johnson
Pvt.	Raymond A. Landua
Pvt.	Walter D. Mason
*Pvt.	Frank J. Schidlowski
*Pvt.	Luigi Salinetti

Headquarters Company

*Sgt.	Peachy G. Sanders
Cpl.	William L. Bowling
Cpl.	Joseph A. Connelly
*Cpl.	Ander J. Fredson
Mus. 3 cl.	Ray A. Master
Mus. 3 cl.	Ross A. Wilcox
Pvt.	Alfred W. Anderson
Pvt.	Angus Armel
Pvt.	Gilbert Bond
Pvt.	Chester P. Buescher
Pvt.	James Burns
Pvt.	Richard O. Burns
Pvt.	Alberis Callewaert
Pvt.	Bertenius Christianson
*Pvt.	Peter Defoe
*Pvt.	Lieu S. Dixon
Pvt.	Connie L. Dodd
Pvt.	Paul Dupuy
Pvt.	Ferris L. Eckis
*Cpl.	Edward W. D. Luedtke
Cpl.	George W. McCaffrey
Cpl.	Harry P. Smith
Cpl.	Eugene Trottier
*Pvt. 1 cl.	Erza Curtis
*Pvt. 1 cl.	Ebert Hall
Pvt.	Guy W. Hughes
*Pvt.	Charles Lepczyk
Pvt.	Harry G. Lyerla
*Pvt.	Elmer E. Maurer
Pvt.	Michael Matsco
Pvt.	William D. Mayo
Pvt.	Thomas McHugh
Pvt.	Lloyd S. Ripley
Pvt.	John Robey
Pvt.	John Rowe
Pvt.	Raymond Stake
Pvt.	Ethan Stone
Pvt.	William O. Thompson

MACHINE GUN COMPANY

Sgt.	Harry Berson	Cpl.	George Vannett
Pvt. 1 cl.	John E. Carlson	Pvt.	Goerge Marquardt
*Pvt.	Oliver L. Bedell	Pvt.	Lloyd R. Mason
Pvt.	Harry E. Davis	Pvt.	Emit E. Mitchell
Pvt.	Laurus Fagerland	*Pvt.	John T. Morton
Pvt.	George W. Gillund	Pvt.	Clinton W. Parker
Pvt.	Edgar J. Hartman	Pvt.	Jay Reber
Pvt.	Walter Hecklin	Pvt.	Paul W. Schroeder
Pvt.	Gustaf Lindberg	Pvt.	Albert C. Seiler
Pvt.	Frank M. Lisse	Pvt.	Claude B. Wilcox
Pvt.	Willie Lucas		

SUPPLY COMPANY

*Wag.	Frank Nelson	Wag.	Phillip Sherlock

MEDICAL DETACHMENT

*Pvt	Jesse E. Garretson	Pvt.	Herbert S. Rowden
Pvt.	John G. Owen		

59TH INFANTRY REGIMENT

Officers

Captain. . . . . Howard G. McCall
1st. Lieutenant. . . . . David M. Barry
1st. Lieutenant. . . . . Leoline O. Crane
1st. Lieutenant. . . . . Donald Ghent
1st. Lieutenant. . . . . Oliver T. McKeown
*1st. Lieutenant. . . . . Arthur McAllister
*1st. Lieutenant. . . . . Edmund V. Walsh
*1st. Lieutenant. . . . . Harold G. Wasgatt
1st. Lieutenant. . . . . Alexander D. Wilson
2nd. Lieutenant. . . . . Joseph Brewster
*2nd. Lieutenant. . . . . James J. Boyle
2nd. Lieutenant. . . . . John E. Creeden
2nd. Lieutenant. . . . . Walter W. Craig
2nd. Lieutenant. . . . . Mark E. Heller
*2nd. Lieutenant. . . . . Louis L. LeConte
2nd. Lieutenant. . . . . Forbes Rickard, Jr.
2nd. Lieutenant. . . . . Ennis Stillwell
*2nd. Lieutenant. . . . . Alexander Williams

Enlisted Men

COMPANY A

Sgt.	Homer J. Bleau	Sgt.	Grover Rockerhausen
Sgt.	George Greepa	Sgt.	Lum C. Stringer
*Sgt.	Arthur Paulson	*Sgt.	John Suba

Cpl.	Arthur Bruck	Cpl.	Walter Siegel
Cpl.	Ernest Fasnes	Cpl.	Edward Sullivan
Cpl.	Edward Hampton	Cpl.	Marion Soinski
Cpl.	Robert Langenstein	Bug.	Harold L. Hall
Pvt. 1 cl.	Chris W. Boudouris	Pvt. 1 cl.	Paul Lombardi
*Pvt. 1 cl.	Lonnie Childers	Pvt. 1 cl.	Claude Shankland
*Pvt.	Sam Angos	Pvt.	Patrick McLyneaux
Pvt.	Guy Boardman	Pvt.	Charles Nagle
Pvt.	James F. Booth	Pvt.	Chris Schaflein
Pvt.	Edgar Bulka	Pvt.	Patrick Waters
Pvt.	Angelo Cannizzaro	Pvt.	Barney Wichlacz
Pvt.	Thomas H. Crowley	Pvt.	E. Wilburn
Pvt.	Thomas Cusack	Pvt.	Anthony M. Kling
*Pvt.	Hugh Carney	Pvt.	Nelson LaFrance
*Pvt.	John Cielaskerioicz	*Pvt.	Clarence Logsdon
Pvt.	John Doles	*Pvt.	John Monterastell
Pvt.	Dominic Farrara	Pvt.	Jay Mann
Pvt.	Albert Gelonek	Pvt.	Nicola Martucci
Pvt.	Seldon R. Hickox	Pvt.	Lawrence McCaffrey
Pvt.	John Harvey	Pvt.	William Rautenberg
Pvt.	Stanley Hugunine	Pvt.	Hobart See
*Pvt.	Charles Jacobucci	Pvt.	Charles M. Woodward
Pvt.	Albert L. Jones	Pvt.	Aren Whitcop
*Pvt.	William Knoll	*Pvt.	Hal Whittington

Company B

Sgt.	Thomas E. Lamont	*Sgt.	Frank Brown
Cpl.	Sheldon A. Axelson	*Cpl.	Lewis C. Gray
Cpl.	Domina Caterina	*Cpl.	Herbert E. Macon
Cpl.	Louis W. Hamblen	*Cpl.	Dominic DeAngelo
Cpl.	Frank Kowalkowski	*Pvt. 1 cl.	Steve E. Christian
Pvt. 1 cl.	Felix Burnett	*Pvt. 1 cl.	Bert Elliot
Pvt.	James Brady	Pvt.	William G. Perry
Pvt.	Nicholas W. Burkhart	*Pvt.	Carl J. Pearson
*Pvt.	William Best	Pvt.	George N. Reiss
Pvt.	Harry E. Bassett	Pvt.	Frank Angus Ross
Pvt.	Alva G. Grover	Pvt.	Lyman M. Smith
Pvt.	Hayden Gullet	Pvt.	William Summers
Pvt.	Ottis P. Kahle	*Pvt.	John Strausbough
Pvt.	William F. Kelley	Pvt.	Henry Tehark
Pvt.	Otto Lange	*Pvt.	Howard S. Thomas
Pvt.	Selmer Lee	Pvt.	Charles Vetri
Pvt.	Bruno Monterosso	Pvt.	Franklin W. Vincent
		Pvt.	Benjamin Wiggins

Company C

Sgt.	Peter Edwards	Sgt.	Wade El Mulford
Cpl.	Harry F. Bernard	Cpl.	James Malkowski
Cpl.	Henry Deloy	Cpl.	Charles H. Sells
Cpl.	Joseph J. Forstner	Mech.	Mike V. Saner

Pvt. 1 cl.	James C. Haizlip
Pvt. 1 cl.	Walter J. Halstead
Pvt. 1 cl.	William Leslie
Pvt.	George A. Aldinger
Pvt.	Mike Amodia
Pvt.	Louis Barker
Pvt.	John D. Black
Pvt.	Harold E. Bellmer
Pvt.	Clarence L. Berry
Pvt.	Philip S. Baldridge
Pvt.	George Cerveny
Pvt.	John W. Clerkin
Pvt.	Edgar Conner
Pvt.	Arthur G. Cornwell
*Pvt.	Thomas A. Grubbs
*Pvt.	Samuel N. Graham
Pvt.	Quantrell C. Hart
Pvt.	Aaron O. Holt
Pvt.	John Luczyk
*Mech.	William Knotts
Pvt. 1 cl.	Ruffin C. Lynch
Pvt. 1 cl.	James Paull
Pvt. 1 cl.	John H. Rommerden
Pvt.	Morris Miller
Pvt.	Stanley M. Murray
Pvt.	Leonard Ostrowski
Pvt.	Antonio Porta
Pvt.	Anton Presalowic
Pvt.	Angel Solari
Pvt.	Benjamin A. Stevens
Pvt.	Christoc Theodorow
Pvt.	Peter VanDerHeide
Pvt.	George Wathen
*Pvt.	Iver N. Wean
Pvt.	John W. Wendland
Pvt.	Frank Zeman
Pvt.	Ignatz Zeman

Company D

*Sup. Sgt.	Pierce B. Atwood
Pvt. 1 cl.	Irwin C. Franklin
Pvt. 1 cl.	Harry R. Schumaker
Pvt.	Nicola Carabillo
Pvt.	Arthur J. Clark
Pvt.	Henry D. Cook
Pvt.	Fred Folmer
Pvt.	George D. Gessa
*Pvt.	James J. Gorman
Pvt.	Alois Heuk
Pvt.	Melvin Lane
Pvt.	John Marlo
Col.	Frank Jungers
Cpl.	John Rybarozyk
Pvt.	Carl Panzarella
Pvt.	Otis B. Robinson
Pvt.	Albert Rusnak
Pvt.	John Sameska
Pvt.	Walter E. Sayer
Pvt.	Reno Schmidt
Pvt.	Adolph M. Schueler
Pvt.	Stanley C. Swift
*Pvt.	Hugh A. Squires
Pvt.	Howard W. Thomas

Company E

*Sgt.	Bert Lee
Cpl.	Morgan W. Evans
Cpl.	Frank Goodran
Pvt.	Eugene F. Amundsen
Pvt.	Herbert L. Armstrong
Pvt.	Burton Arnold
Pvt.	Eugene T. Bellar
Pvt.	William K. Birsgal
Pvt.	Paul Cherney
Pvt.	Thomas M. Conner
*Pvt.	Herman Fichbaum
Pvt.	Arthur Gokey
Pvt.	Dona Geoffrain
Pvt.	William Hunter
Pvt.	Herbert Hovard
Cpl.	Anton Holin
Cpl.	Samuel Rubenstein
Cpl.	Harvey Wilcox
*Pvt.	Joseph Janecek
Pvt.	Emile Kolin
Pvt.	Frank Komski
Pvt.	Edward F. Lloyd
Pvt.	Toney Lotz
Pvt.	Reynard Maestas
Pvt.	Joseph H. Murphy
Pvt.	John T. McCarthy
Pvt.	Theophil Memic
Pvt.	Harry Olson
Pvt.	Carl Olson
Pvt.	Russell Parr

Pvt.	Piedro Pisacretta	Pvt.	John Silvia
Pvt.	Joseph Pejka	Pvt.	Thomas Skinner
Pvt.	Ross Reid	Pvt.	Bert Storm
Pvt.	Joseph C. Romanda	Pvt.	Joseph Taylor
Pvt.	Rufie Reese	Pvt.	Nick D. VanDerGrind
*Pvt.	Lloyd H. Stocker	Pvt.	Ernest M. Willgohs
*Pvt.	Woofin H. Sponangle	Pvt.	James Williams
Pvt.	Stephen Shaw	Pvt.	Smith Warren

Company F

Sgt.	Al Edson	Sgt.	David R. James
*Cpl.	Arthur Case	Cpl.	Willard Cooley
Cpl.	Albert W. Freelich	Cpl.	Edward J. Nolan
Cook	Ned J. Cooper	Pvt.	James R. Harlow
Pvt.	Harry W. Aronson	Pvt.	Albert S. Johnson
Pvt.	Roy H. Bailey	Pvt.	Claude W. Jones
Pvt.	William L. Barker	Pvt.	Daniel O. Kelley
Pvt.	Herriel Corneal	Pvt.	Lester F. LaFlair
Pvt.	Edward J. Coll	Pvt.	Paul R. Martin
Pvt.	George I. Conn	Pvt.	Robert McKane
Pvt.	Louis D. Crane	*Pvt.	Ralph Nussdorfer
Pvt.	Delbert Davis	Pvt.	Harry E. Olson
Pvt.	William Dix	Pvt.	Clarence Radke
Pvt.	Garth C. Davis	*Pvt.	Jacob Risch
*Pvt.	Clair DeMott	*Pvt.	Walter Rochester
*Pvt.	Mike Duda	Pvt.	Joseph Sandusky
*Pvt.	Herbert Edgar	Pvt.	Arthur Schmidt
Pvt.	Albert S. Erskon	Pvt.	William Skiff
Pvt.	James Gaston	*Pvt.	Donald Sutherland
Pvt.	Joseph M. Gosz	*Pvt.	Robert J. Swift
Pvt.	Walter Gaulke	Pvt.	Garett W. Vervelde
Pvt.	Joseph Gillis	Pvt.	Harry E. Westover

Company G

Sgt.	Edward Menge	*Sgt.	H. H. Bartz
Cpl.	John Sparagoski	Cpl.	Pearl G. Wireman
Cpl.	Clarence Thibaudeau	Cpl.	Carl F. Hansen
Cpl.	William Walters	*Mech.	Paul A. Sass
Cpl.	Eeward S. Blackman	Pvt. 1 cl.	Carl Smith
*Pvt.	J. Bonar	Pvt.	Clarence L. Gardner
Pvt.	Joseph B. Benoit	Pvt.	Roy Hamilton
Pvt.	Howard O. Besinger	Pvt.	Roll A. Harper
Pvt.	Leonard A. Brewer	Pvt.	Daniel Jones
Pvt.	Michael Brogen	Pvt.	Daniel E. Kempf
Pvt.	Camiel Dedobdelaero	Pvt.	Anton Losconszkc
Pvt.	James Dignan	Pvt.	Arthur A. Krahn
Pvt.	Clifford Dye	*Pvt.	Edward J. Lex
Pvt.	Percy C. Fagin	Pvt.	James C. Lampton
Pvt.	Freddie J. Finn	*Pvt.	William A. Pierce
Pvt.	Joseph E. Flickinger	Pvt.	James Remedes

*Pvt.	Horace R. Raper	Pvt.	Robert W. Stephens
Pvt.	Willis P. Rinehart	Pvt.	Joseph D. Truhler
Pvt.	Roy Sampson	Pvt.	Harry Utic
Pvt.	Clarence Sheppard	Pvt.	Ernest L. Walker
*Pvt.	Arthur Standisch		

Company H

Sgt.	Patrick Burns	Sgt.	Earl Conners
Cpl.	Arthur C. Beattie	Cpl.	Anthony Wizbicki
Cpl.	Charles F. Kersey	*Cpl.	Edward Martel
Cpl.	Angelo Mustc	Bug.	Mark Campbell
Pvt. 1 cl.	Arthur Finnersen	Pvt. 1 cl.	Arthur D. York
Pvt.	Guard W. Allen	Pvt.	William C. McDonald
Pvt.	Edward C. Bakula	Pvt.	Richard Miller
Pvt.	Tony Battisto	Pvt.	Walter F. Monsin
Pvt.	Noel L. Bearer	*Pvt.	Ellis Peterson
Pvt.	Michael J. Brice	Pvt.	Fred Roberts
Pvt.	Petro Capponi	Pvt.	William H. Roberts
Pvt.	Merle DeLine	Pvt.	James E. Sanders
*Pvt.	Charles Fletcher	Pvt.	Steve Torcheak
Pvt.	Charles Hofer	*Pvt.	Robert Wright
Pvt.	Carson Marchi		

Company I

Sgt.	Walter R. Koerth	*Sgt.	Herbert L. Payne
Cpl.	Carl L. Anderson	Cpl.	Herbert R. Reese
Cpl.	Harry F. Bernard	Cpl.	Luite Severson
Cpl.	John A. Kell	Cpl.	Clarence T. Sutcliffe
Mech.	Alfred Hansen	Pvt. 1 cl.	Don Londagin
Pvt. 1 cl.	Phillip Baldridge	Pvt. 1 cl.	John A. Sundstrom
Pvt. 1 cl.	John A. Heinatz	Pvt.	Harvey L. Nelson
Pvt.	William Baleski	Pvt.	Alvin A. Porter
Pvt.	Frank Bean	*Pvt.	Enoch S. Parker
Pvt.	Clarence L. Berry	*Pvt.	Russell Ramsey
Pvt.	Albert C. Carlson	Pvt.	John Romocky
Pvt.	Dana B. Chambers	Pvt.	Norman Sheritts
Pvt.	Seigel F. Cook	Pvt.	William H. Smith
*Pvt.	Jim M. Doffin	*Pvt.	Claude Smoot
Pvt.	William E. Dennehy	Pvt.	Herbert Togan
Pvt.	Daniel L. Ester	Pvt.	Tracy Torrey
Pvt.	Marion Floram	Pvt.	Rudolph Totty
Pvt.	Roy E. Foraker	Pvt.	Ellis C. Westfall
Pvt.	Rudolph Miks	*Pvt.	Henry F. Walker
*Pvt.	Anton Morganthaler		

Company K

Cpl.	Joseph E. Grubb	Cpl.	William Ruth
Cpl.	Daniel O'Connor	Cpl.	Virgil D. Smith
Cpl.	John Prochaska	Mech.	Harold F. Cole

Bug.	John Hunter
Pvt. 1 cl.	Park S. Chapin
Pvt.	Major Anderson
Pvt.	Arch Barnes
*Pvt.	Frank Bardons
*Pvt.	Rube Brock
Pvt.	LeRoy Cotton
Pvt.	Wyatt D. Covell
Pvt.	Earl W. Green
Pvt.	Gleason Johnson
Pvt.	Andreas Kappelin
Pvt.	William Kayton
Pvt. 1 cl.	Henry G. Kenney
*Pvt. 1 cl.	Milford H. Bates
Pvt.	Charles Love
*Pvt.	Kasper J. Levejewski
*Pvt.	Andrew Matya
*Pvt.	Clem Oglethorpe
Pvt.	Joe Peterson
Pvt.	Charles H. Sonntag
*Pvt.	Amos Witt
Pvt.	Pearl B. Waldsmith
Pvt.	Joseph H. Zielinski

Company L

Sgt.	Richard Henson
Cpl.	Van D. Jones
Cpl.	George L. Slagle
Pvt.	Frank Bendowski
Pvt.	Michael Chesum
Pvt.	Glen O. Clevenger
*Pvt.	Achillo Caputo
*Pvt.	John G. Dougherty
Pvt.	Harry J. Devaney
Pvt.	Frederick P. Dickson
Pvt.	Douglas Dixon
*Pvt.	Edward D. Fisher
Pvt.	Albert L. Forsythe
*Pvt.	Morris W. Gray
Pvt.	Sam Grapsas
Pvt.	William A. Guin
Pvt.	William H. Hall
Pvt.	Dan P. Hankins
Pvt.	Goerge Harry
Pvt.	Charlie C. Hayslip
Pvt.	George F. Hooper
Pvt.	Alfred Housefield
Pvt. 1 cl.	Elmer W. Kinsey
Pvt. 1 cl.	Martin Sunflot
*Pvt. 1 cl.	Henru Hancock
Pvt.	Zid Hutson
Pvt.	Arthur L. Lehman
Pvt.	Charles W. Lowman
Pvt.	Thomas F. Mahon
Pvt.	Howard S. Maine
Pvt.	Avak Manzooian
Pvt.	Walter L. Meador
Pvt.	David L. Midgley
Pvt.	William Miller
Pvt.	Jean Millet
Pvt.	Clinton C. Moore
Pvt.	Michael F. Murphy
*Pvt.	William D. Mazey
Pvt.	Manuel S. Mennenger
Pvt.	Josiah B. Nickerson
Pvt.	Roy Noel
Pvt.	Paul B. Palsmountain
Pvt.	Joseph Saliteris
Pvt.	Mike Tedesco

Company M

Sgt.	Carl A. Chilson
Cpl.	Panagiotes Malavazas
*Cpl.	Sterling Baker
Pvt. 1 cl.	Alexander Miller
*Pvt. 1 cl.	Andrew P. Jedynek
*Pvt.	H. Davies
Pvt.	Raymond Goodson
*Pvt.	Victor Green
Pvt.	Carr E. Hector
Pvt.	Herman F. Huske
*Cpl.	John Francisco
*Cpl.	Jack D. Raymond
Mech.	Clint Miller
*Pvt. 1 cl.	John E. Seaton
Pvt.	Paul M. Keibler
Pvt.	William H. Lawson
Pvt.	Charlie F. Shultz
Pvt.	John Tiberii
Pvt.	William E. Wegner
*Pvt.	Alfred Wiers

Headquarters Company

Sgt.	Thorwald Peterson	*Sgt.	Basil Myers
Sgt.	Andy L. Snow	*Mus. 3 cl	Howard Mehl
Pvt.	Albert Bushey	Pvt. 1 cl.	Nathaniel Harrison
Pvt.	Frank Burtch	Pvt.	Arthur Thayer
*Pvt.	Thomas Miller	Pvt.	Raymond A. Warn

Machine Gun Company

Sgt.	Stanley V. Decker	*Sgt.	George Wright
Cpl.	Fred W. Rapp	*Cpl.	Victor Handley
*Pvt.	Edwin Beebe	Pvt.	George B. Johnson
Pvt.	Brace Briggs	Pvt.	John Kelly
*Pvt.	Oswald Bonholzer	Pvt.	Clyde Kuntz
Pvt.	Wells C. Cochran	Pvt.	William Luska
Pvt.	George H. Croft	Pvt.	Alfred Martin
*Pvt.	Ernest P. Faulkner	Pvt.	Frank Partyka
*Pvt.	Ralph Gilcher	Pvt.	Louis Ring
Pvt.	Carleton Harlow	Pvt.	Raymond M. Schmidt
Pvt.	Willis Harper	Pvt.	Gilden Springer
Pvt.	Arthur Herbert	*Pvt.	Lloyd Shoemaker
Pvt.	Charles Herron	Pvt.	Byrd White
Pvt.	Christopher Johnson		

Supply Company

Sup. Sgt.	James P. England	*Pvt.	Burton Stockwell
Pvt.	Russell Greeder		

Medical Detachment

Pvt.	Grover M. Thorn	*Pvt.	Charles L. Brown

12TH MACHINE GUN BATTALION

Officers

Captain........................Harold C. Hoopes
*1st. Lieutenant..................Earl R. Fretz

Enlisted Men

Company A

*Sgt.	Charles F. Hyatt	Pvt. 1 cl.	Carl Johnson
*Pvt. 1 cl.	Carl H. Leichty	Mech.	John G. Vishnia
Pvt.	James Brown	Pvt.	Francis J. Jaeger
Pvt.	William Dowling	Pvt.	Ralph Rombough
*Pvt.	Vickers Hall	Pvt.	Norman Roberts

Company B

Sgt.	Edward W. McAndrews	Sgt.	Charles A. May
*Sgt.	Samuel H. Hanna	Bug.	Ward B. Morrison

Pvt.	William Conners	Pvt.	Thomas Maher
Pvt.	Carl Dretske	*Pvt.	Lewis Mills
*Pvt.	Otto M. Kunert	*Pvt.	Richard Taylor
Pvt.	Benjamin F. Lair	Pvt.	Elias Williams
Pvt.	Harry E. Meade		

Company C

1st. Sgt.	Henry Pietzmeyer	Sgt.	Allen F. Stone
Mess Sgt.	Roscoe C. Hutt	Sgt.	John E. Norton
Sgt.	John R. Hughes	Cpl.	Charles Perrine
Pvt.	Roy Hamilton	Pvt.	Louis A. Roberts
Pvt.	Gordon E. Jackson		

Company D

Cpl.	Frank V. Grinnon	Cpl.	Morrison Hayes
Pvt. 1 cl.	George J. Jung	Pvt. 1 cl.	Albert G. Perkins
Pvt.	John H. Buchlein	Pvt.	William Hovious
*Pvt.	William J. Bermele	Pvt.	John E. Patrick
Pvt.	Frank Cooney	Pvt.	Emmet M. Woodward
Pvt.	Leon W. Gordon		

Medical Detachment

*Sgt.	Harry E. Delaney	Pvt.	Donald H. Pegg

4TH FIELD ARTILLERY BRIGADE

Brigade Headquarters

Enlisted Men

*Pvt. John Bruno

13TH FIELD ARTILLERY REGIMENT

Enlisted Men

Battery A

*Cpl.	Glynn C. Martin	*Pvt.	Clifford L. Lantis
Pvt.	Olaf Caesar		

Battery B

*Sgt.	Willie Hawkins	Pvt.	Joseph P. Hoeller

Battery C

Pvt. 1 cl.	Henry P. Furtado	Mech.	Frank J. Vavrussa
Pvt.	Charles Bulgrin	Pvt.	Creighton M. Hutchins
Pvt.	Carl Driver	Pvt.	Oscar P. Jensen

BATTERY D

Pvt.	Axel G. Broden	Pvt.	Frank Pawloski
Pvt.	Bernard Farrell		

BATTERY E

*Pvt.	James I. Reed

BATTERY F

Cpl.	Glenn C. Sharrow	*Pvt. 1 cl.	Marcell L. Phillips
*Pvt. 1 cl.	Andrew Anderson	*Pvt.	Samuel P. Collins
*Pvt. 1 cl.	Edward P. Bell	*Pvt.	Melvin J. Nunley
Pvt.	Raphael L. Carlos	Pvt.	Vaughn M. Willets

HEADQUARTERS COMPANY

*Pvt.	Chester E. Bohn	*Pvt.	Frederick Clevinger

SUPPLY COMPANY

Wag.	Arthur L. Weaver	Pvt.	Shelton C. Luttrell

MEDICAL DETACHMENT

Pvt. 1 cl.	Charles L. Charlton	*Pvt. 1 cl.	Martin J. Clements

16TH FIELD ARTILLERY REGIMENT

Officers

Captain. . . . . .Joseph S. Sweeney
1st. Lieutenant. . . . . .Edward J. Wolff
*1st. Lieutenant. . . . . .Clarence A. Knudtson
*1st. Lieutenant. . . . . .Winfield M. Putnam
2nd. Lieutenant. . . . . .William B. Merselis

Enlisted Men

BATTERY A

Cpl.	George Mattick	*Pvt. 1 cl.	Clarence E. Geyer
*Cpl.	Robert F. Smith	Pvt.	Arthur F. Thart
*Pvt.	John H. Jones		

BATTERY B

Sgt.	James F. Healy	Cpl.	Fred V. Milham
Pvt. 1 cl.	Arthur S. Shane	*Pvt. 1 cl.	Robert B. Taylor
Pvt.	Virgil L. Adams	*Pvt.	Joe Laucjs
Pvt.	Antonio Battaglio		

BATTERY C

Pvt. 1 cl.	Harry V. Plessinger	Pvt. 1 cl.	James R. Stover
*Pvt.	Henry Bloomer	Pvt.	Tony Pritchett
Pvt.	John Deitrich	Pvt.	Julius Zanette
Pvt.	James Kramer		

BATTERY D

Cpl.	Clark Moore	Cook	Lon Fiala
*Bug.	Marel H. Roberts	*Pvt. 1 cl.	Carnelius E. Prohl
*Pvt.	Philip Hassinger	Pvt.	Tofil Roginsky
Pvt.	Fred J. Lamphere	Pvt.	Wilbur G. Temperly
Pvt.	George Pecheta		

BATTERY E

Sgt.	John Gilmore	*Cpl.	Errel S. Fellows
Pvt. 1 cl.	James T. Rowan	Pvt.	John Bonegar

BATTERY F

Cpl.	Bertram Pickhardt	Mech.	Sterling R. Hunt
*Ch.Mech	George H. Eckels	Mech.	James Morris
*Pvt. 1 cl.	Lynn E. Davis	Pvt.	Joseph F. Nichols
Pvt.	Merrill C. Bloss	Pvt.	Alfred S. Rice

HEADQUARTERS COMPANY

Sgt.	James R. Coughlin	Pvt. 1 cl.	Ralph C. Beyer
Pvt.	Franklin T. Burns	*Pvt.	James O. McKenny
Pvt.	Albert F. Lertora		

77TH FIELD ARTILLERY REGIMENT

Officers

*Captain........................Charles B. Duncan

Enlisted Men

BATTERY A

Pvt. 1 cl.	Joseph J. Kuhar, Jr.	Pvt.	Alexander Johnston
Pvt.	Parmer C. Karns	*Pvt.	Carl W. Tielsch

BATTERY B

Sgt.	George W. Engelking	*Cook	Herman Rifflert

BATTERY C

1st. Sgt.	John E. Polinaszek	Sgt.	George H. Pearson
Cook	Wladyslaw Kempski	*Cpl.	Elmer Lindback
Pvt.	Frank Ebner	*Pvt.	Charles R. Prather
*Pvt.	Keefer J. Gray		

BATTERY D

*Sgt.	Verren C. Parr	Cpl.	Phillip J. Cavan
Pvt.	Clarence Shiplock	*Pvt.	Clarence O. Collins
*Pvt.	Edward A. Beuford	*Pvt.	William J. Rupert

BATTERY F

*Ch.Mech	Leonard E. Thoma	*Pvt.	Morris Elkan, Jr.

SUPPLY COMPANY

Wag.	Stanley Butvilowich	Wag.	Samuel C. Shawn

UNITS NOT BRIGADED

10TH MACHINE GUN BATTALION

Officers

Captain..........................Henry B. Keep
1st. Lieutenant....................Julius Love
2nd. Lieutenant....................Harols S. Baxter
*2nd. Lieutenant...................Mervin R. Fox
2nd. Lieutenant....................James J. Porter

Enlisted Men

COMPANY A

*Pvt. 1 cl.	Lester R. Griffith	Pvt. 1 cl.	Fred Haddox
Pvt. 1 cl.	Floyd Gray	Pvt. 1 cl.	Carl Johanson
*Pvt.	Walter Baranowski	Pvt.	William E. Bateman
Pvt.	John Laughlin	Pvt.	James O'Brian
*Pvt.	Harvey M. Gooch		

COMPANY B

Sgt.	Bernard W. Baker	*Pvt.	Carl W. Brown
Pvt.	Roy Brown	Pvt.	Joseph T. England
Pvt.	Harry Harris	Pvt.	Anthony Millette
*Pvt.	John P. Ambuchl		

HEADQUARTERS DETACHMENT

Sgt.	Otis K. Norton	Pvt.	Victor Cimei
Pvt.	Ben H. Jones	Pvt.	Clarence McGlynn

4TH ENGINEER REGIMENT

Officers

Captain..........................Kirby B. Sleppy
1st. Lieutenant...................Phillip S. Stafford

Enlisted Men

COMPANY A

Sgt.	Roy S. Butzerin	*Cpl.	William C. Ashby
Cpl.	Lowell F. Coleman	Pvt. 1 cl.	Darrell M. Dunkle
Pvt.	Samuel J. Chiarelli	Pvt.	Fred T. Falk
Pvt.	Anton Nerad		

COMPANY B

*Sgt.	Verner B. McWhorter	Cpl.	Andrew C. Copeman
Cpl.	Robert E. Clark	Cpl.	William E. Fallin
Cpl.	Orville J. Miller	Pvt. 1 cl.	Clyde M. Biles
Pvt.	Carl Farnsworth	Pvt.	Ezra B. Cattell
Pvt.	Walter S. Douglas	Pvt.	Francisco Galtieri
Pvt.	Albert Mausner	Pvt.	George I. Reding

COMPANY C

Cpl.	Marshal N. Braden	Cpl.	Lee R. Griffin
Cpl.	David A. Rhone	Pvt. 1 cl.	Walter E. Drissler
Pvt. 1 cl.	Leonard Heinrich	Pvt.	Bernard J. Howe
Pvt.	Elwood Kindle	*Pvt.	Paul C. Roseman
*Pvt.	Pasquale Servino	*Pvt.	Paul O. Springer
Pvt.	William E. Webb		

COMPANY D

Cpl.	Tom F. Barto	Pvt. 1 cl.	James E. Fowler
*Pvt.	Victor L. Klein	Pvt.	Carl A. Lundstrom
*Pvt.	James P. McDonnell	Pvt.	Domenico Laino

COMPANY E

Sgt. 1 cl.	Christian Luhs	Sgt.	Abraham L. Morris
Cpl.	Emery C. Reynolds	*Pvt. 1 cl.	Thomas E. Duncan
Pvt. 1 cl.	John W. Knowles	Pvt.	Stanton H. Marshall

COMPANY F

*Cpl.	William R. Terrill	Pvt.	Tony Dordas
Pvt.	Arthur W. Fritch		

HEADQUARTERS DETACHMENT

ColorSgt.	Arthur G. Smith	Pvt.	Adolphus G. Busby

MEDICAL DETACHMENT

Pvt.	Charles C. Rismiller

8TH FIELD SIGNAL BATTALION

Officers

*2nd Lieutenant. . . . . .Harold T. Schreiner

Enlisted Men

COMPANY A

*Sgt.	George B. Wilson

COMPANY B

*Pvt.	Philip S. Bowling

COMPANY C

*Msr. Sig. El.	Clifton H. Clark	Cpl.	Leonard W. Vaughn
*Cpl.	Nelson C. Waterman	*Pvt.	Otto C. Bletzer
*Pvt.	Earl C. Chadwick	*Pvt.	Joseph N. Costello
Pvt.	Haskell P. Kendig	*Pvt.	Reinhard Lassanske
Pvt.	William C. Martin	*Pvt.	Gale L. Moore
*Pvt.	Paul F. Richardson		

SUPPLY DETACHMENT

*Chauf.	Thomas L. Freestone	Chauf.	Loy E. Hammer

4TH AMMUNITION TRAIN

Enlisted Men

COMPANY B

*Cpl.	Fred B. Bowers	*Pvt.	Otto E. Moberg

COMPANY C

*Pvt.	Tony Mautino	Pvt. 1 cl.	Leonard Schoenwetter

COMPANY G

*Sgt.	Aaron J. Oppenheim

4TH SUPPLY TRAIN

Enlisted Men

COMPANY B

*Cpl.	Russell J. Bayer

4TH SANITARY TRAIN

Enlisted Men

FIELD HOSPITAL 21

Wag.	Herman Coats	Pvt. 1 cl.	Elza E. Denny

AMBULANCE COMPANY 19

Sgt.	Jean H. Pearl	Pvt. 1 cl.	George L. Marshall

AMBULANCE COMPANY 21

*Pvt. 1 cl.	Michael Jereczkek	Pvt. 1 cl.	Charles H. Reese

AMBULANCE COMPANY 28

Sgt.	Nelson F. Cole	Wag.	Elwyn S. Wheeler
Pvt. 1 cl.	George L. D. Parrish	Pvt.	Louis Sandler
Pvt.	Harry F. Voorhees		

AMBULANCE COMPANY 33

Sgt.	Henry C. Stevens	Pvt.	John A. Dean

4TH MILITARY POLICE COMPANY

Enlisted Men

*Pvt. Donald L. Rose

DIED FROM CAUSES OTHER THAN WOUNDS RECEIVED IN ACTION

HEADQUARTERS TROOP, 4TH DIVISION

Enlisted Men

Regt. Sgt. Maj.	Robert A. Herkert	Pvt.	John W. Burnes
Pvt. 1 cl.	Burns W. Bailey		

7TH INFANTRY BRIGADE

39TH INFANTRY REGIMENT

Officers

2nd. Lieutenant. . . . . Edward A. Selway

Enlisted Men

COMPANY A

Cook	Almer Plouffe	Pvt.	William P. Hutchins
Pvt.	Albert R. Davis	Pvt.	Henry A. Johnson

COMPANY B

Pvt.	James L. Cannon		

COMPANY C

Cpl.	Fred Staton		

COMPANY D

Pvt.	Magnes R. Christensen	Pvt.	Cosmo Molnio
Pvt.	Bert Powell	Pvt.	Philip Tomanio

COMPANY E

Pvt.	Richard R. Kelly	Pvt.	Archie Milmine

COMPANY F

Pvt.	Wm. H. Wemberly		

COMPANY G

Pvt. 1 cl.	William G. Kodada	Pvt.	Harold M. Ward

COMPANY H

Pvt.	Steve A. Diafokeros	Pvt.	Clast B. McDonald

COMPANY I

Pvt.	Peter K. Anderson	Pvt.	Ross C. McConnell
Pvt.	Joseph J. Bourdeau		

COMPANY K

Sup. Sgt.	Ray A. Wilson	Cpl.	Van Kelley
Cpl.	Arthur Gilley		

COMPANY L

1st. Sgt.	Thomas J. Heatherly		

HEADQUARTERS COMPANY

Mus. 2nd cl.	Roscoe Ross	Pvt.	Ralph V. Helmer
Pvt.	Alphonso Dicerbo		

MACHINE GUN COMPANY

Cpl.	George A. Brand	Pvt.	William Seward

SUPPLY COMPANY

Wag.	Tommy O. Anderson	Pvt.	Grover C. Mattuschek

MEDICAL DETACHMENT

Pvt.	Manuel Ormonde

47TH INFANTRY REGIMENT

Officers

2nd. Lieutenant.................... Sidney Beardsley

Enlisted Men

COMPANY A

Pvt.	Corbett J. Jenkins	Pvt.	Young M. Pope

COMPANY B

Cpl.	Peter J. Nee	Pvt.	Harvey Dillon

COMPANY C

Pvt.	Ralph M. Burdett	Pvt.	Willie Reusch

COMPANY F

Cpl.	Daniel K. Iopa	Pvt.	Cleo C. Heuchan
Mech.	John R. Veary	Pvt.	William Krensing
Pvt.	Wilbur Fulton		

COMPANY G

Pvt.	Charles Musar

COMPANY H

Pvt.	Lewis Prokop	Pvt.	Henry Wittmeier

COMPANY I

Pvt.	Leon E. Hale

COMPANY K

Pvt.	Joseph A. O'Keefe

COMPANY L

Pvt. 1 cl.	William P. Gould	Pvt. 1 cl.	Fernand J. Miguez

COMPANY M

Cpl.	Ernest V. Blacet	Pvt.	Morris Van Der Jagt
Cpl.	Richard J. Maybour		

HEADQUARTERS COMPANY

Pvt. 1 cl.	Carl H. Gerfen	Pvt.	Henry Baumgardt
Pvt.	Charles J. Aul	Pvt.	Arthur John

11TH MACHINE GUN BATTALION

Enlisted Men

Mech.	James C. Wallace	Pvt.	Zellen Haught

8TH INFANTRY BRIGADE

BRIGADE HEADQUARTERS

Enlisted Men

Pvt.	Paschel Ashmore

58TH INFANTRY REGIMENT

Officers

1st. Lieutenant.................Raymond F. Wehrle

Enlisted Men

COMPANY A

Cpl.	Martin F. Swanty	Pvt.	Bruno M. Calabretta

COMPANY B

Cpl.	George S. Prince	Pvt.	Rexford H. Stevens

COMPANY C

Cook	William Hafski	Pvt.	Franklin F. Randolph
Pvt.	Louis J. Baum		

COMPANY D

Bug.	Nicola Di Pietro	Cook	Leo J. Schoner

COMPANY E

Sgt.	George W. Benney	Pvt.	Hughey A. Weatherford
Pvt.	Walter R. Monetto		

COMPANY F

Pvt.	George B. Evenson	Pvt.	Charles Suhocki
Pvt.	Lish C. Rutledge		

COMPANY G

Pvt.	Horace K. Duffy	Pvt.	Guy V. Welch
Pvt.	Howard Livingston		

COMPANY H

Pvt.	Edward J. Cosgrove	Pvt.	Augustus W. McCarthy
Pvt.	James F. Cummisky	Pvt.	Arthur L. Trottier
Pvt.	Leo H. Klotten		

COMPANY I

Pvt.	Loren H. Hodges

COMPANY K

Sgt.	Leo E. McDaniel	Cpl.	Hanford Hanson
		Pvt.	Willis Fields

COMPANY L

Pvt.	William E. Gaus

COMPANY M

Pvt.	George H. Keese	Pvt.	I. T. Singleton

HEADQUARTERS COMPANY

Bn. Sgt. Maj.	Arthur Willett	Sgt.	James J. Dorney
St. Sgt.	Giles Clyde Borden	Pvt. 1 cl.	Paul E. Wolfe
Pvt.	Dennis Breen	Pvt.	Howard R. Spaulding
Pvt.	John Blain Kuhn		

MACHINE GUN COMPANY

Cpl.	Wilbert G. Huffman	Pvt.	James H. Powers

SUPPLY COMPANY

Wag.	Walter D. Rodgers

59TH INFANTRY REGIMENT

Enlisted Men

COMPANY A

Pvt.	Herbert Sage	Pvt.	Charles Yanacek

COMPANY B

Pvt.	Claude V. Watson	Pvt.	John Weakley

COMPANY C

Sgt.	Andrew Farley	Pvt.	Noy Tickle

COMPANY D

Pvt. Thomas C. Cunningham

COMPANY E

Cook Walter Gajek

COMPANY H

Sgt.	Mike Kwasneske	Pvt.	John W. Moore

COMPANY I

Pvt.	Pedro Lucero	Pvt.	Virgil Perrine

COMPANY K

Pvt. Homer L. Hodge

COMPANY M

Pvt. Isaac Moser

MACHINE GUN COMPANY

Pvt. George M. Gault

MEDICAL DETACHMENT

Pvt. Charles L. Brown

12TH MACHINE GUN BATTALION

Officers

Major. . . . . John B. Duckstad

Enlisted Men

Wag.	Thomas J. Cleary	Wag.	Charlie E. Crumrine
Wag.	Robert C. Deemer	Wag.	Ralph Kuhl
Pvt. 1 cl.	James M. Patterson	Pvt.	George A. Saur
Pvt.	D. H. Hughson	Pvt.	John Urey
Pvt.	Wilbur D. Potter		

4TH FIELD ARTILLERY BRIGADE

BRIGADE HEADQUARTERS

Enlisted Men

Regtl. Sgt. Maj. Eugene Frankel

13TH FIELD ARTILLERY REGIMENT

Officers

1st. Lieutenant................Joseph W. Del Alamo
2nd. Lieutenant................Edward K. Pentecost

Enlisted Men

Battery A

Pvt. 1 cl.	Guy R. Cobel	Pvt.	Willie Sharp
Pvt.	Henry Heyer		

Battery B

Cpl.	Lester W. Loomis	Pvt.	Taylor S. Dearth
Pvt. 1 cl.	Cardiff E. Cozier		

Battery C

Cpl.	Ray Park

Battery D

Pvt.	Jim V. Martin

Battery E

Mech.	Don H. Wadleigh	Pvt.	Allwyn W. Harrison
Pvt. 1 cl.	Baldwin F. Deevers	Pvt.	Paul E. Settle

Battery F

Sgt.	Harvey E. Love	Pvt. 1 cl.	Thomas J. Masterson

Headquarters Company

Cpl.	Phillip W. Parkinson	Pvt.	Louis E. Andrus
Wag.	Albert Maurone		

16TH FIELD ARTILLERY REGIMENT

Enlisted Men

Battery A

Sgt.	Charles Montgomery

Battery B

Sgt.	Edward C. Bergner	Sgt.	John Sutherland
Pvt.	Gus Fulkerson	Pvt.	John W. Mulhollen

BATTERY C

Cpl.	Fravel E. Alleman	Cpl.	Fred Pelaske
Pvt.	George L. Spencer		

BATTERY D

Cpl.	James H. Metcalfe	Pvt.	Charles B. Burke

HEADQUARTERS COMPANY

Pvt.	Raymond A. Lee		

SUPPLY COMPANY

Wag.	Victor Danguy	Wag.	Lee T. Prescott
Pvt.	Lester E. Kittering		

77TH FIELD ARTILLERY REGIMENT

Enlisted Men

BATTERY A

Sgt.	Albert O. Swift	Horseshoer	Frank E. Wiles

BATTERY B

Pvt.	Walter J. Jones	Pvt.	Fred J. Hamilton

BATTERY C

Pvt. 1 cl.	Feodore K. Semensow	Pvt.	Bernard A. Carr
Pvt. 1 cl.	Joseph A. Willi		

BATTERY E

Pvt.	Theodore Rochester	Pvt.	David Moser

BATTERY F

Pvt. 1 cl.	John Henry	Pvt.	Robert J. Bruns

HEADQUARTERS COMPANY

Cpl.	Warren S. Percival	Cpl.	Roy De Camp
Pvt.	Ray E. Nipper		

SUPPLY COMPANY

Cook	Chauncey Windrum		

10TH MACHINE GUN BATTALION

Enlisted Men

COMPANY A

Pvt. Ray Mallo

4TH ENGINEER REGIMENT

Officers

Captain.......................Burton E. Hocker

Enlisted Men

COMPANY B

Pvt. 1 cl. John B. Clark

COMPANY C

Pvt. Konstanti Sulesky

COMPANY D

Sgt.	Cornelius J. O'Brien	Pvt.	John P. McCartin
Pvt.	Burl Hicks	Pvt.	Frank L. Shingldecker

COMPANY F

Pvt. 1 cl.	Ivan J. Johnson	Pvt.	John H. Ehni

HEADQUARTERS DETACHMENT

Pvt. 1 cl. Ralph G. Steele

8TH FIELD SIGNAL BATTALION

Enlisted Men

COMPANY A

Cpl.	Carl Christensen	Pvt.	John W. Hansard
Cook	Russell L. Coe		

COMPANY B

Pvt. Florian C. Von Kaenel

COMPANY C

Sgt.	Elmer E. Wilson	Cook	John G. Hildebrand
Cpl.	John F. Schall	Cook	Syl P. Hodges
Cpl.	John R. Souther		

4TH AMMUNITION TRAIN

Officers

Captain. ....William C. George
Captain.......................Leonard K. Hart
2nd. Lieutenant...............Robert Eckenroad

Enlisted Men

COMPANY A

Pvt.	James J. Dugan		

COMPANY B

Pvt.	Francisco Duemez	Pvt.	Chris A. Kruger

COMPANY D

Pvt. 1 cl.	Andre H. Elliot	Pvt.	Philip Epstein

COMPANY E

Mech.	Rube T. Cressey		

COMPANY F

Pvt.	Ira C. McConnell	Pvt.	Nicholas Roth

4TH SANITARY TRAIN

Officers

Captain.......................George H. Hockett

Enlisted Men

FIELD HOSPITAL 33

Wag.	Fred P. Eberle		

AMBULANCE COMPANY 28

Sgt.	George D. Raines	Pvt.	Jov Orechva
Pvt.	Patrick H. Gallagher		

Index

Compiled by H. S. Parsons, Library of Congress

Note: Numbers following the entries in this index refer to the pages of the text.

THE COUNTRY LIFE PRESS, GARDEN CITY, NEW YORK

CPSIA information can be obtained
at www.ICGtesting.com
Printed in the USA
BVHW070617271222
654995BV00001B/3

9 781944 193553